Global Urban Agriculture

Global Urban Agriculture

Edited by

Antoinette M.G.A. WinklerPrins

Johns Hopkins University

CABI is a trading name of CAB International

CABI
Nosworthy Way
Wallingford
Oxfordshire OX10 8DE
UK

CABI
745 Atlantic Avenue
8th Floor
Boston, MA 02111
USA

Tel: +44 (0)1491 832111
Fax: +44 (0)1491 833508
E-mail: info@cabi.org
Website: www.cabi.org

Tel: +1 (617)682 9015
E-mail: cabi-nao@cabi.org

© CAB International 2017. All rights reserved. No part of this publication may be reproduced in any form or by any means, electronically, mechanically, by photocopying, recording or otherwise, without the prior permission of the copyright owners.

A catalogue record for this book is available from the British Library, London, UK.

Library of Congress Cataloging-in-Publication Data

Names: WinklerPrins, Antoinette M. G. A., editor.
Title: Global urban agriculture : convergence of theory and practice between North and South / edited by Antoinette M.G.A. WinklerPrins.
Description: Boston, MA : CABI, [2017] | Includes bibliographical references and index.
Identifiers: LCCN 2016045662 | ISBN 9781780647326 (hbk : alk. paper) | ISBN 9781780647340 (epub)
Subjects: LCSH: Urban agriculture.
Classification: LCC S494.5.U72 G56 2017 | DDC 630.9173/2--dc23 LC record available at https://lccn.loc.gov/2016045662

ISBN-13: 978 1 78064 732 6

Commissioning editor: David Hemming
Editorial assistant: Emma McCann
Production editor: Tim Kapp

Typeset by SPi, Pondicherry, India
Printed and bound in the UK by CPI Group (UK) Ltd, Croydon, CR0 4YY

Contents

Contributors vii

Preface ix
Nathan McClintock

Acknowledgements xv

1 **Defining and Theorizing Global Urban Agriculture** 1
Antoinette M.G.A. WinklerPrins

2 **A View from the South: Bringing Critical Planning Theory to Urban Agriculture** 12
Stephanie A. White and Michael W. Hamm

3 **North American Urban Agriculture: Barriers and Benefits** 24
Leslie Gray, Lucy Diekmann and Susan Algert

4 **A Survey of Urban Community Gardeners in the USA** 38
Tammy E. Parece and James B. Campbell

5 **Gardens in the City: Community, Politics and Place in San Diego, California** 50
Fernando J. Bosco and Pascale Joassart-Marcelli

6 **'Growing food is work': The Labour Challenges of Urban Agriculture in Houston, Texas** 66
Sasha Broadstone and Christian Brannstrom

7 **The Marketing of Vegetables in a Northern Ghanaian City: Implications and Trajectories** 79
Imogen Bellwood-Howard and Eileen Bogweh Nchanji

8 **Hunger for Justice: Building Sustainable and Equitable Communities in Massachusetts** 93
Timothy F. LeDoux and Brian W. Conz

9	**Sustainability's Incomplete Circles: Towards a Just Food Politics in Austin, Texas and Havana, Cuba** *Jonathan T. Lowell and Sara Law*	106
10	**A Political Ecology of Community Gardens in Australia: From Local Issues to Global Lessons** *Jason A. Byrne, Catherine M. Pickering, Daniela A. Guitart and Rebecca Sims-Castley*	118
11	**Urban Agriculture as Adaptive Capacity: An Example from Senegal** *Stephanie A. White*	134
12	**Intersection and Material Flow in Open-space Urban Farms in Tanzania** *Leslie McLees*	146
13	**Relying on Urban Gardens for Survival within the Building of a Modern City in Colombia** *Colleen Hammelman*	159
14	**Regreening Kibera: How Urban Agriculture Changed the Physical and Social Environment of a Large Slum in Kenya** *Courtney M. Gallaher*	171
15	**Farm Fresh in the City: Urban Grassroots Food Distribution Networks in Finland** *Sophia E. Hagolani-Albov and Sarah J. Halvorson*	184
16	**The Appropriation of Space through 'Communist Swarms': A Socio-spatial Examination of Urban Apiculture in Washington, DC** *Lauren Dryburgh*	196
17	**Urban Agriculture and the Reassembly of the City: Lessons from Wuhan, China** *Sarah S. Horowitz and Juanjuan Liu*	207
18	**The Contribution of Smallholder Irrigated Urban Agriculture Towards Household Food Security in Harare, Zimbabwe** *Never Mujere*	220
19	**Community Gardens as Urban Social–Ecological Refuges in the Global North** *Joana Chan, Bryce B. DuBois, Kristine T. Nemec, Charles A. Francis and Kyle D. Hoagland*	229
20	**Global Urban Agriculture into the Future: Urban Cultivation as Accepted Practice** *Antoinette M.G.A. WinklerPrins*	242

Index 249

Contributors

Susan Algert, University of California Agriculture and Natural Resources Cooperative Extension, San Francisco, San Mateo and Santa Clara Counties, California, USA (retired). E-mail: salgert@gmail.com
Imogen Bellwood-Howard, Institute of Social and Cultural Anthropology, Göttingen University, Göttingen, Germany. E-mail: ibellwoodh@gmail.com
Eileen Bogweh Nchanji, Institute of Social and Cultural Anthropology, Göttingen University, Göttingen, Germany. E-mail: eileen-bogweh.nchanji@sowi.uni-goettingen.de
Fernando J. Bosco, Department of Geography, San Diego State University, San Diego, California, USA. E-mail: fbosco@mail.sdsu.edu
Sasha Broadstone, Stokes Nature Center, Logan, Utah, USA. E-mail: sashabroadstone@gmail.com
Christian Brannstrom, Department of Geography, Texas A&M University, College Station, Texas, USA. E-mail: cbrannst@geos.tamu.edu
Jason A. Byrne, Environmental Futures Research Institute, Griffith University, Gold Coast, Queensland, Australia. E-mail: jason.byrne@griffith.edu.au
James B. Campbell, Department of Geography, Virginia Polytechnic Institute and State University, Blacksburg, Virginia, USA. E-mail: jayhawk@vt.edu
Joana Chan, University of Nebraska, Lincoln, Nebraska, USA and Cornell University, Ithaca, New York, USA. E-mail: jchan@huskers.unl.edu
Brian W. Conz, Department of Geography and Regional Planning, Westfield State University, Westfield, Massachusetts, USA. E-mail: bconz@westfield.ma.edu
Lucy Diekmann, Environmental Studies and Sciences, Santa Clara University, Santa Clara, California, USA. E-mail: ldiekmann@scu.edu
Lauren Dryburgh, School of International Service, American University, Washington, District of Columbia, USA. E-mail: dryburgh.lauren@gmail.com
Bryce B. DuBois, Cornell University, Ithaca, New York, USA and City University of New York, New York, USA. E-mail: bd333@cornell.edu
Charles A. Francis, University of Nebraska, Lincoln, Nebraska, USA and Norwegian University of Life Sciences, Ås, Norway. E-mail: cfrancis2@unl.edu
Courtney M. Gallaher, Departments of Geography and Women's, Gender and Sexuality Studies, Northern Illinois University, DeKalb, Illinois, USA. E-mail: cgallaher@niu.edu
Leslie Gray, Environmental Studies and Sciences, Santa Clara University, Santa Clara, California, USA. E-mail: lcgray@scu.edu

Daniela A. Guitart, Griffith School of Environment, Griffith University, Gold Coast, Queensland, Australia. E-mail: daniela.guitart@griffithuni.edu.au

Sophia E. Hagolani-Albov, Department of Agricultural Sciences, University of Helsinki, Helsinki, Finland. E-mail: sophia.hagolani-albov@helsinki.fi

Sarah J. Halvorson, Department of Geography, University of Montana, Missoula, Montana, USA. E-mail: sarah.halvorson@umontana.edu

Michael W. Hamm, Department of Community Sustainability, Michigan State University, East Lansing, Michigan, USA. E-mail: mhamm@msu.edu

Colleen Hammelman, Culinaria Research Centre, University of Toronto-Scarborough, Toronto, Ontario, Canada. E-mail: c.hammelman@utoronto.ca

Kyle D. Hoagland, University of Nebraska, Lincoln, Nebraska, USA. E-mail: khoagland1@unl.edu

Sarah S. Horowitz, Community-Based Conservation and Development Research Center, Guizhou Normal University, Guiyang, Guizhou Province, China. E-mail: sarah.s.horowitz@gmail.com

Pascale Joassart-Marcelli, Department of Geography, San Diego State University, San Diego, California, USA. E-mail: pmarcell@mail.sdsu.edu

Sara Law, Sustainable Food Center, Austin, Texas, USA. E-mail: selaw09@gmail.com

Timothy F. LeDoux, Department of Geography and Regional Planning, Westfield State University, Westfield, Massachusetts, USA. E-mail: tledoux@westfield.ma.edu

Juanjuan Liu, School of Geographic and Environmental Sciences, Guizhou Normal University, Guiyang, Guizhou Province, China. E-mail: liujuanj@hotmail.com

Jonathan T. Lowell, Department of Geography and the Environment, University of Texas, Austin, Texas, USA. E-mail: jonathan.t.lowell@utexas.edu

Nathan McClintock, Toulon School of Urban Studies and Planning, Portland State University, Portland, Oregon, USA. E-mail: n.mcclintock@pdx.edu

Leslie McLees, Department of Geography, University of Oregon, Eugene, Oregon, USA. E-mail: lmclees@uoregon.edu

Never Mujere, Department of Geography, University of Zimbabwe, Harare, Zimbabwe. E-mail: nemuj@yahoo.co.uk

Kristine T. Nemec, University of Northern Iowa, Cedar Falls, Iowa, USA and University of Nebraska, Lincoln, Nebraska, USA. E-mail: kristine.nemec@unl.edu

Tammy E. Parece, Department of Social and Behavioral Sciences, Colorado Mesa University, Grand Junction, Colorado, USA. E-mail: tammyparece@gmail.com

Catherine M. Pickering, Environmental Futures Research Institute, Gold Coast, Queensland, Australia. E-mail: c.pickering@griffith.edu.au

Rebecca Sims-Castley, Independent scholar, Gold Coast, Queensland, Australia. E-mail: rebecca.sims8@bigpond.com

Stephanie A. White, Center for Regional Food Systems, Michigan State University, East Lansing, Michigan, USA. E-mail: whites@msu.edu

Antoinette M.G.A. WinklerPrins, Environmental Sciences and Policy Program, Johns Hopkins University, Washington, DC, USA. E-mail: winklerprinsconsulting@gmail.com

Preface

During the late 1800s, vegetable crops blanketed the slopes above Portland, Oregon's Tanner Creek, where it coursed through Goose Hollow, a few blocks west of the city's burgeoning downtown (Fig. P.1). Tended by a legion of Cantonese immigrants, many of whom had come to the American West a decade or two earlier to build railroads and mine for gold, the bounty from these gardens fed the citizens of the booming young city. By the late 19th to early 20th centuries, however, much of this agricultural landscape had been tilled under and graded for the construction of a mainstay of modernist urbanism, a municipal stadium (Wong, 2004). Construction of the stadium – indeed, of the city as a whole – was made possible first by the removal of indigenous people, and later by a state constitution and a suite of exclusion acts that prevented non-whites from owning the land they stewarded and tilled. Half a century later, Elvis would perform onstage in the stadium, and today, another half-century later, thousands of soccer fans cheer for the Portland Timbers as they kick a ball over manicured turf growing from the soil that the Chinese gardeners once cultivated to eke out a modest living.

Far across a continent and an ocean to the east, in Bamako, Mali, a green patchwork sprawls across a vast terrain in the Hippodrome neighbourhood (Fig. P.2). The abandoned racetrack for which the neighbourhood is named was closed in the 1970s by government decree in the wake of the Sahelian drought and the ensuing famine. Mali's leaders felt that feeding the fledging nation's people should take precedence over feeding its racehorses. In the years since, the site has been tilled and cultivated by scores of market gardeners struggling to earn enough to feed their families. Like the horse racing that preceded the gardening, the horticultural techniques used to produce vegetables in the square, raised beds are a legacy of the French colonial era. Around the same time that Cantonese gardeners were feeding Portlanders, Bamako boomed from a colonial outpost of a few hundred to a bustling urban centre of several thousand people and colonial capital of the French Sudan (Roberts, 1987, p. 140). The handful of French officers and their cadre of Senegalese administrators and soldiers, along with a small number of French, Lebanese and Syrian merchants, brought with them a taste for fresh vegetables and the seeds and gardening practices adopted and adapted to meet local preferences and growing conditions, and which today remain inscribed across the entirety of Bamako's urban landscape (Zallé, 1999, pp. 4–5). Today, the Chinese Embassy towers over the vast expanse of Hippodrome gardens, a reminder of changing tides of political economic power in the region, while the garden beds themselves serve as a testament to the gardens' vital contribution to urban livelihoods and food security.

Together, the seemingly disparate vignettes of urban agriculture (UA) speak to UA's ubiquity and obduracy, both spatially and temporally. Indeed, UA as a phenomenon is as old as urbanization itself. People have *always* grown food in cities, both because cities grew up around farms and because

Figure P.1. Chinese vegetable gardens in Goose Hollow, Portland, Oregon, USA, *c.* 1909. Photographer unknown. (Source: Oregon Historical Society, OrHi 12176.)

people brought their farming techniques to the nascent towns to feed themselves and their fellow denizens (Steel, 2013). But with the development of capitalism, which both fuelled and was fuelled by the interdependent engines of industrialization and urbanization, agriculture began to play new and diverse roles in the city. Raising crops and livestock not only served as a coping mechanism and social safety valve in times of economic and political crisis (Lawson, 2005), but also persisted in cities in times of abundance. Indeed, it has often been during such periods of economic growth that agriculture's place in the city has been the most hotly contested (Brinkley and Vitiello, 2014; Moore, 2006; Schmelzkopf, 2002).

Clearly, both vignettes speak to UA's prominent role in feeding city dwellers, perhaps tempting us to take a side – whether on Team Pollyanna or Team Dubious – in the ongoing debate over UA's ability to feed a rapidly urbanizing world. But I would caution us to be wary of the productivist myopia that dominates these tiresome discussions. Not only are such arguments often couched in Malthusian terms, but indeed, they simply miss the diverse contributions that UA makes at various scales, contributions that have less to do with feeding people, and more to do with social cohesion, political mobilization, education and de-alienation (Battersby and Marshak, 2013; McClintock, 2010; Sbicca, 2014), actions that together prove central to production of urban space (Eizenberg, 2012; Lefebvre, 1992; Purcell and Tyman, 2014; Shillington, 2013).

Indeed, these two vignettes shed light on UA's imbrications in often contradictory processes of urbanization, processes that may be at once radical and reformist, liberatory and neoliberal, depending on the scale of analysis (McClintock, 2014). Whether those growing in the rain-soaked topography of the US Pacific Northwest or the sun-parched Niger floodplain, these gardens can open our eyes to processes of colonialism and nation-building. To survival in times of economic hardship and

Figure P.2. Hippodrome gardens as seen from La Terrasse. Bamako, Mali, 2006. The Chinese Embassy can be seen in the distance. (Photo by Nathan McClintock.)

political transition. To the ways in which spaces of food production are often racialized. Indeed, to understand and explain UA, we must do more than repackage the anodyne tales of locavores and heirloom tomatoes celebrated in the media cybersphere today, but must also recognize that, like other tales of accumulation and dispossession, the history of UA is often 'written in the annals of mankind in letters of blood and fire' (Marx, 1976, p. 875).

In Bamako, recently ranked as the fastest-growing city in Africa, urbanization is fuelled by an influx of rural migrants eager to escape the fluctuations and instability of markets and a changing climate. The outward sprawl of the city reshapes the landscape, pitting use values against one another and superimposes new exchange values, enclosing and effacing spaces where gardens grew and livestock grazed. But more ignominious forces are also upending the lives of the city's 2 million people. A 2012 coup d'état toppled Mali's democratically elected government, for the past two decades the poster child of Western development agencies. At the same time, Salafist militants took over the northern half of the country, spurring massive internal displacement, much of it an exodus toward Bamako. But the violence has not been restricted to the north. In March 2015, Al-Qaeda-affiliated terrorists killed five people in a hail of grenades and machine gun fire at La Terrasse, a restaurant and nightclub overlooking the Hippodrome gardens (BBC, 2015). The conflict and socio-economic upheaval of the coup, rebellion and terrorism are new forces that are dynamically reworking both space and place in Bamako. With a small handful of exceptions (e.g. Corcoran and Kettle, 2015; Maconachie *et al*., 2012), scholars have yet to fully examine UA's various functions in conflict and post-conflict zones, but based on our understanding of UA's perennial function as a coping strategy, we might expect to see new roles for gardens and livestock, as internally displaced migrants head to the city.

Meanwhile, back in Portland, a paradigmatic Sustainable City, the contradictions of capitalist urbanization have come to a head. Housing costs are skyrocketing. The mayor recently declared a housing state of emergency, given the rise of homelessness. The city has earned the shameful crown of 'fastest gentrifying city in America' (Smith, 2015). Like other green infrastructure and amenities for which Portland is lauded, UA is deeply entangled in processes of gentrification here. The story of UA often unfolds in a way well documented by scholars of New York's gardens during the Giuliani era (Schmelzkopf, 2002; Smith and Kurtz, 2003; Staeheli et al., 2002). In these instances, developers and gardeners are pitted against one another in an iconic standoff between use and exchange value, a sad story where the bulldozer usually wins, and condos sprout in place of collards. But in Portland – like in Vancouver and Detroit (Safransky 2014; Quastel, 2009; Walker, 2015) – this tension is less clear cut. Indeed, more salient is UA's contribution to forces leading to the condo construction in the first place (Goodling et al., 2015). In a city renowned for sustainability, the ecotopian, localist discourse often framing UA practice is part and parcel of the eco-gentrification and displacement underway in Portland. At the same time, some UA practitioners have begun to recognize their unwitting role in this process and are allying with affordable housing and anti-displacement campaigns. The Greeley Forest Garden, for example, has partnered with an adjacent homeless encampment called Hazelnut Grove and the non-profit Sisters of the Road Café, not only to provide campers with fresh produce in exchange for volunteering in the garden, but also to provide them with a safe space to stay and to advocate for their right to stay (McClintock et al., forthcoming).

How then do we theorize UA in a way that makes sense of both its historical and newly emergent forms and functions? Moreover, how might we weave these disparate narratives together? How might we use a 'relational comparative approach' (Ward, 2010) that emphasizes UA's embeddedness within 'sticky, history-laden contexts' (Jacobs, 2012, p. 414) to understand the emergence of productive agricultural spaces in cities as different as Portland and Bamako? Such an approach to comparative research not only takes into account the similarities and differences of individual cases, but also emphasizes the ways in which they are defined by their relationship to historically and geographically distinct and contingent political economic networks and flows—and to each other (Brenner, 2001; McFarlane, 2010; Robinson, 2011). Central to such an epistemological project is what Ananya Roy (2011) describes as the 'worlding' of urban theory. Such an approach 'is fundamentally concerned with how cities – of both the global North and global South – are represented in the canon of urban studies and its archives of knowledge' (McCann et al., 2013, p. 584). It places urban spaces in the North and South on equal footing 'to recover and restore the vast array of global strategies that are being staged at the urban scale around the world' (Roy, 2011, p. 10), strategies that may be 'closely tied to elite aspirations and the making of world-class cities' or to the 'laboring bodies circulat[ing] in search of survival, livelihood and hope' (Roy, 2011, p. 10).

Which brings us to the book you're holding now. With *Global Urban Agriculture*, Antoinette WinklerPrins presents Roy's concept as a point of embarkation on our journey to better understand UA, its similarities and differences, as well as its connectedness. She brings together a talented crowd of UA scholars to engage with urban political ecology, critical urbanism and sustainable livelihoods, as they examine UA's emergence in relation to a diversity of sticky contexts that are at once site-specific and global in scale. From the Gold Coast to Ghana, San Jose to Senegal, Massachusetts to Medellín, the fields, gardens and livestock described herein are linked in time and space by flows of material, capital, people, animals, plants and ideas. And it is via these flows – and the various friction points that slow them (Lawhon, 2013) – that UA and urbanization proceed in tandem, shaping the globe in new, uneven and contradictory ways. One of the first of its kind to bring Global North and Global South UA into the same theoretical bailiwick, this book contributes beautifully both to the worlding of UA theory and to the worlding of urban practices that such theory attempts to understand.

Nathan McClintock
Portland State University, Portland, Oregon, USA

References

Battersby, J. and Marshak, M. (2013) Growing communities: integrating the social and economic benefits of urban agriculture in Cape Town. *Urban Forum* 24(4), 447–461.

BBC (2015) Mali bar attack kills five in Bamako. *BBC News* (7 March). Available at: http://www.bbc.com/news/world-africa-31775679 (accessed 26 October 2016).

Brenner, N. (2001) World city theory, globalization and the comparative-historical method – reflections on Janet Abu-Lughod's interpretation of contemporary urban restructuring. *Urban Affairs Review* 37(1), 124–147.

Brinkley, C. and Vitiello, D. (2014) From farm to nuisance: animal agriculture and the rise of planning regulation. *Journal of Planning History* 13(2), 113–135.

Corcoran, M.P. and Kettle, P.C. (2015) Urban agriculture, civil interfaces and moving beyond difference: the experiences of plot holders in Dublin and Belfast. *Local Environment* 20(10), 1215–1230.

Eizenberg, E. (2012) Actually existing commons: three moments of space of community gardens in New York City. *Antipode* 44(3), 764–782.

Goodling, E.K., Green, J. and McClintock, N. (2015) Uneven development of the sustainable city: shifting capital in Portland, Oregon. *Urban Geography* 36(4), 504–527.

Jacobs, J.M. (2012) Urban geographies I: still thinking cities relationally. *Progress in Human Geography* 36(3), 412–422.

Lawhon, M. (2013) Flows, friction and the sociomaterial metabolization of alcohol. *Antipode* 45(3), 681–701.

Lawson, L.J. (2005) *City Bountiful: A Century of Community Gardening.* University of California Press, Berkeley, California, USA.

Lefebvre, H. (1992) *The Production of Space.* Wiley Publishers, Hoboken, New Jersey, USA.

Maconachie, R., Binns, T. and Tengbe, P. (2012) Urban farming associations, youth and food security in post-war Freetown, Sierra Leone. *Cities* 29(3), 192–200.

Marx, K. (1976) *Capital: A Critique of Political Economy.* Penguin Classics, London, UK.

McCann, E., Roy, A. and Ward, K. (2013) Assembling/worlding cities. *Urban Geography* 34(5), 581–589.

McClintock, N. (2010) Why farm the city? Theorizing urban agriculture through a lens of metabolic rift. *Cambridge Journal of Regions, Economy and Society* 3(2), 191–207.

McClintock, N. (2014) Radical, reformist, and garden-variety neoliberal: coming to terms with urban agriculture's contradictions. *Local Environment* 19(2), 147–171.

McClintock, N., Miewald, C. and McCann, E. (forthcoming) The politics of urban agriculture: governance, policy-making, and contestation. In: Jonas, A. and Ward, K. (eds) *Handbook on Spaces of Urban Politics.* Sage Publications, London, UK.

McFarlane, C. (2010) The comparative city: knowledge, learning, urbanism. *International Journal of Urban and Regional Research* 34(4), 725–742.

Moore, S. (2006) Forgotten roots of the Green City: subsistence gardening in Columbus, Ohio, 1900–1940. *Urban Geography* 27(2), 174–192.

Purcell, M. and Tyman, S.K. (2014) Cultivating food as a right to the city. *Local Environment* 20(10), 1132–1147.

Quastel, N. (2009) Political ecologies of gentrification. *Urban Geography* 30(7), 694–725.

Roberts, R.L. (1987) *Warriors, Merchants, and Slaves: The State and the Economy in the Middle Niger Valley, 1700–1914.* Stanford University Press, Stanford, California, USA.

Robinson, J. (2011) Cities in a world of cities: the comparative gesture. *International Journal of Urban and Regional Research* 35(1), 1–23.

Roy, A. (2011) Urbanisms, worlding practices and the theory of planning. *Planning Theory* 10(1), 6–15.

Safransky, S. (2014) Greening the urban frontier: race, property, and resettlement in Detroit. *Geoforum* 56, 237–248.

Sbicca, J. (2014) The need to feed: urban metabolic struggles of actually existing radical projects. *Critical Sociology* 40(6), 817–834.

Schmelzkopf, K. (2002) Incommensurability, land use, and the right to space: community gardens in New York City. *Urban Geography* 23(4), 323–343.

Shillington, L.J. (2013) Right to food, right to the city: household urban agriculture, and socionatural metabolism in Managua, Nicaragua. *Geoforum* 44 (1), 103–111.

Smith, C.M. and Kurtz, H.E. (2003) Community gardens and politics of scale in New York City. *Geographical Review* 93(2), 193–212.

Smith, D.M. (2015) The 10 U.S. cities that are gentrifying the fastest. *Grist*, 6 February. Available at: http://grist.org/cities/the-10-u-s-cities-that-are-gentrifying-the-fastest (accessed 27 January 2016).

Staeheli, L.A., Mitchell, D. and Gibson, K. (2002) Conflicting rights to the city in New York's community gardens. *GeoJournal* 58(2/3), 197–205.

Steel, C. (2013) *Hungry City: How Food Shapes Our Lives*. Vintage, London, UK.

Walker, S. (2015) Urban agriculture and the sustainability fix in Vancouver and Detroit. *Urban Geography*, X, 1–20.

Ward, K. (2010) Towards a relational comparative approach to the study of cities. *Progress in Human Geography* 34(4), 471–487.

Wong, M.R. (2004) *Sweet Cakes, Long Journey: The Chinatowns of Portland, Oregon*. University of Washington Press, Seattle, Washington, USA.

Zallé, D. (1999) Stratégies politiques pour l'agriculture urbaine, rôle et responsabilité des autorités communales : le cas du Mali. In Smith, O.B. (ed.) *Agriculture Urbaine en Afrique de l'Ouest / Urban Agriculture in West Africa*, IDRC , Ottawa, Canada, pp. 1–18.

Acknowledgements

I thank the many people who cultivate urban spaces around the world and whose practices, thoughts, and ideas about it we were able to share.

1 Defining and Theorizing Global Urban Agriculture

Antoinette M.G.A. WinklerPrins*
Johns Hopkins University, Washington, District of Columbia, USA

> Urban agriculture (UA) is sprouting up in empty spaces of post-industrial landscapes throughout the industrialized world – in vacant lots, road medians, parks – reminiscent of the patchwork of vegetable gardens and livestock enclosures that are part of the urban streetscape in much of the Global South.
>
> (McClintock, 2010, p. 191)

1.1 Introduction

Time has come to rethink and theorize urban agriculture (UA) at a global scale as its importance continues to rise in a world that is becoming ever more urban, and perhaps more importantly, a world in which the differences between the Global North (GN) and the Global South (GS) regarding the practice and motivations for urban agriculture are lessening. The objective of this volume is to bring together research that focuses on productive cultivation in urban spaces from around the world and to place these empirics in a theoretical context to provide cohesion. The motivation for compiling this book and titling it as I have come from years of research on home gardens and urban agriculture in the Global South (e.g. Winkler-Prins, 2002, 2006; Murrieta and Winkler-Prins, 2003; WinklerPrins and de Souza, 2005,

2009, 2010; Gallaher et al., 2013a, b, 2015) while also advising students on the topic (Egger, 2007; Gallaher, 2012; White, 2014). Years ago, while working with my Amazonian collaborator on our home-garden project, she asked what I grew in mine. Although I do indeed cultivate some vegetables and fruits, this launched us into a conversation about how most home gardens in the Global North contain plants primarily for landscaping (aesthetic) purposes. This baffled her, as it seemed a waste of potential utilitarian plants. This exchange provided me with thoughts and insights about the role of plants about us. This, combined with an awareness of the surging movement in urban agriculture in the Global North through teaching on sustainable food systems at American universities, has propelled me to interrogate the divide between the GN and the GS. In reading about these practices in the various places, I sensed differences in the cited literature, semantics, and the approach between case materials from the GS and the GN, with literatures rarely crossing over. Yet the trends in practice that are occurring point to a seeming convergence in practice. On the one hand, for example, urban agriculture in Detroit, Michigan (e.g. White, 2011; Colasanti et al., 2012; Safransky, 2014) and other rust-belt cities of the USA has become a survival

---

*E-mail: winklerprinsconsulting@gmail.com

strategy for the disenfranchised and marginalized left behind in that city's tumultuous deindustrialization and is, in many ways, becoming similar to the self-help survival strategies witnessed in many cities of the GS (Zezza and Tasciotti, 2010; Opitz *et al.*, 2016). On the other hand, as wealth has increased in the GS, middle-income women are gardening in the cities of Senegal (White, 2015; Chapter 11, this volume) for reasons that have less to do with their immediate need for food and is more in line with gardening as a recreational and time-filling activity, reminiscent of urban gardening in the GN.

Despite the seeming convergence in practice, the literatures on UA in the GN and GS remain impressively separated, with researchers working on case material in the GS and rarely referencing work on the GN, and vice versa. The moment is here to think about UA at a global scale and focus on shared experience. My intention with this volume is to move towards greater interaction and engagement across this divide, as this will enrich both focus areas of inquiry. I refer here to urban theorist Ananya Roy and her invocation of the term 'worlding', which refers to 'alternative modernities that produce multiple urban sites and experiences and can speak to and inform one's analysis of other places' (Roy, 2011, p. 828).

In addition to the increasing convergence of motivations and practice, there is the potential for convergence in theorizing UA as well, and this book makes an attempt to do so, both in some of the individual chapters and *in toto*. This can be done by engaging with a broadly defined urban political ecology, especially its attention to UA as a way of healing the metabolic rift, as well as attention to the idea of 'urban assemblage' and new ideas from critical urban studies. The food relocalization movement in the GN has focused its attention on UA as a way of reconnecting people and their food sources, as well as the numerous environmental benefits such as UA's role in greening the city and contributing to urban sustainability. The livelihoods framework and its attention to the five capitals that the poor have access to – physical, natural, human, financial and social – is also a helpful way of framing empirical studies. More on the theoretical approaches used in this volume is elaborated below.

Geographer Tom Bassett (1981) was quite prescient when he stated in his conclusions on the history of community gardening in America that 'what unites these groups [those that instigated gardening efforts] is their collective effort to make living in the city a more palatable experience'. This indeed remains the case, whether we are talking about self-help community organizations, development non-governmental organizations (NGOs), or formal governmental and international institutions in the GN or GS. The 'palatable experience' I see emerging as a unifying theme in global UA, and by extension a convergence of theory and practice, is the very active rethinking of the role and purpose, and even conceptualization of nature in the city, and of efforts to 'green' the city, not just to improve aesthetics and people's quality of life, but because a green city is a more sustainable city. UA contributes to a greening of the city by converting this green to productive spaces that nourish the city in more than aesthetic ways and also provide necessary ecosystem services.

1.2 Defining Urban Agriculture (UA)

Defining urban agriculture is not an easy task. Many definitions exist, and I settle here on a variation derived from Pearson *et al.* (2010, p. 7) which itself is an amalgam of other sources. For the purpose of this volume, *urban agriculture is the production, processing and marketing of food and related products in urban and peri-urban areas, usually through intensive cultivation and for consumption in the same urban or peri-urban area*. The existing literature covers a wide range of practice that some call 'gardening' and some call 'agriculture'. Gardening usually connotes leisure, aesthetics and recreation, small scale, and in some parts of the world is women's domain. In contrast, farming typically connotes production for subsistence or commercial purposes. It refers to a livelihood, a way of life, and is usually practised on a relatively larger scale than gardening. In many parts of the world, farming is associated with men and is considered a male domain. The reality is that, in practice, much of what we have traditionally talked about in UA is gardening, but it has taken on elements of farming and there is semantic fluidity between the two. Neither term is entirely satisfactory for encompassing what actually occurs, and therefore I suggest that instead of using the terms 'gardening' and/or 'farming' that we refer to this suite of activities as 'urban cultivation' and refer to the people who practise it as 'urban cultivators'. This is difficult to do given the deep embeddedness of the term 'urban agriculture', which is why in this volume there will be a mix.

Urban cultivation encompasses plants being grown for some utility, but also includes activities that involve animals. This may range from home gardeners keeping or enabling bee foraging in their yards through the planting of appropriate flowers to the keeping (legally or not) of chickens or other fowl, to the maintenance of cows (usually for milk) or even horses and other animals. Although more common in the GS, the keeping of poultry as part of the home-garden system is gaining traction in many cities in the GN as the health benefits of free-range and locally produced eggs and meat have become clear. Activists in cities large and small in the GN are working on the legal issues of keeping poultry, while those in the GS work to keep such activities from becoming illegal.

One of the characteristics of urban cultivation is its great diversity of practice. Nathan McClintock provides us with an excellent typology of urban agriculture (2014, p. 150) and I borrow from his work, as modified by Gray et al. (2014 and Chapter 3, this volume) to summarize the various forms of UA in Table 1.1. Pearson et al. (2010) also provide a very helpful organization of UA typology, and they add to McClintock and Gray et al.'s typologies a discussion of the scale of the UA production (micro, meso and macro). I have incorporated their elements into Table 1.1 as well. The range of UA practice ranges from individual household gardens, organized allotment and community gardens, and the use of interstitial spaces (Galt et al., 2014) such as berms and public rights of way to macro-scale urban (hydroponic) farming on the ground or on rooftops, and even in the vertical dimension (Despommier, 2010). Used spaces range from the officially public to the intimately private.

In the GN, the focus of UA research has been on how it empowers local communities and how it contributes to the relocalization of a food system that has become disconnected from the community. It is usually conceptualized as something organic that arises from the bottom up, from the community. It is often enveloped in the discourse of social justice that gives voice to marginalized people and empowers them to take control of their lives and communities. UA in the GN is often seen as a solution to many urban challenges, including addressing social woes and efforts to 'green' the city.

In contrast, in the GS the focus of UA research has been on how it assists the transition to urban subjects for newly arrived rural migrants and provides food security for those new arrivals, however marginal it may be. It is usually conceptualized as something that is a necessary process on the way to more 'modern' ways of urban living, including the purchasing of food in supermarkets. In the long run it should be eliminated. It is usually enveloped in developmentalist discourse and undergirded with top-down efforts to 'aid' locals (often by NGOs). UA in the GS is often seen as a necessary problem that needs to be dealt with as a city urbanizes.

Additionally, UA is seen as part of Alternative Food Networks (AFNs) (Jarosz, 2008) which capture 'a wide array of new linkages between agricultural production and food consumption that differ from "conventional" processes and routes' (Galt et al., 2014, p. 134). Many UA practices are part of these networks, although in many places around the world, especially in the GS, they are seen as much more 'conventional' than in the GN. Additionally, aspects of a broadly defined UA are encompassed by what Galt et al. (2014) termed 'SIFS', or Subversive and Interstitial Food Spaces, a phrase that is meant to point to the fact that many activities encompassed by UA subvert the usual use of spaces and places, and are meant to challenge this normative use.

The essential similarity between UA in the GS and GN is that it increases social capital – that food production is important, but not as important as what comes with the process of cultivation. Research to date, very difficult to do, is that the amount of total food produced via UA is not enough to feed the cities of the world, no more than about 15–20% (Pearson et al., 2010; Ackerman et al., 2014; Thebo et al., 2014). But as the shift in discourse in the GN, from UA being for recreational purposes to its greater role in urban sustainability and resilience, there is a convergence with what is closer to the focus of UA in the GS, where UA is a form of social resilience. The

Table 1.1. Range of urban cultivation. (Based on McClintock, 2014, as modified by Gray, 2015; also Pearson et al., 2010.)

Type	Organized	Management	Location	Purpose	Scale
Home gardens; yards	Sometimes	Individual or household	Backyards, front yards, containers, sacks	Household food production, landscaping, recreation	Micro-Meso
Community gardens; allotments	Usually	Municipality or non-profit programme	Vacant lots, parks, open land	Food production, cultural reproduction, recreation	Meso
Non-profit urban farms	Yes	Non-profit organization	Vacant lots, rooftops	Education, food access, vocational training, youth and children's programming	Macro
For-profit urban farms	Yes	For-profit company (individual or individuals)	Vacant lots, warehouses, client yards, greenhouses	Food production, garden installation	Macro
Institutional gardens	Yes	Hired staff or volunteers	Schools, churches, prisons, hospitals	Education, rehabilitation	Micro-Meso
Interstitial food spaces (e.g. guerilla gardening, gleaning and foraging)	Sometimes	Individuals or group	Berms, traffic circles, alleys, parks, forests, backyards, front yards	Reclaiming urban spaces, food production and consumption, urban greening	Micro

overarching unity, and its strength as a social movement, is that UA is a series of processes through which communities gain greater resilience and thereby are more sustainable, with positive social and environmental implications. The small but marginally important food production aspects of UA, common in the GS, are becoming a more visible part of the portfolio of the urban poor in under-resourced cities of the GN, and expose the fundamental similarities in UA's importance to urban survival globally. Likewise, the rise of the middle class in parts of the GS and their desire and ability to garden for recreational purposes converges with elements of UA in the GN. Everywhere UA is seen as good for human health, including getting people moving and eating healthier foods, but also for their mental health, especially for the elderly and economically or culturally dislocated, as UA provides something to do that is meaningful and allows all to feel they are a productive part of society (Airress and Clawson, 1994; Egger, 2007; WinklerPrins and de Souza, 2009).

Globally, the growth in attention to UA also reflects a change in values and priorities. There is a long tradition of productive gardening in the GN as responses to crises (Bassett, 1981; Moore, 2006), but once the crisis has subsided, gardening loses its productive focus and becomes a way of creating aesthetic spaces, i.e. by demonstrating that you can afford to not have your garden be productive. In the GS, one can argue, there is a perpetual crisis for the majority of those in cities, hence the need for continuous UA and for it to be productive. Today in the GN, the movement towards making gardens, yards and other 'unproductive' spaces productive is motivated by different values that reflect not so much that you cannot afford to buy food, but that people want to be in control of the source of their food. This movement is part of the 'relocalizing' of food movement and urban sustainability. From the perspective of critical social science, this can be seen as a privileged position not available to the poor and marginalized who are closer to the majority that practise UA in the GS.

Another shift in values and aesthetics is one towards UA and its ability to 'green' the city: both literally, as UA provides green space in the city, and figuratively, as UA contributes to sustainability. It has always had that function, but this has not been seen as a valuable component of its existence in the past. Today, UA is looked to, in both the GN and GS, as a way of making the city a *productive* green. UA thus can be seen as contributing to ecosystem services that make the city more pleasant and sustainable. Beyond the potential agricultural output is the acceptance of foraging as part of the cityscape, and a rethinking of nature in the city (McClain et al., 2014). Green and productive cities have the potential of being more sustainable (and resilient) than those that are not, and there is an emerging global focus on a desire for the sustainable city in this era of global environmental change and rapid urbanization.

1.3 Theoretical Framing

Despite their differences in expression, there is an increasing convergence between UA in the GS and GN in practice, and that, by extension, there is the potential for convergence in theorizing. Work on UA is a convergence of early research in cultural ecology on home gardens and their biodiversity and spatial configurations (e.g. Christie, 2004; Kimber, 2004) combined with urban studies and urban planning, and a more recent turn toward critical theory. Much of the literature is highly applied and remains descriptive and at the case study level as a result. Where possible I have asked authors of this volume to engage with at least one of the following theoretical frameworks in order to emerge above the case study, emphasizing not just the uniqueness of their findings but the lines of similarity and convergence with their case and others in order to contribute to an emerging global UA.

Urban political ecology (UPE) asks questions about who produces what kind of socio-ecological configuration and for whom (Keil, 2003, 2005; Swyngedouw and Heynen, 2003; Heynen et al., 2006; Wachsmuth, 2012; Angelo and Wachsmuth, 2014). It assumes that urbanization is a process of socio-ecological change and foregrounds the urban condition as fundamentally a socio-environmental process. UPE pays close attention to existing unequal power relations, the social construction of nature, and especially its attention to the 'metabolic rift' that has occurred between people and land in urban spaces. There are three dimensions that UA can contribute to in terms of healing the metabolic rift (McClintock, 2010). First there is the ecological – the rescaling, or closing, of the nutrient cycle that occurs as spaces in the city are increasingly cultivated, ideally using materials already in place and engaging in waste (nutrient) recycling in the process of cultivation and reconstituting urban soil for cultivation. This is linked to the movement toward greening the city. Then there is the social, the building of community and social capital, which I have already outlined above and is the fundamental piece of the success of UA as a global effort as it reconnects dispossessed people to each other and helps build and support community at various scales. UA stands to be able to heal this social disconnect through its ability, not only to provision people with direct need, but also to empower the marginalized in urban places, which can lead to many other actions that improve their lives. Lastly, there is the healing at the individual level, in part also already mentioned above: UA as a means of reconnecting people to nature that improves their individual health, but also connecting them to work that makes people feel a productive part of society.

Nathan McClintock (2014) has elaborated on urban agriculture's necessary contradiction or tension, that it is at once both radical and neoliberal. It is radical in that its activities often stand in opposition to the accepted norms of what a city should look like and how its people should behave, but it also enables a neoliberal agenda to be pursued. He points out that what is common in all UA is that it can be seen as a subsidy from self-provisioning (self-exploitation), which lets the state get 'off the hook', as it were, from providing for its citizens. This is particularly strong in places where the state has failed, which has happened in both the GN and GS, as the state is often absent in its support for marginalized people. McClintock argues that UA enables the state to leave people behind because they do take care of themselves.

There are also forms of UA that are corporate and seek to profit from a new sector of production, and this vision sees little in unity with the small-scale and organic movements of much of UA to date. This contrast and contradiction, and what I call McClintock's 'tension', is well illustrated by the stops on a 2015 field excursion to see 'Urban Farms and Gardens' at the Metropolitan Solutions Congress in Berlin, Germany. The title in itself is quite telling: farms and gardens are treated separately and were not placed under one category. And the excursion demonstrated the two extremes of the way UA is practised and conceptualized. The first stop was to a brand new urban farm, ECF FarmSystems, a demonstration farm for a company that will help others develop high tech aquaponic farm systems (including fish tanks and hydroponically grown vegetables) (Fig. 1.1). Not traditionally viewed as the norm of sustainable (urban) farming, although possibly its future, highly integrated and technologically sophisticated farming such as this is spreading to both rural and urban spaces. Hydro- and aquaponic systems in greenhouses permit year-round production in rural areas and permit a degree of intensification in urban spaces that might result in the volume needed to farm in the city in a self-sufficient manner, something that is questioned as a possibility. The second stop was at the community gardens at the abandoned Tempelhof airport (the 'air bridge' airport when Berlin was a divided city), *Stadtteilgarten Schillerkiez*, a very organic (figuratively as well as literally) and grassroots effort to reclaim green space in the city and build community (Fig. 1.2). It is a classic community garden, including the challenges of maintaining land tenure, as the Tempelhof development plans are uncertain at this time and are an ongoing tension between the Berlin city government (who want housing and development) and its residents (who want to leave it as open green space). West Berlin, due to its existence as a virtual island during the Cold War, has long focused on being able to sustain itself, and though it is now reconnected with its hinterland, the principles of self-sufficiency pervade thinking about urban planning there. Both the urban farm and the community garden seek this self-sufficiency in very different ways. Their contrast embodies what is happening in many parts of the world.

From the field of *critical urban studies* comes the *urban assemblage framework* that focuses on the role UA plays in the assemblages of urban life and its multiple flows (food, people, knowledge, materials, etc.) (Brenner *et al.*, 2011; McFarlane, 2011a, b; Shillington, 2013). This approach seeks to view quite holistically the entirety of the urban system in a new light, and those using

Fig. 1.1. An urban demonstration farm in Berlin, Germany: ECF FarmSystems, a company that will help develop high-tech aquaponic farm systems in the city. (Photo by author, May 2015.)

this approach can understand the important component UA plays in urban assemblages today. This approach helps shift the discourse about UA from a developmentalist and problem-based narrative to a solutions-focused one, that emphasizes the creative powers of marginalized people in cities everywhere and focuses on their ability to make the city work, despite seemingly chaotic conditions. By extension I link here to the work of Roy (2011), her *urban 'worlding' theory*, in which she emphasizes the need to consider reversing the flow (or at least consider its potential for bi-directionality) from the usual urban theory flowing from models in the GN to applications in the GS – to treat ALL cities together, erasing the differences between them to have productive conversations. There is a great need to think creatively of global city-regions that can be compared, considering their extra-territoriality and the increasing reliance of mega-cities on a super exurban, post-border hinterland that is global in scope and scale. The GN can learn from the experiences of the GS as much as the GS can from the GN.

The *sustainable livelihoods framework* has been effectively used by a number of investigators working primarily in the GS (Carney, 1998; Rakodi, 2002; Gallaher *et al.*, 2015). It started as a tool in rural development planning, but it has been adapted for use in urban settings. Livelihoods are the 'capabilities, assets and activities required for a means of living'. This approach is a tool that helps identify the main factors affecting livelihoods and the relationships between them, and places targeted households at the centre of the development process. It starts with considering household capabilities and assets, rather than just their problems, and it elaborates on the various 'capitals' the poor have – physical, natural, human, financial and social. Given the strong connection between UA and the urban poor in many studies of UA in the GN, this tool has the potential to be effectively applied in the GN as a framing of UA.

Fig. 1.2. The *Stadtteilgarten Schillerkiez* community gardens at the former Tempelhof airport in Berlin, a grassroots effort to reclaim green space in the city and build community. (Photo by author, May 2015.)

1.4 Summary

In reviewing the state of research on UA it is clear that it is a field that has moved well beyond a description of food production systems to the role of UA in producing spaces of community, resistance and empowerment. UA is a way to claim a right to the city by those not necessarily seen as central to its function by those in power. This is not the way UA in the GS has been framed, but in practice it does work in this manner, and this approach has been used as a way of framing UA in the GN. UA is a way of exercising people's right to the city because of its metabolic role, especially its way of healing metabolic rifts and of being a part of an urban assemblage. UA represents a different framing and a making visible of the way an economy works, through social and other livelihood capitals, gifting, informal volunteer labour, salvage, guerilla gardening, foraging, and the growing of things in places where the norm has not placed them. In moving forward the key is to focus on processes more than traits and treat cities as more similar than different (Roy, 2011). Lastly, UA is a rich field of interdisciplinary inquiry for developing ideas and ideals for urban sustainability. This richness will be ensured if investigators work towards erasure of the GN/GS divide, truly 'worlding' this field of inquiry by informing each other. The intent of this book is to make a contribution towards that end.

1.5 Outline of the Book

This chapter serves as an introduction to the volume, providing definitions and framing of the theoretical work. From here we move to two general chapters that provide focused overviews. In Chapter 2, White and Hamm discuss the role that urban agriculture plays in the broader context of urban food systems, especially in the Global South, arguing that UA should be an accepted urban livelihood and fully integrated into urban processes. Gray et al. in Chapter 3 consider some of the barriers and benefits of UA, particularly in the form of community gardens in the Global North. Chapter 4 deepens our knowledge about community gardens, as Parece and Campbell discuss a survey they conducted in the US. The next few chapters develop a particular theme, illustrating it through rich case material. In Chapter 5, Bosco and Joassart-Marcelli, via an empirical study of community gardens in San Diego County, unpack and challenge the broadly accepted idea of 'community', which is often a highly contested notion, though taken for granted in the UA literature. Broadstone and Brannstrom, in Chapter 6, illustrate the challenges and difficulties in consistently mobilizing labour for UA through an empirical study in Houston. In Chapter 7, Bellwood-Howard and Nchanji take a close look at how products from urban agriculture are marketed in greater Accra, Ghana, via complex networks. Through a comparison of different community gardens in Greater Springfield, Massachusetts, LeDoux and Conz in Chapter 8 demonstrate the enabling power of community gardening as a form of social justice. In Chapter 9, Lowell and Law consider how the concept of sustainability is or is not fully engaged in UA with a comparative study of Austin, Texas and Havana, Cuba. In Chapter 10, Byrne et al. consider how UA maintains ecosystem services in Australian community gardens, while in Chapter 11 White illustrates how UA creates greater resilience in urban systems with a case from Senegal.

The next few chapters illustrate less expected elements of UA. McLees in Chapter 12, building on material from Tanzania, situates UA as urban process, considering the intersections, movements and flows in the continual processes of remaking that which defines and makes different. Hammelman, in Chapter 13, links UA as a survival strategy for refugee women who at the same time challenge the concept and notion of the ideals of a world city in Medellin, Colombia. Gallaher, in Chapter 14, shows how just a small amount of UA in one of the most densely populated urban places on earth can provide green respite for its residents. In Chapter 15, Hagolani-Albov and Halvorson share a highly innovative variation on urban community supported agricultural practices in Finland. In Chapter 16, Dryburgh, in a creative application of Foucault, deliberates the role of bees in the city, specifically in Washington DC. Horowitz and Liu in Chapter 17 take us to the challenging urban scape of China, where UA is practised on apartment balconies of the very newly urbanized residents, while structural limitations make UA illegal, yet so critical. Mujere, in Chapter 18, illustrates how perceived political affiliation can undermine the success of UA with a case from Zimbabwe. In Chapter 19, Chan et al. elaborate further on the idea of UA as a form of resilience by considering how community gardens in three very different locations are a form of socio-ecological refugia. The book concludes with a chapter that brings together the various elements illustrated in the chapters and poses some considerations for the future.

References

Ackerman, K., Conrad, M., Culligan, P., Plunz, R., Sutto, M.P. and Whittinghill, L. (2014) Sustainable food systems for future cities: the potential of urban agriculture. *The Economic and Social Review* 45(2), 189–206.

Airress, C.A. and Clawson, D.L. (1994) Vietnamese market gardens in New Orleans. *The Geographical Review* 84(1), 16–31.

Angelo, H. and Wachsmuth, D. (2014) Urbanizing urban political ecology: a critique of methodological city-ism. *International Journal of Urban and Regional Research*. DOI: 10.1111/1468-2427.12105.

Bassett, T.J. (1981) Reaping on the margins: a century of community gardening in America. *Landscape* 25(2), 1–8.

Brenner, N., Madden, D. and Wachsmuth, D. (2011) Assemblage urbanism and the challenges of critical urban theory. *City* 15, 225–240.

Carney, D. (ed.) (1998) Sustainable rural livelihoods: what contribution can we make? Department for International Development, London, UK.

Christie, M.E. (2004) The cultural geography of gardens. *The Geographical Review* 94(3), iii–iv.

Colasanti, K.J., Hamm, M.W. and Litjens, C.M. (2012) The city as an 'agricultural powerhouse'? Perspectives on expanding urban agriculture from Detroit, Michigan. *Urban Geography* 33(3), 348–369.

Despommier, D. (2010) *The Vertical Farm: Feeding the World in the 21st Century*. Picador, New York, USA.

Egger, M.C. (2007) Cultivating social capital: community gardens in Lansing, Michigan. MA thesis, Michigan State University, East Lansing, Michigan, USA.

Gallaher, C.M. (2012) Livelihoods, food security and environmental risk: sack gardening in the Kibera slums of Nairobi, Kenya. PhD dissertation, Michigan State University, East Lansing, Michigan, USA.

Gallaher, C.M., Kerr, J., Njenga, M., Karanja, N. and WinklerPrins, A.M.G.A. (2013a) Urban agriculture, social capital, and food security in the Kibera slums of Nairobi, Kenya. *Agriculture and Human Values* 30, 389–404.

Gallaher, C.M., Mwaniki, D., Njenga, M., Karanja, N. and WinklerPrins, A.M.G.A. (2013b) Real or perceived: the environmental health risks of urban sack gardening in Kibera slums of Nairobi, Kenya. *EcoHealth* 10, 9–20.

Gallaher, C.M., WinklerPrins, A.M.G.A., Njenga, M. and Karanja, N.K. (2015) Creating space: sack gardening as a livelihood strategy in the Kibera slums of Nairobi, Kenya. *Journal of Agriculture, Food Systems and Community Development* 5(2), 155–173. DOI: 10.5304/jafscd.2015.052.006.

Galt, R.E, Gray, L.C. and Hurley, P. (2014) Subversive and interstitial food spaces: transforming selves, societies, and society-environment relations through urban agriculture and foraging. *Local Environment* 19(2), 133–146.

Gray, L.C. (2015) Urban agriculture's benefits and barriers in California's Silicon Valley. Paper presentation, annual meeting of the Association of American Geographers, Chicago, Illinois, USA.

Gray, L., Guzman, P., Glowa, K.M. and Drevno, A.G. (2014) Can home gardens scale up into movements for social change? The role of home gardens in providing food security and community change in San Jose, California. *Local Environment* 19, 187–203.

Heynen, N., Kaika, M. and Swyngedouw, E. (eds) (2006) *In the Nature of Cities: Urban Political Ecology and the Politics of Urban Metabolism*. Routledge, New York, USA.

Jarosz, L. (2008) The city in the country: growing alternative food networks in metropolitan areas. *Journal of Rural Areas* 24, 231–244.

Keil, R. (2003) Urban political ecology. *Urban Geography* 24(8), 723–738.

Keil, R. (2005) Progress report – urban political ecology. *Urban Geography* 26(7), 640–651.

Kimber, C. (2004) Gardens and dwelling: people in vernacular gardens. *The Geographical Review* 94(3), 263–283.

McClain, R.J., Hurley, P.T., Emery, M.R. and Poe, M.R. (2014) Gathering 'wild' food in the city: rethinking the role of foraging in urban ecosystem planning and management. *Local Environment* 19(2), 220–240.

McClintock, N. (2010) Why farm the city? Theorizing urban agriculture through a lens of metabolic rift. *Cambridge Journal of Regions, Economy and Society* 3, 191–207.

McClintock, N. (2014) Radical, reformist, and garden-variety neoliberal: coming to terms with urban agriculture's contradictions. *Local Environment* 19(2), 147–171.

McFarlane, C. (2011a) The city as assemblage: dwelling and urban space. *Environment and Planning D: Society and Space* 29, 649–671.

McFarlane, C. (2011b) Encountering, describing and transforming urbanism. *City* 15, 731–739.

Moore, S. (2006) Forgotten roots of the green city: subsistence gardens in Columbus Ohio, 1910–1935. *Urban Geography* 27(2), 174–192.

Murrieta, R.S.S. and WinklerPrins, A.M.G.A. (2003) Flowers of water: homegardens and gender roles in a riverine Caboclo community in the Lower Amazon, Brazil. *Culture and Agriculture* 25(1), 35–47.

Opitz, I., Berges, R., Piorr, A. and Krisker, T. (2016) Contributing to food security in urban areas: differences between urban agriculture and peri-urban agriculture in the Global North. *Agriculture and Human Values* 33, 341–358.

Pearson, L.J., Pearson, L. and Pearson, C.J. (2010) Sustainable urban agriculture: stocktake and opportunities. *International Journal of Agricultural Sustainability* 8, 7–19.

Rakodi, C. (2002) A livelihoods approach: conceptual issues and definitions. In: Rakodi, C. and Lloyd-Jones, T. (eds) *Urban Livelihoods: A People-Centred Approach to Reducing Poverty*. Earthscan, London, UK, pp. 3–22.

Roy, A. (2011) Urbanisms, worlding practices, and the theory of planning. *Planning Theory* 10(6), 6–15.

Safransky, S. (2014) Greening the urban frontier: race, property, and resettlement in Detroit. *Geoforum* 56, 237–248.

Shillington, L.J. (2013) Right to food, right to the city: household urban agriculture, and socionatural metabolism in Managua, Nicaragua. *Geoforum* 44, 103–111.

Swyngedouw, E. and Heynen, N.C. (2003) Urban political ecology, justice and the politics of scale. *Antipode* 35(5), 898–918.

Thebo, A., Drechsel, P. and Lambin, E. (2014) Global assessment of urban and peri-urban agriculture: irrigated and rainfed croplands. *Environmental Research Letters* 9, 114002.

Wachsmuth, D. (2012) Three ecologies: urban metabolism and the society–nature opposition. *The Sociological Quarterly* 53, 506–523.

White, M.M. (2011) D-town farm: African American resistance to food insecurity and the transformation of Detroit. *Environmental Practice* 13(4), 406–417.

White, S.A. (2014) Cultivating the city: exploring production of place and people through urban agriculture. Three studies from M'Bour, Senegal. PhD dissertation, Michigan State University, East Lansing, Michigan, USA.

White, S.A. (2015) A gendered practice of urban cultivation: performing power and well-being in M'Bour, Senegal. *Gender, Place and Culture* 22(4), 544–560.

WinklerPrins, A.M.G.A. (2002) Linking the urban with the rural: house-lot gardens in Santarém, Pará, Brazil. *Urban Ecosystems* 6, 43–65.

WinklerPrins, A.M.G.A. (2006) Urban house-lot gardens and agrodiversity in Santarém, Pará, Brazil: spaces of conservation that link urban with rural. In: Zimmerer, K.S. (ed.) *Globalization and New Geographies of Conservation*. University of Chicago Press, Chicago, Illinois, USA, pp. 121–140.

WinklerPrins, A.M.G.A. and de Souza, P.S. (2005) Surviving the city: urban home gardens and the economy of affection in the Brazilian Amazon. *Journal of Latin American Geography* 4, 107–126.

WinklerPrins, A.M.G.A. and de Souza, P.S. (2009) House-lot gardens as living space in the Brazilian Amazon. *FOCUS on Geography* 52, 31–38.

WinklerPrins, A.M.G.A. and de Souza, P.S. (2010) The diversity and circulation of plants in urban homegardens, Santarém, Pará, Brazil. *Boletim do Museu Paraense Emílio Goeldi – Ciências Humanas* 5(3), 493–507.

Zezza, A. and Tasciotti, L. (2010) Urban agriculture, poverty, and food security: empirical evidence from a sample of developing countries. *Food Policy* 35(4), 265–273.

2 A View from the South: Bringing Critical Planning Theory to Urban Agriculture

Stephanie A. White* and Michael W. Hamm
Michigan State University, East Lansing, Michigan, USA

2.1 Introduction: Re-framing Urban Agriculture

This chapter discusses the role that urban agriculture (UA) plays in urban food systems and how theoretical framings of UA that draw attention to it as an 'urbanistic practice', both constituting and constituted by urban assemblages, offer new directions for research. We argue that studies of urban agriculture can be 'put to work' to:

1. develop more accurate understandings of regional and city food provisioning and exchange, especially in relation to informality;
2. shed light on urban socio-ecological processes and relationships, including those that reproduce food insecurity, poverty and social marginalization;
3. provide case-study accounts of peripheral livelihoods that challenge 'conventional understandings about how the city is put together' (Simone, 2010, p. 39). More nuanced and deeper appreciations of UA, in turn, should inform more conscientious and place-based efforts to link food and income security to urbanization processes in the Global South, which in many cases are 'de-linked from the growth of formal economies and employment opportunities' (Duminy et al., 2014, p. 8; McClintock, 2014).

Urban agriculture studies in the Global South have tended to focus on its material output and its remedial effects on urban material deprivation (Slater, 2001; Webb, 2011). Such studies have been critiqued based on the idea that they construe UA as 'space into which development projects can be inserted in the future' (Page, 2002, p. 41) or, according to Webb (1998a), 'a case of the development discourse creating a niche for itself' (p. 201). In addition, UA is generally studied in isolation from other urban processes, which tends to obscure its *urbanness*, thus fostering a perception of it as a misplaced rural activity. Such approaches to UA have inevitably led to overly simplistic prescriptions of UA as a pathway out of poverty, but as Hovorka (2006) notes, 'outright promotion of UA . . . is not necessarily simple, nor desirable' (p. 52). Indeed, such outright promotion in the context of 'bootstraps' neoliberal development thinking may work to relieve municipal officials of their responsibilities to serve urban populations (Rakodi, 1985; Sanyal, 1987; Hovorka, 2006).

In Chapter 1 of this volume, WinklerPrins asserts that a more global understanding and theorizing of urban agriculture as one that recognizes the convergence between the Global North (GN) and Global South (GS) – can serve a

---

*Corresponding author; e-mail: whites25@msu.edu

project to generate more sustainable cities. Urban agriculture, because of its potential to green the city in productive ways, seems to be particularly well suited to a sustainability agenda. But, there is a danger to conferring sustainability credentials to UA without working out how and why it exists in particular places and whom it serves. In other words, it is necessary to qualify its role in particular places and in relation to urban assemblages to determine its role in 'actually existing sustainabilities' (Evans and Jones, 2008, p. 1417).

Postcolonial and critical urban planning theory offer promising theoretical frames for putting UA to work by improving researchers' ability to connect it to 'the assemblages of urban life and its multiple flows' (Chapter 1, this volume). Among critical planning theorists who bring southern perspectives to debates about the production of urban space, there is a unifying recognition that before cities can be remade to serve the goals and aspirations of all their inhabitants, we must remake them in our own imaginations. Roy (2011) notes the importance of theory that meaningfully engages with 'the human condition of much of the contemporary world', which is marked by 'the materiality of late capitalism and its political closures and openings' (p. 7). As an *urban* livelihood practised in many cities by many different kinds of people, studies of urban agriculture have the potential to contribute to this reimagination by shedding light on the numerous ways people navigate and survive the city. Reimagining UA also means recognizing how Robinson's 'regulating fictions' and Roy's 'referenced urbanisms' are used to privilege some claims to space, while criminalizing others, and to more fully engage with informality as a mode of development that is likely to characterize the emergence and workings of cities well into the future (Roy, 2005; Robinson, 2006; Roy, 2011, p. 10). By regarding UA research as engagements with 'actually existing metropolitan complexities and ambitions', one can begin to see UA, not as temporary pathology or universally beneficial, but, rather, as an urban *livelihood* that produces, and is produced by, cities that are 'at once heterogeneously particular and . . . irreducibly global' (Ong, 2011, pp. 4 and 8). In practical terms, putting UA to work means employing methods that 'permit a degree of flexibility and innovation in the choice and use of analytical concepts, the use of very different types of data, as well as a high degree of contextual adaptability' (Duminy *et al.*, 2014, pp. 8–9). In short, it means discerning the contemporary *urban-ness* of urban agriculture in ways that have the potential to pull UA beyond developmentalist fictions of self-sufficiency and neoliberal imaginaries of modernity, and into contemporary 'ordinary' and lived realities (Robinson, 2006, p. 11).

The remainder of this chapter provides a foundation for regarding UA differently based on recent research on urban food systems and food security. Much of the discussion is based on recent scholarship in sub-Saharan Africa, but the same broad urbanization/globalization dynamics that influence locally specific articulations in African cities also shape those in Asia, Europe and the US. It may be useful to think of city dynamics in relation to ideas proposed by Comaroff and Comaroff's (2012) notion of 'living politics', which posits that the South is the vanguard of the new capitalist world order, or Simone (2010), who refers to an 'anticipatory urban politics', whereby making a life in the city requires practising the 'art of staying one step ahead of what might come' (p. 62). Such observations and reframing of what contemporary urbanity entails provide the impetus for closer attention to shared experience; urban agriculture, as a widely practised global–local phenomenon, provides one entry point for improving understanding of 21st-century city life.

2.2 Urban Agriculture as a Component of Urban Food Systems

Urban agriculture is a persistent feature of cities everywhere in the world and often occurs informally and opportunistically, in the 'in-between spaces' of towns and cities. Thebo *et al.* (2014) estimate that 456 million hectares, an area about the size of the European Union, are devoted to urban and peri-urban agriculture, and that 80% of this cropped area occurs in low- to middle-income countries. UA has received enormous interest from development organizations, specifically in relation to improving urban food security and income-generation for the poorest urban residents. However, recent research has urged caution towards this generally celebratory view of UA based on a number of urban realities:

1. poor people do not often have access, or only insecure access, to productive resources, such as land and water, in an urban environment;
2. municipal officials are often antagonistic towards urban agriculture for a number of reasons (e.g. issues of hygiene and safe production of food, a perceived backwards use of city space, urban cultivator opportunist use of land), and, therefore, the livelihood security of urban cultivators is always tenuous;
3. UA produces only a small amount of the *aggregate* food needs of urban populations;
4. urban food insecurity is not due to a lack of food, but the inability to *access* food (Binns and Lynch, 1998; Crush *et al.*, 2011, 2012; Webb, 2011).

From a wholly 'productivist' framing, such critiques weigh heavily against urban agriculture and associated advocacy efforts. Recent research from southern Africa argues that a more enlightened approach to food insecurity would focus on the factors that prevent people from accessing food, which are economic, political and social, and may range from issues related to inequality to geographic location to cooking methods (Ellis and Sumberg, 1998; Crush *et al.*, 2010, 2011, 2012; Satterthwaite *et al.*, 2010; Webb, 2011; Battersby, 2012a). As Jane Battersby (2012b) notes, 'In urban areas, the need to consider food accessibility, stability, and utilization is ultimately as great as the need to consider food availability' (p. 38). In addition, an over-reliance on urban agriculture, which is often interpreted as a measure to improve 'self-sufficiency', particularly in relation to women, risks relieving urban officials of their duties to respond to the needs of the most marginalized urban residents by paying attention to and addressing urban processes that hinder people and their abilities to access food (Rakodi, 1985; Hovorka, 2006).

Such cautions are important because they temper the expectations that often surround UA. But, rather than wholly eliminating UA from discussions on urban food security, such cautions should prompt advocates to clearly link it to more complicated and grounded understandings of urban food environments. In much of the Global South, and in contrast to the highly capitalized and consolidated food provisioning systems in the Global North, urban food provisioning and exchange is an idiosyncratic urbanistic practice, comprising market and non-market sourcing strategies. It is highly contingent on individual household constraints and opportunities, as well as on claims to urban space (Battersby-Lennard and Haysom, 2012; Battersby and Crush, 2014). Urban agriculture is but one component of a complex food system, practised in various ways, at various scales depending on the goals, opportunities and constraints of urban cultivators. Understanding UA as an *integrated* dimension of an urban food system suggests that research or policy efforts regarding UA must be done *in relation to* the socio-ecological processes that comprise food systems. Several concepts that have proved helpful in framing food production and exchange in relation to urban socio-ecological processes can provide starting points for examining how urban agriculture is relevant to daily urban life.

2.2.1 Definition: urban and peri-urban agriculture

For the sake of simplicity, urban agriculture (UA) and peri-urban agriculture (PUA) are collectively referred to as UA. Peri-urban agriculture generally refers to agriculture that takes place on the outskirts of a town or city, but what counts as UA or PUA is fluid. As cities grow, the particular factors that influence how UA is practised will change and will be contingent on how place and people urbanize. In general, it is helpful to think of UA as a distinctly *urban* livelihood (rather than a rural livelihood 'misplaced') that takes shape as a result of the socio-ecology in which it is found. This means that urban cultivators integrate and shape their production and exchange practices in ways contingent on urban processes as well as personal preferences. For example, in order to distribute their produce, city farmers may develop relationships with multiple and diverse agents of distribution, and, as a result, may be able to access various types of markets, as Chapter 7, this volume, illustrates. As an informal urban livelihood, it may be one income-generating activity among several that the cultivator practises (Owusu, 2007). In addition, as a practice that 'integrates' with the environment, urban cultivators may often use urban 'natural resources', such as compost from city dumps or nutrient-rich (but bacteria-laden) effluent, as production inputs.

2.2.2 Definition: food systems

A conventional definition of food systems conceives of them as a *set of activities* involving food, from production to consumption. Ericksen (2008) suggests a broader definition that draws attention to the political, social and environmental dimensions of food systems based on the idea that food systems are *embedded* within societies and environments, and are thus shaped by political, social and ecological factors. Therefore, she proposes a definition that includes the following.

- The relationships between and within social and ecological environments that comprise food provisioning processes and practices, as well as the practices themselves.
- The outcomes of these processes and practices, and their effect on social and ecological environments, such as improved food security, pollution and social welfare, including economic development.
- Other determinants of food security, such as policies that permit raising urban live stock, deal with food safety, and/or the quality of infrastructure relative to the food systems' processes and practices.

Such a definition compels attention to gender relationships, cultural food beliefs and practices, spatial dimensions of the city that may inhibit or enable access, infrastructure, policies and practices that determine the quality of food, and policies or other relationships that govern production and exchange practices within the urban environment. This definition also suggests the need to address factors at multiple scales, such as national and international policies that govern trade, and across space, such as rural–urban transfers, or the location of urban markets.

2.2.3 Definition: food environments

The idea of 'food environments' draws attention to the idea that peoples' eating practices and their capacity to acquire food are *situated in* and *affected by* the places, or environments, in which they live. For example, the recent neologism, 'obesogenic environments' is increasingly used to draw attention to the multiple factors that promote obesity (Egger and Swinburn, 1997; Wooten *et al.*, 2015). The terms 'food deserts' and 'food swamps', which are more often used to explore food environments in the GN, help to perceive how space and economy play a constitutive role in people's food access (Beaulac *et al.*, 2009; Rose *et al.*, 2009). Battersby and Crush (2014) adapt the concept of food deserts to explore the social and political complexities of food provisioning in an urban African context. Such conceptual frames enable analyses that link eating practices to health, and show how political and economic marginalization is the result of certain food regimes. Such a concept helps to open analyses to considering a broad range of factors that prevent or enable access to food, as well as how bodies are *created by* the social and ecological environments in which they reside (Guthman, 2011).

2.3 Urban Agriculture as an Urban Process

Leveraging urban agriculture towards improved food security requires understanding obstacles and opportunities in the context of food environments. The ways in which people experience food environments are highly variable and depend on the intersection of multiple factors, such as spatial location, political and economic dynamics at multiple scales, livelihood, gender, socio-economic status, age, and many other factors. Without such an understanding, there is a real danger that the productivist approach to food security that dominates rural development will be uncritically adopted in urban areas (Crush *et al.*, 2011).

What is required, rather, is attention to the social, economic and ecological dimensions of food provisioning *in relation to* production, which can enable analyses that embed UA within the larger food environment. Such a perspective helps to move UA beyond simplistic advocacy efforts that present its only value in materialistic terms, and reframes it as a *social process* that is characterized and influenced by the same issues that make surviving the city challenging. In the following sections, a number of frames are presented that can help to 'unpack' urban agriculture in order to better understand how it integrates with urban processes.

2.3.1 Gender and women in urban agriculture

Hovorka (2013) highlights the value of conceptualizing gender and power through 'feminist foodscapes', and notes 'the urgent need to generate insightful, comprehensive findings on how food "works" in African cities' (p. 124). Perhaps because food is so mundane, the ways in which its production, exchange, preparation and eating are gendered enactments of power and privilege have gone unexamined. But, shedding light on these relationships and practices of power 'tells us things about what is valued in society and how that becomes embedded in relation to food' (Hovorka, 2013, p. 125).

In much of the celebratory literature about UA, authors often note the importance of women in urban food production. However, as Hovorka (2006) and Crush *et al.* (2011) point out, this is not the same as understanding the factors that may cause women to take up urban agriculture in the first place, and whether or not it represents a desperate measure of survival or an emancipatory measure of self-reliance. A number of authors caution against assuming that farming always produces positive outcomes for women, and suggest that unless it is accompanied by an emancipatory agenda, urban agriculture projects targeted at women may serve only to reproduce the conditions that limit their opportunities and, thus, perpetuate their oppression (Rakodi, 1985; Hovorka, 2005, 2006). As Hovorka *et al.* (2009) note, 'women are in the majority among urban farmers around the world, but they tend to predominate in subsistence farming, whereas men play a greater role in urban food production for commercial purposes' (p. 5). An emancipatory agenda, for example, might seek to improve the commercial opportunities for women by leveraging and improving their existing skills in the subsistence sector or addressing prejudicial land tenure laws.

In both the GN and GS access to food is more problematic for those who are socially and structurally disempowered. Livelihoods that improve social standing and social networks may not only improve food security, but also fundamentally change the structural factors that reproduce poverty and marginalization (Slater, 2001; Twyman and Slater, 2005; WinklerPrins and de Souza, 2005; Battersby-Lennard and Haysom, 2012; Gallaher *et al.*, 2013; LeDoux and Conz, Chapter 8, this volume). Hovorka (2006) explores some of the ways in which urban agriculture can enable political and social empowerment of women. For example, one subsistence chicken producer was able to take advantage of a government programme that allocated land for chicken production, and enrolled in courses in agricultural production at the Botswana College of Agriculture. As a result, she expanded her production and eventually began earning a good income that allowed her to live in a middle-income bracket. Gallaher *et al.* (2013) demonstrate how sack gardening improved social capital, especially if carried out collectively, which enabled a measure of resistance to food insecurity by poor women in the Kibera slums of Nairobi.

2.3.2 The heterogeneity of urban agriculture reflects the variability in cities

In many cities of the Global South people typically pursue multiple livelihoods (Owusu, 2007), and farming for income or food is one livelihood that can be practised in association with other livelihoods. How one practises and what one grows is contingent on a range of factors that determine how a cultivator can benefit (see Table 1.1 in Chapter 1): the location of the cultivated space, other obligations, market demand (which is seasonally, socially and spatially contingent), access to inputs (the poorer one is, the more difficult it is to access inputs, land and information), social connections, and many other factors. People may grow on plots within or close to the home either on public or private land (Ellis and Sumberg, 1998). To a great extent, urban farming is done on land that has not yet been developed, or which cannot be developed due to various environmental concerns, such as recurrent flooding. Farmers very often do not have legal land access, so that urban agriculture as a livelihood is often precarious and limited (Crush et al., 2011). In other cases, farmers may make arrangements with landowners or public institutions to farm portions of land that have yet to be developed. Limited access to water may prevent year-round cultivation activities. Where one cultivates may raise associated public health or land-tenure issues, both of which are potentially important to municipal decision makers and planners.

Although recent work shows that urban agriculture is not as important to food security as is sometimes portrayed, it is important to realize that there is wide variation among cities (Battersby, 2011; Crush and Frayne, 2011; Crush et al., 2011, 2012; Battersby, 2012a). In some places, UA appears to be critical to mitigating food insecurity, even if it does not solve it. Table 2.1 demonstrates variation among 11 cities. It also shows that in cities where individual rural–urban food transfers are more important, urban agriculture tends to be less important. Rural–urban food transfers refers to the private exchange of food between individual households, and represents an important source of food, especially among the food insecure (WinklerPrins and de Souza, 2005; Frayne et al., 2010).

Variability of urban agriculture is also the result of the farmers' objectives. Webb (1998b) notes that most urban agriculture studies mention urban crops in only the most cursory manner, and therefore miss a critical dimension of the ways in which cultivators ascribe importance to particular plants, and how they strategize and plan production or distribute their time and effort. To remedy such an oversight, Webb developed an index that he used to analyse the relative importance of various crops to household diets, the monetary value of those crops, and how the savings achieved by home growing was diverted to other household needs (1998b). The index revealed the underlying reasons for growing particular plants, which were not always related to economic issues but were nonetheless important to urban cultivators. Such indices, because they are 'more intimately connected with household welfare' might be used, for example, to better assess crops of local importance and to develop technical support measures.[1]

2.3.3 The impact of global processes on city food provisioning

Most studies of urban agriculture do not theorize or address how urban food provisioning is affected by global processes, though these processes will become critically important to understand as city food systems become increasingly affected by global economic and environmental dynamics. For example, the prevalence of urban agriculture appears to increase in times of economic distress in both the GN and GS, such as those created by war (e.g. victory and depression-era gardens in the US), urban decay (e.g. the bourgeoning urban farms in Detroit), economic embargoes (e.g. urban agroecology in Cuba) and structural adjustment (Maxwell, 1995; Page, 2002; Hansen and Vaa, 2004; Owusu, 2007; Koont, 2009; McClintock, 2010). Battersby-Lennard and Haysom (2012) argue that urban food production may be important to dealing with the effects of economic uncertainty, financialization of commodity markets, and climate change, especially for the urban poor, and that preserving and protecting agricultural land in close proximity to the city is preferential to land development that is more conventionally urban. Such perspectives draw attention to the ways in which use of urban space is eminently political and tied to factors in the global political economy.

Table 2.1. Urban agriculture and rural–urban food transfers (% of households). (Derived from Crush et al., 2012, p. 286.)

	Urban agriculture	Rural–urban transfers of food
Windhoek	3	72
Gaborone	5	70
Msunduzi	30	15
Johannesburg	8	24
Cape Town	4	14
Manzini	9	53
Maseru	47	49
Blantyre	63	38
Harare	60	37
Maputo	22	23
Lusaka	3	39

Thus, the pattern to emerge from the research is that urban agriculture articulates with city processes in diverse ways. These processes are not spatially confined to the geographic borders surrounding the city. The reasons that people engage in urban agriculture, as well as the reasons underlying food security, will have explanations that lie well outside the spatial borders of the city. Globalization and economic liberalization, for example, affect the ways in which people can obtain food, the price of land, or access to inputs. Although spatial boundaries and urban spatial management are clearly important to how UA can be carried out and to what degree it can contribute to food security, analyses must also consider processes beyond the city to explain and understand the value of UA in particular places. For example, many urban citizens rely on food remittances from relatives in rural areas. The food-security status of a household that relies heavily on these remittances may differ substantially from their next-door neighbour, who relies more heavily on food produced in the city. As regions become more urbanized, the close relationships with rural families will decrease, which is likely to increase urban food insecurity. Addressing this changing food provisioning dynamic will require attention, and may be at least partially addressed through increased proximate food production that aims to meet the needs of poor and marginalized urban citizens.

2.4 Linking Food Security and Urban Agriculture Differently

For many people in the world, in both the GS and the GN, food security should be thought of as a dynamic process or a continuum, along which households are constantly shifting according to numerous variables, such as season or household members' employment status. That is, there are degrees to which households are food insecure, as well as different factors among and between households that shape food insecurity. In environments where food provisioning can be quite an uncertain and tenuous endeavour, small amounts of income or food matter, and there is strong evidence that UA can move people along the continuum to improved food security, while not necessarily wholly eliminating it (Battersby-Lennard and Haysom, 2012; Gallaher *et al.*, 2013). For example, in their study of the Philippi Horticultural area of Cape Town, South Africa, Battersby-Lennard and Haysom (2012) found that though many of the families who worked for or bought food from Philippi could still be considered food insecure, their food insecurity status would be more dire had they not had this source of food.

The complex and multifaceted nature of urban food systems helps to understand urban agriculture differently. It is neither a panacea nor insignificant. Rather, it is one food-provisioning practice among many, occurring within complex and idiosyncratic urban food-provisioning systems that help to mitigate household food insecurity. It also represents a claim to urban space that can be disproportionately beneficial to poorer urban residents. What is produced *can* be critically important to urban residents in terms of income and food. But, as a process that integrates with other urban processes, including other components of the food system, it also has social and political dimensions that can shed light on the structural and spatial dimensions of food security. As such, including the widespread practice of urban agriculture in analyses of food insecurity can provide insight into the factors that cause people to farm the city, and into how food systems are failing or succeeding. In other words, in addition to investigating the instrumental aspects of urban agriculture and how they can be strengthened to support the food needs of urban residents, research could use urban agriculture as an entry point for understanding wider food-provisioning issues and food insecurity, which, in turn, can be used by municipal governments to inform city planning processes.

2.4.1 An urban planning research agenda: making sense of urban food systems

Well-functioning urban food systems are an important part of enabling citizen well-being and functional cities. Despite this, food concerns do not generally occupy a spot on the agendas of urban planners or municipal officials in the global North or South, and much of the research on food systems has been limited to economic analyses and the set of activities to get food from farm to bowl, i.e. growing, processing, transporting and marketing. While such analyses are important, they do not give a good and *grounded* sense of how people navigate food environments from day to day, and how, over space and time, they leverage both informal and formal practices, relationships, and markets to their advantage.

A survey conducted in 11 cities throughout southern Africa showed that 70% of households obtained their food from informal sources, with 31% doing so on a daily basis, higher than for any other food source (Frayne *et al.*, 2010). Due to the numerous avenues for acquiring food, Riley and Dodson (2014) observe that food insecurity can be 'set apart' from urban poverty 'because of the broad range of entitlements that can allow low-income households to become food secure in different contexts' (p. 228). A better understanding of non-consolidated and emergent food systems might be used to support the efforts of urban municipalities to develop planning practices that enable food security in local contexts and which focus on improving the resilience and responsiveness of the system in relation to existing entitlements.

The spatial and infrastructural dimensions of cities affect to varying degrees the ways in which people access food. For example, open-air food markets remain central to urban economic, social and cultural life in the Global South and play multiple roles in the GN. Such markets have enjoyed a renaissance and a reinvention in the UK and US, where they are referred to as farmers' markets, and persist as a vital part of social life in many parts of Europe. The location and design of these markets can have a major impact on how they are used and who is able to benefit from them, which should be considered by policy makers as cities continue to grow and become increasingly dense. In a study of spontaneous vs planned markets in Cali, Colombia, Ray Bromley (1980) found that the reasons a city government erects a market are quite different from the reasons that consumers and vendors use a market. Without an understanding of community needs and preferences, markets can fail, and thus represent a missed opportunity for enhancing the urban civic and economic environment.

The addition of supermarkets to urban spaces is seen as one way to offer lower prices and higher quality on a range of products. One effect of supermarkets in the Global North has been to centralize food access, which has resulted in the evolution of 'food deserts'. In the process of deciding where to locate, supermarket developers typically use models to determine locations offering the best profit-maximizing potential (Battersby, 2012a).[2] Such locations do not generally benefit the urban poor, who are often severely limited in their ability to travel or cannot afford regular trips on public transportation. Although consumers might pay higher prices at small neighbourhood shops, they offer at least two advantages over supermarkets in addition to convenient location:

1. they regularly offer credit to their patrons;
2. they sell things in smaller quantities than supermarkets.

This second point is of critical importance to urban consumers who typically must buy small quantities of food on a daily basis because of small incomes and lack of, or unreliable access to, electricity and/or refrigeration to preserve food for longer periods of time.

As noted earlier, *access* to food is more of a problem than overall *availability*, which draws attention to the role of poverty in food security (Frayne *et al.*, 2010; Crush *et al.*, 2012). However, the hard conceptual distinction between access and availability may mask important nuances of food provisioning that vary among cities. For example, urbanization is rapidly removing farmland from production, which threatens overall availability of healthy, accessible food. In many areas, food production occurs in cities or within very close proximity to urban areas. Because urban planning and development approaches generally do not consider food production as a concern or objective, this food production capacity may become severely constrained as urbanization proceeds.

Lastly, increasing resource scarcity, inequality and climate change will require cities and individuals to reassess current food-provisioning and exchange practices, and adapt them to changing circumstances, which Kevin Morgan and Roberta Sonnino (2010) refer to as the New Food Equation (NFE). Many 'modern' food systems, such as those that predominate in the GN, are highly energy-intensive and rely on cheap inputs of oil, and are therefore incompatible with the NFE. Such incompatibility results in 'burgeoning prices for basic foodstuffs and growing concerns about the security and sustainability of the agri-food system' (p. 209). Mitigating this instability, the authors assert, requires a fundamental reorganization of the relationships that govern food-related labour, land and capital.

2.5 Conclusion – Resiliency as a Guide

The food environments of today and tomorrow do not match the food environments of yesterday. The trajectory in 'modern' food systems has been to develop extensive food systems that rely heavily on petrochemicals, and which are based on the principle of competitive or comparative advantage, economies of scale and liberalization of markets. In an environment of cheap energy and relative indifference to the environmental effects of energy-intensive agri-food systems, this kind of extensive food system, governed by economies of scale, was more tenable. What happens when it becomes untenable? It may be that a sea-change in the objectives of the food system, which moves it from the province of 'globalization' to 'regionalization and localization in a global context', needs consideration. When considering the future of food production and access, it may be that urban agriculture makes increasingly more sense as part of a *resilient* food system. Resiliency as a guide would be more likely to call for spatially and procedurally diverse food-provisioning protocols, which would include urban agriculture. Indeed, food-sourcing strategies guided by the principle of resilience would call for increased diversity at all levels of food systems.

There is much research to be done to better understand the existing and potential role of urban agriculture in relation to urban processes and urban food insecurity. From our perspective, the following are major points to emerge from the literature and provide guidance for future research; some of these points are further developed in the chapters that follow.

1. Move research beyond a focus on UA's instrumental value. Research must better contextualize the practice of UA with other food-provisioning practices, within the entire food system, and in relation to social, political and environmental dynamics at multiple scales. In discussions of urban food security, better theorizing of 'urban' is needed, and how urban dynamics in any particular place influence food security. As Battersby (2012a) notes, in most conventional approaches, which focus only on the household, the wider structural and spatial context is written out. That is, contemporary understandings of food security remove 'the urban as an actor' (p. 146). Examining the *urbanness* of urban agriculture can provide insight into how people navigate urban places. For example, what urban 'natural' resources do urban cultivators use and what relationships do they draw on to access them?

2. The role of markets and multiple food practices and their diverse intersection with UA needs to be better understood. The concepts of resiliency and vulnerability may help to make sense of idiosyncratic, diverse and decentralized food practices in relation to the needs and entitlements of city residents. This kind of information can be used to better balance city management and the evolution of a 'modern' space with the needs and socio-economic status of urban citizens. For example, research might examine how UA is leveraged by individuals and communities to weather the uncertainties that result from climate change and global economic fluctuations. Research and theoretical approaches found in the resiliency and vulnerability literature may be especially helpful in understanding how food and agriculture dynamics across spatial and temporal scales affect the efforts of urban farmers and food access.

3. UA should be thought of as an entry point into the broader food system. Urban food provisioning and exchange is carried out through diverse practices that depend on the circumstances of individuals and households. 'Following' urban agriculture can lead to other areas of urban food networks and shed light on the relationships and practices that people use to access food.

4. Mixed research methods can shed light on the complexities and contingencies of urban food provisioning and exchange. Researchers should employ methods that analyse problems/issues from a number of scales (e.g. citywide, neighbourhood, household, individual), and a number of perspectives (e.g. disaggregated by gender, education, ethnicity, income, spatial location), and both qualitative and quantitative methods should be used (e.g. surveys, interviews, GIS, income data). Rigorous mixed methods can help to develop new understandings of, and approaches to, urban food security. For example, how do space and infrastructure combine to produce or prevent food insecurity?

5. The power dimensions of food security and urban agriculture need to be better articulated. Gaining access to city space (both physical and economic) can be a highly contested process, but is necessary for well-being. For example, how does gender intersect with the ability to practise urban agriculture, and how do men and women differentially experience or practise urban agriculture in a given place? Hovorka (2005) notes 'there are few systematic, comprehensive investigations regarding the effects of men's and women's differential socio-economic status on the net outcomes of urban agricultural systems' (p. 295). Understanding these dynamics better can provide directions for more targeted food-security planning.

To plan and implement food systems that generate improved food security, research must engage with pressures wrought by the challenges we face. Rapid urbanization and climate change will require a diversity of adaptation and mitigation strategies. Various forms of urban agriculture are likely to play an important role, but it will require moving beyond conventional understandings that are largely concerned with material output, and into more political and collaborative terrain that deals better with the challenges that poor and marginalized citizens face in their efforts to survive the city.

Notes

[1] One cautionary note – Webb's index does not disaggregate by gender or other social difference, possibly obscuring the different importance that men and women ascribe to certain crops and how they are able to access urban space to grow those crops.
[2] See Brown's Super Stores in Philadelphia for example of counter-trends.

References

Battersby, J. (2011) Urban food insecurity in Cape Town, South Africa: an alternative approach to food access. *Development Southern Africa* 28, 545–561.

Battersby, J. (2012a) Beyond the food desert: finding ways to speak about urban food security in South Africa. *Geografiska Annaler: Series B, Human Geography* 94, 141–159.

Battersby, J. (2012b) Urban food security and climate change: a system of flows. *Climate Change, Assets and Food Security in Southern African Cities*. Earthscan, Abingdon, UK.

Battersby, J. and Crush, J. (2014) Africa's urban food deserts. *Urban Forum* 25, 143–151.

Battersby-Lennard, J. and Haysom, G. (2012) *Horticultural*. AFSUN and Rooftops Canada Abri International, Cape Town, South Africa.

Beaulac, J., Kristjansson, E. and Cummins, S. (2009) A systematic review of food deserts, 1966–2007. *Preventing Chronic Disease* 6(3), A105.

Binns, T. and Lynch, K. (1998) Feeding Africa's growing cities into the 21st century: the potential of urban agriculture. *Journal of International Development* 10, 777–793.

Bromley, R. (1980) Municipal versus spontaneous markets? A case study of urban planning in Cali, Colombia. *Third World Planning Review* 2, 205.

Comaroff, J. and Comaroff, J.L. (2012) Theory from the South: or, how Euro-America is evolving toward Africa. *Anthropological Forum* 22(2), 113–131.

Crush, J.S. and Frayne, G.B. (2011) Urban food insecurity and the new international food security agenda. *Development Southern Africa* 28, 527–544.

Crush, J.S., Hovorka, A.J. and Tevera, D. (2010) *Urban Food Production and Household Food Security in Southern African Cities*. African Food Security Urban Network (AFSUN), University of Cape Town, Cape Town, South Africa.

Crush, J., Hovorka, A. and Tevera, D. (2011) Food security in Southern African cities. *Progress in Development Studies* 11, 285–305.

Crush, J., Frayne, B. and Pendleton, W. (2012) The crisis of food insecurity in African cities. *Journal of Hunger and Environmental Nutrition* 7(2/3), 271–292.

Duminy, J., Watson, V. and Odendaal, N. (2014) Introduction. In: Duminy, J., Andreasen, J., Lerise, F., Odendaal, N. and Watson, V. (eds) *Planning and the Case Study Method in Africa: The Planner in Dirty Shoes*. Palgrave Macmillan, London, UK, pp. 1–17.

Egger, G. and Swinburn, B. (1997) An 'ecological' approach to the obesity pandemic. *British Medical Journal* 315, 477–480.

Ellis, F. and Sumberg, J. (1998) Food production, urban areas and policy responses. *World Development* 26, 213–225.

Ericksen, P.J. (2008) Conceptualizing food systems for global environmental change research. *Global Environmental Change* 18, 234–245.

Evans, J. and Jones, P. (2008) Rethinking sustainable urban regeneration: ambiguity, creativity, and the shared territory. *Environment and Planning A* 40, 1416–1434.

Frayne, B., Pendleton, W., Crush, J., Acquah, B., Battersby-Lennard, J., Bras, E., Chiweza, A., Dlamini, T., Fincham, R. and Kroll, F. (2010) The state of urban food insecurity in southern Africa. *Urban Food Security Series No. 2*. Queen's University and AFSUN Cape Town, Kingston and Capetown, South Africa.

Gallaher, C., Kerr, J., Njenga, M., Karanja, N. and WinklerPrins, A.M.G.A. (2013) Urban agriculture, social capital, and food security in the Kibera slums of Nairobi, Kenya. *Agriculture and Human Values* 30(3), 1–16.

Guthman, J. (2011) *Weighing In: Obesity, Food Justice, and the Limits of Capitalism*. University of California Press, Berkeley, California, USA.

Hansen, K.T. and Vaa, M. (2004) *Reconsidering Informality: Perspectives from Urban Africa*. Nordic Africa Institute, Uppsala, Sweden.

Hovorka, A.J. (2005) The (re) production of gendered positionality in Botswana's commercial urban agriculture sector. *Annals of the Association of American Geographers* 95, 294–313.

Hovorka, A.J. (2006) Urban agriculture: addressing practical and strategic gender needs. *Development in Practice* 16, 51–61.

Hovorka, A.J. (2013) The case for a feminist foodscapes framework: lessons from research in urban Botswana. *Development* 56(1), 123–128.

Hovorka, A., Zeeuw, H.D. and Njenga, M. (eds) (2009) *Women Feeding Cities: Mainstreaming Gender in Urban Agriculture and Food Security*. Practical Action Publishing, Rugby, UK.

Koont, S. (2009) The urban agriculture of Havana. *Monthly Review* 60(1), 63–72.

Maxwell, D.G. (1995) Alternative food security strategy: a household analysis of urban agriculture in Kampala. *World Development* 23, 1669–1681.

McClintock, N. (2010) Why farm the city? Theorizing urban agriculture through a lens of metabolic rift. *Cambridge Journal of Regions, Economy, and Society* 3(2), 191–207.

McClintock, N. (2014) Radical, reformist, and garden-variety neoliberal: coming to terms with urban agriculture's contradictions. *Local Environment* 19(2), 147–171.

Morgan, K. and Sonnino, R. (2010) The urban foodscape: world cities and the new food equation. *Cambridge Journal of Regions, Economy and Society* 3, 209–224.

Ong, A. (2011) Introduction: worlding cities or the art of being global. In: Roy, A. and Ong, A. (eds) *Worlding Cities: Asian Experiments and the Art of Being Global*. Blackwell-Wiley, Chichester, West Sussex, UK.

Owusu, F. (2007) Conceptualizing livelihood strategies in African cities: planning and development implications of multiple livelihood strategies. *Journal of Planning Education and Research* 26, 450–465.

Page, B. (2002) Urban agriculture in Cameroon: an anti-politics machine in the making? *Geoforum* 33, 41–54.

Rakodi, C. (1985) Self-reliance or survival? Food production in African cities with particular reference to Zambia. *African Urban Studies* 21, 53–63.

Riley, L. and Dodson, B. (2014) Gendered mobilities and food access in Blantyre, Malawi. *Urban Forum* 25, 227–239.

Robinson, J. (2006) *Ordinary Cities: Between Modernity and Development*. Routledge, New York, USA.

Rose, D., Bodor, N., Swalm, C., Rice, J., Farley, T. and Hutchinson, P. (2009) *Deserts in New Orleans? Illustrations of Urban Food Access and Implications for Policy*. University of Michigan National Poverty Center/USDA Economic Research Service Research, Ann Arbor, Michigan, USA.

Roy, A. (2005) Urban informality: toward an epistemology of planning. *Journal of the American Planning Association* 71, 147–158.

Roy, A. (2011) Urbanisms, worlding practices, and the theory of planning. *Planning Theory* 10(1), 6–15.

Sanyal, B. (1987) Urban cultivation amidst modernization: how should we interpret it. *Journal of Planning Education and Research* 6, 197–207.

Satterthwaite, D., McGranahan, G. and Tacoli, C. (2010) Urbanization and its implications for food and farming. *Philosophical Transactions of the Royal Society B: Biological Sciences* 365, 2809–2820.

Simone, A. (2010) *City Life from Jakarta to Dakar: Movements at the Crossroads*. Routledge, New York, USA.

Slater, R. (2001) Urban agriculture, gender and empowerment: an alternative view. *Development Southern Africa* 18, 635–650.

Thebo, A., Drechsel, P. and Lambin, E. (2014) Global assessment of urban and peri-urban agriculture: irrigated and rainfed croplands. *Environmental Research Letters* 9, 114002.

Twyman, C. and Slater, R. (2005) Hidden livelihoods? Natural resource-dependent livelihoods and urban development policy. *Progress in Development Studies* 5, 1–15.

Webb, N.L. (1998a) Urban agriculture. *Urban Forum* 9, 95–107.

Webb, N.L. (1998b) Urban cultivation: food crops and their importance. *Development Southern Africa* 15, 201–213.

Webb, N.L. (2011) When is enough, enough? Advocacy, evidence and criticism in the field of urban agriculture in South Africa. *Development Southern Africa* 28, 195–208.

WinklerPrins, A.M.G.A. and de Souza, P.S. (2005) Surviving the city: urban home gardens and the economy of affection in the Brazilian Amazon. *Journal of Latin American Geography* 4(1), 107–126.

Wooten, K.J., Blackwell, B.R., McEachran, A.D., Mayer, G.D. and Smith, P.N. (2015) Airborne particulate matter collected near beef cattle feedyards induces androgenic and estrogenic activity in vitro. *Agriculture, Ecosystems and Environment* 203, 29–35.

3 North American Urban Agriculture: Barriers and Benefits

Leslie Gray,[1]* Lucy Diekmann[1] and Susan Algert[2]
[1]*Santa Clara University, Santa Clara, California, USA;* [2]*UC ANR Cooperative Extension, San Francisco, San Mateo and Santa Clara Counties, California, USA (retired)*

3.1 Introduction

Urban agricultural research in the Global North (GN) and Global South (GS) has been fairly unconnected, despite overlapping themes of food security, social capital, nutrition, urban greening and access to land (Chapter 1, this volume). One reason might be that much of urban agriculture in the GN has resulted from urban agricultural projects initiated to solve a host of social and economic ills, resulting in formalized movements and organized spaces for people to engage in urban agriculture (Bassett, 1981; Lawson, 2005).[1] In contrast, gardens in the GS are commonly planted in informal public spaces or in homes by urban dwellers seeking household food security or opportunities for income generation (Zezza and Tasciotti, 2010). Despite these differences, urban agriculture in the GN and GS are similar in that they are frequently transitory in nature. In the GN, organized projects have typically waxed and waned with economic and social crises. Lawson (2005) contends that after each successive crisis passed, urban agricultural programmes have been difficult to sustain. The interstitial and frequently illegal nature of urban agriculture in the GS also results in tenure insecurity, particularly in expanding urban areas (Lynch et al., 2001; Bryld, 2003).

Our chapter examines the recent renaissance of urban agriculture in North America, asking whether this is leading to enduring landscapes and socio-economic changes or, like past movements, is transitory in nature. Recent North American urban agricultural movements began in the 1960s with the creation of community gardens and now encompass diverse forms of production and organization with different goals and outcomes. The various spaces and places of urban agriculture reflect the environmental, health and economic justice concerns of North America's expanding urbanized population. Urban agricultural research covers a wide variety of themes, from civic engagement, community development, and food justice to ecosystem services. After a discussion of these broad themes of North American urban agriculture, we consider the impacts of and barriers to various forms of urban agriculture in Santa Clara County, the heart of Silicon Valley, an area of Northern California known for its innovative technology-based economy, but where significant economic disparities are reflected in access to different forms of urban agriculture.

---

*Corresponding author; e-mail: lcgray@scu.edu

3.2 The Varied Forms of Urban Agriculture's Resurgence

The history of North American urban agriculture is sprinkled with examples of public works ventures in vacant lots and open spaces in cities. During the First World War, for example, Liberty Gardens promoted patriotism and responded to social unrest due to lack of sufficient food. This movement spurred 5 million residents to garden (Hynes, 1996). Relief gardens during the Great Depression helped mitigate unemployment, providing both food and purpose to those who were unemployed (Bassett, 1981). Liberty Gardens were reborn during the Second World War with the development of Victory Gardens that provided 40% of American's produce by 1944 (Bassett, 1981). The modern urban agricultural movement was born in the 1960s and 1970s when the USDA developed an urban agricultural programme to address recession woes in urban areas (Hynes, 1996). Many urban agricultural movements focus on community gardens that benefit both individuals and communities rather than home gardens. This focus on community is a theme that links agriculture in the Global North to the Global South, where the social aspect of gardening is also important (Winkler-Prins and de Souza, 2005).

Urban agriculture is defined as 'the growing, processing, and distribution of food and other products through intensive plant cultivation and animal husbandry in and around cities' (Brown and Carter, 2003, p. 3). Contemporary forms of urban agriculture have multiple functions, operate at multiple scales in both public and private spaces, and involve a variety of different actors: individuals or households, collective enterprises, city departments, non-profits, and for-profit businesses (McClintock, 2014). Although the extent of various forms of urban agriculture in the US is unknown, several studies show that it is a significant land use in urban areas. A survey by the National Gardening Association (2009) found that 36 million households – slightly less than one-third of all US households – participated in food gardening in 2008 (see also Chapter 4, this volume). Of these households, 91% had gardens at home and 3% had plots in a community garden. Taylor and Lovell (2012) identified and mapped urban food gardens in Chicago, locating a total of 4648 urban agriculture sites with a combined production area of 65 acres. Urban agricultural sites can account for a significant portion of urban green space; in a study of several UK cities, gardens comprised 35–47% of total green space (Goddard et al., 2010).

Of the multiple forms of urban agriculture (Table 3.1), the most commonly studied is the community garden, a semi-public space (see, for example, Armstrong, 2000; Glover, 2004; Kingsley and Townsend, 2006; Firth et al., 2011; Surls et al., 2015; Taylor and Lovell, 2014). Community gardens are generally shared lots where urban residents can grow their own produce, usually in individually tended plots. Community gardens appear to be a persistent form of urban agriculture, with some gardens operating continuously since the 1970s. However, one of the challenges for community gardens can be securing available land or keeping land for existing gardens, particularly in cities with high property values, where there is a strong economic incentive to develop the land or convert it to other uses (e.g. Schmelzkopf, 1995; Irazabal and Punja, 2009; Drake and Lawson, 2014; see also Chapters 6 and 13, this volume). Bassett (1981) maintains that community gardens have been successful because they benefit not just individuals but entire communities through cooperation that strengthens both social and economic relationships.

Home gardens are the most common form of urban agriculture. Home gardens are individual or household gardens that generally produce for household and family consumption. While most home gardens are individually initiated, they have recently become part of the portfolio of non-profit organizations as a strategy to fight urban poverty and enhance community food security, particularly among low-income households (Gray et al., 2014).[2] Among the benefits of home gardens the proximity and ease of growing food close to home leads to higher vegetable consumption and financial savings. Kortright and Wakefield (2011) demonstrate that home gardens can enhance food security by encouraging a more nutritious diet and more healthful lifestyle, though garden skills and finding suitable land can be a barrier to growing food in urban areas. The scale of home gardens can be significant, though the number and impact of home and backyard gardens is difficult to assess given their private and often enclosed nature.

Table 3.1. Urban agriculture's multiple forms. (Adapted from McClintock, 2014.)

	Organized	Location	Purpose	Management
Home gardens	Sometimes	Backyards, front yards, containers	Household food production, landscaping, recreation	Individual or household
Community gardens	Yes	Vacant lots, parks	Food production, cultural reproduction, recreation	Municipality or programme
Non-profit urban farms	Yes	Vacant lots, rooftops	Education, food access, vocational training, youth and children's programming	Non-profit organization
For-profit urban farms	Yes	Vacant lots, warehouses, client yards, greenhouses	Food production, garden installation and maintenance	For-profit company (individual or individuals)
Institutional gardens	Yes	Schools, prisons, hospitals	Education, rehabilitation	Hired staff or volunteers
Interstitial food spaces (e.g. guerilla gardening, gleaning and foraging)	Sometimes	Sidewalk/pavement strips, alleys, city forests, backyards, front yards	Reclaiming urban spaces, food production and consumption, urban greening	Individuals or group

Using remote sensing to quantify the extent of urban agriculture in Chicago, Taylor and Lovell (2012) found that the aggregate area of home gardens far exceeds that of other forms of urban agriculture.

On urban farms, food is grown with the goal of producing for sale, donation or educational purposes. While for-profit urban farms exist, many of them are non-profit organizations with the mission of serving the community. Non-profit urban farms often operate as more collaborative enterprises, where staff and community members grow food, engage in education about agriculture and food systems and enhance community development through vocational training. Organizations frequently have the explicit goal of changing the food system. Vitiello and Wolf-Powers (2014) find that urban farms are better at achieving social goals (e.g. developing human and social capital, supplementing incomes and providing food to the hungry) than acting as a traditional economic development tool.

Of gardens that serve institutions such as schools, prisons and hospitals, school gardens are the most prevalent form. School gardens are defined as intentional plantings for the benefit of students; they range in size from a few garden beds to designated areas of a school yard. A growing body of academic studies has identified the positive effects of school gardens on children. These include improving overall academic performance, providing opportunities for science-based education, increasing nutritional and environmental awareness, and connecting children to nature (e.g. Blair, 2009). School garden programmes and curricula are usually focused more on education than production, partly because most school garden plots are small in size and the main production season does not coincide with the academic calendar. Interest in school gardens, though, has been fostered through the growth of public programmes such as Michelle Obama's Let's Move campaign[3] and concerns about health, science education and children's lack of connection to nature (Louv, 2005; Williams and Brown, 2011).

Urban agriculture also occurs in what Galt et al. (2014) call interstitial and subversive food spaces. Activities such as guerilla gardening, gleaning and foraging often take place in the interstices of the physical urban landscape – for example, in planting strips between the sidewalk and the curb, in alleys, in vacant lots, and in urban parks and forests. These disparate activities are united by the fact that they take place on abandoned or vacant urban properties where use rights are ill-defined. These activities frequently increase the aesthetic appeal of and create value in under-utilized spaces. McLain et al. (2014) describe how urban foragers reconnect urban dwellers to nature, providing products for human use and challenging commonly held ideas about humans' role in urban areas. Some of the activities under this umbrella are explicitly more political, e.g. reclaiming commons areas from urban blight, crime or even gentrification.

3.3 Themes in Urban Agriculture

Overarching themes in urban agriculture vary by academic discipline. For example, public health research on urban agriculture has focused on the benefits to health, well-being and nutrition. Other social scientists such as geographers and sociologists who have written about alternative food networks and food justice approach the issue with a different lens, focusing on issues of race, class and inequality. The discussion below is an attempt to categorize the different themes emerging from urban agricultural research, mostly from North America, but also including research from other global regions.

3.3.1 Food access, economic savings and health

Many scholars and activists in both the Global North and the Global South have approached urban agriculture as a means to increase food security and to improve health. Access to healthy food has been widely acknowledged as a problem in the US, where more than 50 million people live in households unable to obtain a safe, culturally acceptable, nutritionally adequate diet (Hamm and Bellows, 2003; Coleman-Jensen et al., 2012). Many low-income urban residents have difficulty obtaining fresh produce because it is too expensive or they lack access to full-service grocery stores or other sources of healthy foods (Gottlieb and Joshi, 2010; McClintock, 2010; Alkon and Agyeman, 2011). Urban agriculture is seen as an effective strategy for enhancing food security because producing for personal consumption or with the intent of distributing food to those in need makes it possible for some urban residents to obtain food they could not otherwise afford or find (Smit and Nasr, 1992).

Urban agriculture's ability to feed urban populations depends on its ability to increase food production in a given area. However, the productivity of various forms of urban agriculture varies widely, depending on environmental conditions, crop choice, scale of production and other factors (Duchemin et al., 2008; McClintock, 2014). Studies from different US cities illustrate this variability: community gardeners produced an average of 1.4 lb of vegetables per square foot in Philadelphia (Vitiello and Nairn, 2009), an average of 0.75 lb of vegetables per square foot in San Jose (Algert et al., 2012), and an estimated 0.44 lb of vegetables per square foot in New York City (Farming Concrete, 2012). Both home and community gardeners have reported significant cost savings by substituting garden-grown produce for store-bought foods (Wakefield et al., 2007; Carney et al., 2012; Gray et al., 2014), as well as improved access to fresh foods (Armstrong, 2000; Kortright and Wakefield, 2011; Carney et al., 2012). Increased access to nutritious food often extends beyond the gardeners themselves as many report sharing extra produce with friends, relatives or community organizations (Blair et al., 1991; Vitiello and Nairn, 2009; Gray et al., 2014). This theme of food sharing also emerges from urban agricultural research in the Global South (WinklerPrins and de Souza, 2005).

The most commonly cited health benefits of urban agriculture are increased consumption of fruits and vegetables, increased physical activity, and reduced stress and enhanced psychosocial well-being (Armstrong, 2000; Brown and Jameton, 2000; Twiss et al., 2003; Alaimo et al., 2008; Litt et al., 2011; Carney et al., 2012). Urban agriculture has also contributed to behaviour changes; home gardeners in Toronto reported that their gardening experiences made them more likely to purchase fresh produce than processed foods and produce that was in season (Kortright and Wakefield, 2011).

3.3.2 Education, civic engagement and community development

As sites where gardeners interact with and learn from fellow gardeners and the natural world, urban gardens have been associated with an interest in renewing urbanites' connection with nature, educating urban populations, and promoting democratic self-improvement (Lawson, 2005). The education goals of urban agricultural projects vary from imparting practical agronomic knowledge to building civically engaged citizens (Lawson, 2005). As spaces where the natural world interfaces with the metropolis, urban agricultural projects provide the opportunity to have a hands-on connection with the soil, plants and elements. Many city dwellers are interested in experiential-based classes on topics such as composting with worms, natural pest management, seed-saving and preserving foods.

Gardens can also be democratic spaces for civic engagement (Lawson, 2005). By providing urban residents with the opportunity to learn about the food system and to take part in local decision-making processes, urban agriculture can encourage participants to become more engaged citizens, not just better-informed consumers (Travaline and Hunold, 2010). Urban garden projects have been praised for their ability to bring together people from different cultural, socio-economic or age groups to work for mutual gain. By bringing together diverse groups of people, sharing agricultural and culinary knowledge and creating stronger bonds in the community, these projects can highlight the social relations behind the production of food (McClintock, 2010). Participation in community gardens can empower some members to become more active in their communities (Blair et al., 1991; Armstrong, 2000; Saldivar-Tanaka and Krasny, 2004; Wakefield et al., 2007).

Finally, gardens can serve as important cultural and community spaces. Community gardeners have increased their interactions with other community members, making new social connections and strengthening existing social ties (Glover, 2004; Alaimo et al., 2010). Indeed, some community gardeners value gardens more as sites for social and cultural gatherings than as sites of agricultural production (Saldivar-Tanaka and Krasny, 2004). Community gardens have also been shown to provide participants a connection to their cultural heritage, in particular helping immigrants to maintain farming traditions and uphold traditional food ways in their new communities (Schmelzkopf, 1995; Corlett et al., 2003; Baker, 2004; WinklerPrins and de Souza, 2005; Carter et al., 2013).

3.3.3 Food justice and critical food studies

Do groups involved in urban agricultural projects go on to mobilize politically around broader food and economic justice issues? Alkon and Agyeman's (2011) edited volume *Cultivating Food Justice* provides many case studies that demonstrate how marginalized communities work to create sustainable and just food systems. For example, Morales (2011) examines food justice organizations in Milwaukee that aim to promote racial equality by incorporating an anti-racist agenda in their organizing. The cultivation of food can also be a way to build cultural communities that go beyond food as a nutritional commodity to represent deep social and cultural relationships in the building of place-based identities (Mares and Peña, 2011). White (2011) demonstrates how the Detroit Black Community Food Security Network has used its community farm to create new communal spaces where community is built, services are provided, and collective action and political agency are nurtured. White's case study shows how food activism might link to other forms of activism, including 'access to affordable housing, clean water and decent public education'.

While food-justice scholars often agree that urban agriculture has the potential to produce positive economic, environmental and social changes, they criticize the alternative food movement for overlooking how the food system is implicated in racial and economic injustice. Those in the food-justice movement often employ a structural approach to understanding the racial and economic effects of food and agriculture policies (Alkon and Agyeman, 2011). They foreground the social and economic context in which individual choices are made, the structural barriers low-income people and people of colour face in accessing local and organic food, and the processes that have produced contemporary unequal social, economic and cultural landscapes (Alkon and Agyeman, 2011; Mares and Peña, 2011).

Acknowledging the race and class dimensions of food access and food systems – particularly unequal access to healthy food in low-income communities of colour – has led to the inclusion of food justice approaches in some urban agricultural projects (Alkon and Agyeman, 2011). Community gardens and community farms can be sites of resistance to racism and marginalization through collective work and self-reliance (Mares and Peña, 2011; White, 2011; Chapters 8 and 9, this volume). Many food-justice advocates claim that urban agriculture projects can open doors to more radical social transformation through community empowerment (Gottlieb and Joshi, 2010; McClintock, 2010) and link to broader social issues such as health, social and environmental justice (Levkoe, 2011).

Some scholars have adopted a more critical stance, exploring ways in which urban agriculture projects fall short of their transformative goals. Using examples from urban agriculture, critical geographers (Allen and Guthman, 2006; Guthman, 2008; Pudup, 2008; Alkon and Mares, 2012) have argued that alternative food networks reinforce neoliberalization. They argue that by filling in gaps left by the roll-back of the social safety net, these projects tacitly accept the shifting of services once delivered by the state to community groups and the private sector (Pudup, 2008; Alkon and Mares, 2012). By employing neoliberal discourses about consumer choice, entrepreneurialism and self-improvement, some urban agriculture projects accept the notion of individual and community responsibility for addressing problems of inequality and emphasize personal rather than systemic transformation (Alkon and Mares, 2012; McClintock, 2014). Synthesizing the laudatory and critical approaches to the study of urban agriculture, McClintock (2014) has argued that urban agriculture is both neoliberal and radical; neoliberal because it fills the space left by state retrenchment from service provision and radical because it attempts to reconnect production and consumption that have been made separate by the industrial agro-food system.

3.3.4 Ecosystem services

Urbanization has a variety of negative environmental consequences, including habitat loss, fragmentation, increased temperatures, pollution and changes to the hydrology of urban watersheds because of an increase in impervious surfaces (Goddard et al., 2010). Urban agriculture is one way of 'greening the city', through the mitigation of some of the detrimental impacts of urbanization by revitalizing brownfields, improving air quality, ameliorating the urban heat-island effect, enhancing biodiversity, and transforming urban wastes into resources (compost) for growing food (Tzoulas et al., 2007; van Veenhuizen and Danso, 2007; Mendes et al., 2008; Goddard et al., 2010). In addition, many urban agriculture programmes promote organic or sustainable farming methods, which are associated with better soil quality maintenance, carbon sequestration potential, pollination services, control of pests, weeds and diseases, biodiversity conservation and efficient energy use (Kremen and Miles, 2012). Despite these potential benefits, urban agriculture is not without environmental risks or impacts. Urban agriculture sites and the produce grown there may be at risk of contamination from air, soil and water pollution (Brown and Jameton, 2000; Gallaher et al., 2013). Agricultural practices may also introduce toxics through the application of industrial fertilizers, herbicides and insecticides, although the use of chemicals is often restricted in community gardens (Brown and Jameton, 2000).

Ecosystem services are defined as 'the conditions and processes through which natural ecosystems, and the species that make them up, sustain and fulfill human life' (Daily, 1997, p. 3). Agricultural systems in general have the potential to provide an array of ecosystem services (Colding et al., 2006; Kremen and Miles, 2012). In both the Global North and the Global South, urban agricultural systems are often managed expressly for their provisioning services (Calvet-Mir et al., 2012), such as food, firewood, medicinal resources and ornamentals. But they also offer regulating services (e.g. regulating air quality, climate and pests, and water infiltration), supporting services (e.g. soil formation, pollination and nutrient cycling), and cultural services (e.g. aesthetic enjoyment and recreation).

In cities around the world, the cumulative environmental impacts of urban agriculture may be especially important, given its extent and distribution. Recent surveys have revealed that urban agriculture, especially home gardens, is often quite extensive and may comprise a significant portion of a city's green space (National Gardening Association, 2009; Goddard et al., 2010; van Heezik et al., 2012). Yet governments have often overlooked spaces for urban agriculture when devising strategies to achieve their environmental goals (Colding et al., 2006). Because cities are characterized by small, fragmented habitat patches, understanding the urban environment requires adopting a landscape ecology perspective – one that takes into consideration the spatial arrangement of habitat patches, their size, heterogeneity and connectivity (Goddard et al., 2010).

3.4 Barriers and Benefits to Urban Agriculture in Santa Clara County, California

To illustrate these themes, we now turn to a discussion of our case material. Santa Clara County, California, the geographic heart of Silicon Valley, is located at the southern end of the San Francisco Bay Area. Once a regional and national agricultural centre, this region is now a heavily urbanized hub for technological innovation and development. Pellow and Park (2002) argue that there are two Silicon Valleys: one with the power, wealth and privilege of the high-tech economy, and the other where low-income communities of colour are relegated to low-wage jobs in environmentally and economically marginalized neighbourhoods. With high levels of inequality and poverty, food insecurity is a problem in Santa Clara County. In 2010, one in seven county residents were food insecure and more than one in five children lived in households that were food insecure (Feeding America, 2011). The lack of access to healthy food resources disproportionately affects low-income communities of colour (ChangeLab Solutions, 2010; Food Empowerment Project, 2010).

In Santa Clara County, like other urban communities throughout the US, urban agricultural projects proliferated after the economic downturn of 2008. The results reported here are based on interviews and surveys that we have conducted as part of the Food Systems Alliance of Santa Clara County in creating a food system assessment of the county (Santa Clara

County Food System Alliance, 2014). Urban agriculture activities in Santa Clara County include community gardens, community farms, home gardens, school gardens and for-profit gardens of varying sizes.

3.4.1 Community gardens

Most community gardens in Santa Clara County are located on city or other forms of public land in urban areas and have been established for a relatively long period. As of 2010, Santa Clara County had 28 active community gardens with an estimated 1250 residents gardening (ChangeLab Solutions, 2010). With 18 active gardens, the city of San Jose has the largest community garden programme. It is run by a paid garden manager who works with each garden's volunteer management team. In at least two other cities, gardens are run by volunteers and non-profit organizations, although the land is city-owned.

One of the primary benefits community gardens provide is access to land for urban residents who might otherwise lack space to grow their own food. These gardens provide access to fresh fruits and vegetables and engage urban residents with their local food system. In low-income neighbourhoods, community gardens have the potential to reduce household food expenditures and provide healthy foods in areas with limited access to produce in stores. In addition, gardeners in Santa Clara County have noted that gardening reduces stress, provides exercise and helps immigrants feel more connected to their homelands (ChangeLab Solutions, 2010).

There are several barriers to participation in community gardens, the most significant of which are the long waiting lists for a plot. In Santa Clara County, the average waiting list had 46 people, indicating that the demand for community garden plots is much greater than the available space. Another barrier is the yearly fees, which were an average of US$56 in 2010 and have gone up since. While not exorbitant, the fees represent a significant expense for some residents. Another potential barrier for low-income participants is that most community gardens do not allow the sale of produce, although many do encourage the donation and sharing of produce. Few community gardens offer garden-based programming, such as gardening courses, plant sales or seed exchanges (ChangeLab Solutions, 2010).

3.4.2 Home gardens

In contrast to community gardens, home gardens are private spaces, located in back and front yards of individuals' residences. In Santa Clara County, there are two home-gardening programmes – La Mesa Verde and Valley Verde – that help low-income families grow organic vegetables at home. A programme of Sacred Heart Community Service, La Mesa Verde aims to increase healthy food access for low-income communities, with the goals of promoting self-sufficiency and building community for its participants. Participating families receive nutrition and garden training, as well as all the supplies they need to participate in two seasonal plantings during the year. (Although it is an independent non-profit organization, Valley Verde follows a similar model.) We conducted an evaluation of La Mesa Verde that demonstrated the successes of the programme: 91% of families reported eating more vegetables and 25% of families reported saving over US$720 annually by eating fresh organic produce they grew at home. Having a garden also provided an outlet for physical activity. These and other benefits indicate that the programme is promoting healthy living for families who suffer most from food insecurity.

The La Mesa Verde programme is also structured to promote healthy communities. Surveys revealed that parents, children and grandparents shared time together in the garden, strengthening family ties. Many families reported meeting friends and neighbours through the programme and sharing produce and garden knowledge with neighbours, family and friends. Gardens also helped to create 'reconstructed landscapes' reflecting human/nature interactions (Schmelzkopf, 1995; Corlett et al., 2003). For San Jose's large and diverse Latino migrant communities, urban agriculture can be a way to continue food-consumption practices performed in Mexico, where there is a history of maximizing space for kitchen gardens, communal agricultural lands, and using public open space for horticultural and agricultural production (Mares and Peña,

2010). Gardens can also become sites of resistance against economic injustices through political action associated with garden projects. For example, La Mesa Verde's garden programme is explicitly linked to programmes concerning immigrant rights, access to economic resources and poverty alleviation.

3.4.3 Urban farms

Santa Clara County has both for-profit and non-profit urban farms. Many of the for-profit farms are the remnants of small-scale production that existed when the region was a major orchard-producing region, known as the Valley of Hearts Delight. Family farms have largely been replaced by urban development although some small farmers remain at the peri-urban edge in the southern end of county. Newer for-profit urban farms often have different operating models. For example, Ecopia uses warehouses in urban areas to grow hydroponically produced greens that are sold to high-end restaurants. Ecopia's model is high-tech, with stacked trays of solar-panelled lit plants that use minimal water. Other for-profit companies install and maintain raised bed gardens at urban dwellers' houses and on corporate campuses (e.g. eBay and PayPal).

The largest urban farmers are non-profits. As of 2013, three non-profit community farms operated in Santa Clara County: Veggielution in East San Jose, Full Circle Farm in Sunnyvale and Hidden Villa in Los Altos. All three have educational and vocational programmes and sell produce to the community. These organizations share the mission of growing local food and providing farm-based education. Veggielution is a six-acre farm tucked into an urban park in the city of San Jose. Their mission is to increase food access to the low-income, primarily Latino neighbourhood surrounding the farm, which they do through their low-cost Community Supported Agriculture or CSA (where people buy a share in a farm and receive a box of produce in return), farm stand and donations to food pantries. They also host many farm education programmes that encourage healthy eating and garden education.

Barriers that came out in interviews with community farm managers include the high cost of providing programming and the challenge of retaining skilled agricultural labour. Fundraising is often a burden, as the time spent 'chasing funding' can take away from other potential activities. While they do make money from produce sales, these generally do not provide enough funding for the staffing needs. Non-profits are plagued by unpredictable funding streams that ebb and flow. This, in addition to low non-profit-sector wages, means that consistent labour is a problem for most urban agricultural projects. Projects tend to be staffed by recent college graduates eager to become involved in food-movement-based work; enthusiasm often wears thin with the realization that salaries will not cover basic living expenses in this high-income area. Most of the urban farms have a high rate of employee turnover and a significant amount of self-exploitation (see Galt's 2013 discussion of self-exploitation in community-supported agriculture), as employees work long hours for low pay. Several non-profits rely heavily on volunteers, who provide labour for much-needed tasks, but also tend to be less skilled than trained farm workers. These labour concerns bring into question the sustainability of urban agriculture.

3.4.4 School gardens

At the state level, the California Department of Education has been promoting school gardens across California, since it launched the Garden in Every School Initiative in 1995. In 2002, 41% of the approximately 5800 elementary schools in California had a successful school garden (Graham et al., 2005). In this study, most principals reported that the purpose of their school garden was academic enhancement, and the subjects most frequently taught in school gardens were science, environmental studies, nutrition, language arts and mathematics.

In Santa Clara County, approximately one-third of the county's public schools have school gardens. A recent study by Stewart et al. (2013) found that school gardens were not distributed equally within Santa Clara County. They found that schools serving economically disadvantaged neighbourhoods and neighbourhoods with a high percentage of minority students were less likely to have a school garden. In contrast, they found that elementary-school gardens were more prevalent in the wealthier and ethnically less diverse western and northern communities of Santa Clara County. Part of the difference in the prevalence of gardens between low-income and high-income communities arises from differential availability of labour and funding to undertake school gardens that generally rely on volunteer labour and financial donations from parents. School gardens tend to be largely informal; very few schools support gardens over the long term. Typically, school gardens in Santa Clara County lack secure funding and coordinated planning, relying instead on individual teachers and parent volunteers, and private donations or external grants for support bring into question the sustainability of school garden programmes.

Several other barriers affect school gardens. Teachers are unsure how to expand garden programming that does not align with California State Content Standards, which leaves little time for activities that are not seen to contribute to the state requirements (ChangeLab Solutions, 2010). This report also found that produce from school gardens was rarely integrated into school lunch programmes, often because not enough produce was being grown or because schools lacked the kitchen equipment necessary to prepare locally grown food. Because peak harvest season does not coincide with the school year, produce often went unharvested if teachers, parents or other volunteers did not manage the garden during the summer months.

3.5 Conclusion

Urban agriculture owes its current renaissance to its position at the intersection of many concerns viewed as critical to the sustainability and livability of urban environments (van Veenhuizen and Danso, 2007; Urban Design Lab, 2012). In Silicon Valley, urban agriculture programmes have goals that touch on all three social themes outlined above: they strengthen food security, support good nutrition and health, and build community. They also provide ecosystem services to individuals and the community at large. The two home-gardening programmes that specifically target low-income communities and the non-profit urban farms that offer low-cost produce and provide a range of educational programming for the public are explicitly tackling issues of equity in food access. But some of the barriers to participation in urban agriculture programmes also illustrate the ways in which the environmental and social benefits of urban agriculture are unevenly distributed. These benefits, for instance, are limited in the less privileged urban areas of the region because of forces operating at multiple spatial and temporal scales (e.g. the relatively high cost of inputs for individuals; city and county regulatory frameworks; and the historical political and economic processes that have influenced the uneven patterns of development in Santa Clara County). As demonstrated by Stewart et al.'s (2013) study of the distribution of school gardens in the county, significant social disparities (race, ethnicity and class) exist in access to different forms of urban agriculture. Urban agriculture is more prevalent in high-income communities, but low-income communities are more likely to lack access to fresh produce, green space, and the other benefits that urban agriculture can provide.

In the beginning of the chapter, we asked whether today's urban agricultural resurgence is creating landscapes that result in enduring socio-economic change. In the San Francisco Bay Area, the alternative food movement – which encompasses urban agriculture – has persisted since the 1960s and shows no signs of weakening (Fairfax et al., 2012). This movement has helped to garner institutional support for urban agriculture from the local to the state level. Since 2009, Santa Clara County has a Food System Alliance, similar to Food Policy Councils elsewhere. In the past decade, supporters of urban agriculture have also achieved a number of policy victories that help to expand access to locally grown foods and land for urban agriculture. For instances, recent changes to the San Jose zoning code have allowed for small-scale urban agriculture to occur in commercial and industrial zones of the city. A new state law, the Urban Agriculture Incentive Zones Act (AB 551), provides tax incentives for landowners who use vacant lots in metropolitan areas for urban agriculture. Santa Clara County was the first county in the state to create an urban agriculture incentive zone. This framework of supportive policies and institutions for urban agriculture is a positive development and a necessary, but not sufficient, condition for permanence and lasting change.

Our experience in Silicon Valley shows that challenges remain to the viability and sustainability of all forms of urban agriculture – from home gardens to urban farms to school gardens. Sustainable funding sources are a key challenge for non-profit organizations undertaking urban agriculture work. Providing a living wage for agricultural workers in one of the most expensive regions of the country is another threat to the long-term viability of organized urban agriculture in Silicon Valley. For forms of urban agriculture that rely on volunteer labour, it is important to build robust programmes that can survive changes in leadership and administrative interest or support. It is even more important that volunteer-led efforts do not end up reproducing existing social inequities, by excluding marginalized communities from the beneficial impacts of urban agriculture programmes. Furthermore, scale matters in the creation of social, economic and ecosystem service benefits. More individualized urban agricultural production forms result in benefits at household and neighbourhood levels, while larger-scale enterprises have broader socio-economic and political benefits. But as McClintock (2010) argues, it is easier for urban agriculture to help restore individual connections with food and agriculture, much harder for it to successfully tackle the social disruption that accompanies processes of urbanization. However, addressing these broader social issues is more likely to lead to reconfigured landscapes and social change.

Notes

[1] Examples of gardening movements that have come and gone include Potato Patches (1894–1917), School Gardens (1900–1920), Liberty Gardens (1917–1920), Relief Gardens (1930–1939), and Victory Gardens (1941–1945). A detailed description of these movements can be found in both Bassett (1981) and Lawson (2005).

[2] Some examples (other than the case presented in this paper) of non-profits working with low-income communities to build home gardens are city slickers (Oakland, California), GROWINGGARDENS (Portland, Oregon) and GRuB (Olympia, Washington).

[3] Available at: http://www.letsmove.gov (accessed 17 March 2016).

References

Alaimo, K., Packnett, E., Miles, R.A. and Kruger, D.J. (2008) Fruit and vegetable intake among urban community gardeners. *Journal of Nutrition Education and Behavior* 40, 94–101.

Alaimo, K., Reischl, T.M. and Allen, J.O. (2010) Community gardening, neighborhood meetings, and social capital. *Journal of Community Psychology* 38(4), 497–514.

Algert, S.J., Baameur, A. and Renvall, M.J. (2012) Vegetable output and cost savings of community gardens in San Jose, California. *Journal of the Academy of Nutrition and Dietetics* 114(7), 1072–1076.

Alkon, A.H. and Agyeman, J. (2011) Introduction: the food movement as polyculture. In: Alkon, A.H. and Agyeman, J. (eds), *Cultivating Food Justice: Race, Class, and Sustainability*. MIT Press, Cambridge, Massachusetts, USA, pp. 1–20.

Alkon, A.H. and Mares, T.M. (2012) Food sovereignty in US food movements: radical visions and neoliberal constraints. *Agriculture and Human Values* 29(3), 347–359.

Allen, P. and Guthman, J. (2006) From 'old school' to 'farm-to-school': neoliberalization from the ground up. *Agriculture and Human Values* 23(4), 401–415.

Armstrong, D. (2000) A survey of community gardens in upstate New York: implications for health promotion and community development. *Health and Place* 6(4), 319–327.

Baker, L.E. (2004) Tending cultural landscapes and food citizenship in Toronto's community gardens. *The Geographical Review* 94(3), 305–325.

Bassett, T.J. (1981) Reaping on the margins: a century of community gardening in America. *Landscape* 25(2), 1–8.

Blair, D. (2009) The child in the garden: an evaluative review of the benefits of school gardening. *The Journal of Environmental Education* 40(2), 15–38.

Blair, D., Giesecke, C.C. and Sherman, S. (1991) A dietary, social and economic evaluation of the Philadelphia urban gardening project. *Journal of Nutrition Education* 23(4), 161–167.

Brown, K.H. and Carter, A. (2003) Urban agriculture and community food security in the United States: farming from the city center to the urban fringe. A primer prepared by the Community Food Security Coalition's North American Urban Agriculture Committee. Community Food Security, Venice, California, USA, October.

Brown, K.H. and Jameton, A.L. (2000) Public health implications of urban agriculture. *Journal of Public Health Policy* 21(1), 20–39.

Bryld, E. (2003) Potentials, problems, and policy implications for urban agriculture in developing countries. *Agriculture and Human Values* 20(1), 79–86.

Calvet-Mir, L., Gómez-Baggethun, E. and Reyes-García, V. (2012) Beyond food production: ecosystem services provided by home gardens. *Ecological Economics* 74, 153–160.

Carney, P.A., Hamada, J.L., Rdesinski, R., Sprager, L., Nichols, K.R., Liu, B.Y., Pelayo, J., Sanchez, M.A. and Shannon, J. (2012) Impact of a community gardening project on vegetable intake, food security and family relationships: a community-based participatory research study. *Journal of Community Health* 37, 874–881.

Carter, E.D., Silva, B. and Guzmán, G. (2013) Migration, acculturation, and environmental values: the case of Mexican immigrants in central Iowa. *Annals of the Association of American Geographers* 103(1), 129–147.

ChangeLab Solutions (2010) *Healthy Food Resource Assessment: Santa Clara County*. Public Health Law and Policy.

Colding, J., Lundberg, J. and Folke, C. (2006) Incorporating green-area user groups in urban ecosystem management. *AMBIO: A Journal of the Human Environment* 35(5), 237–244.

Coleman-Jensen, A., Nord, M. and Singh, A. (2012) *Household Food Security in the United States in 2011*. Report number 141. BiblioGov, United States Department of Agriculture, Economic Research Service, Washington, DC, USA.

Corlett, J.L., Dean, E.A. and Grivetti, L.E. (2003) Hmong gardens: botanical diversity in an urban setting. *Economic Botany* 57(3), 365–379.

Daily, G. (1997) *Nature's Services: Societal Dependence on Natural Ecosystems*. Island Press, Washington, DC, USA.

Drake, L. and Lawson, L.J. (2014) Validating verdancy or vacancy? The relationship of community gardens and vacant lands in the US. *Cities* 40, 133–142.

Duchemin, E., Wegmuller, F. and Legault, A.-M. (2008) Urban agriculture: multi-dimensional tools for social development in poor neighbourhoods. *Field Actions Science Reports. The Journal of Field Actions*, 1. Available at: https://factsreports.revues.org/113 (accessed 1 November 2016).

Fairfax, S.K., Dyble, L.N., Guthey, G.T., Gwin, L., Moore, M. and Sokolove, J. (2012) *California Cuisine and Just Food*. MIT Press, Boston, Massachusetts, USA.

Farming Concrete (2012) Harvest Map. Available at: https://farmingconcrete.org/2013/03/11/2012-harvest-report (accessed 1 November 2016).

Feeding America (2011) Map the Meal Gap, Food Insecurity in Your County. Available at: http://feedingamerica.org/hunger-in-america/hunger-studies/map-the-meal-gap.aspx (accessed 11 March 2016).

Firth, C., Maye, D. and Pearson, D. (2011) Developing 'community' in community gardens. *Local Environment* 16(6), 555–568.

Food Empowerment Project (2010) Shining a Light on the Valley of Heart's Delight: Taking a Look at Access to Healthy Foods in Santa Clara County's Communities of Color and Low-income Communities. Food Empowerment Project, San Jose, California, USA. Available at: http://www.foodispower.org/documents/FEP_Report_web_final.pdf (accessed 11 March 2016).

Gallaher, C.M., Mwaniki, D., Njenga, M., Karanja, N.K. and WinklerPrins, A.M. (2013) Real or perceived: the environmental health risks of urban sack gardening in Kibera slums of Nairobi, Kenya. *EcoHealth* 10(1), 9–20.

Galt, R.E. (2013) The moral economy is a double-edged sword: explaining farmers' earnings and self-exploitation in community-supported agriculture. *Economic Geography* 89(4), 341–365.

Galt, R.E., Gray, L.C. and Hurley, P. (2014) Subversive and interstitial food spaces: transforming selves, societies, and society–environment relations through urban agriculture and foraging. *Local Environment* 19(2), 133–146.

Glover, T.D. (2004) Social capital in the lived experiences of community gardeners. *Leisure Sciences* 26(2), 143–162.

Goddard, M.A., Dougill, A.J. and Benton, T.G. (2010) Scaling up from gardens: biodiversity conservation in urban environments. *Trends in Ecology and Evolution* 25(2), 90–98.

Gottlieb, R. and Joshi, A. (2010) *Food Justice*. MIT Press, Cambridge, Massachusetts, USA.

Graham, H., Beall, D.L., Lussier, M., McLaughlin, P. and Zidenberg-Cherr, S. (2005) Use of school gardens in academic instruction. *Journal of Nutrition Education and Behavior* 37(3), 147–151.

Gray, L., Guzman, P., Glowa, K.M. and Drevno, A.G. (2014) Can home gardens scale up into movements for social change? The role of home gardens in providing food security and community change in San Jose, California. *Local Environment* 19, 187–203.

Guthman, J. (2008) Neoliberalism and the making of food politics in California. *Geoforum* 39(3), 1171–1183.

Hamm, M.W. and Bellows, A.C. (2003) Community food security and nutrition educators. *Journal of Nutrition Education and Behavior* 35(1), 37–43.

Hynes, H.P. (1996) *A Patch of Eden: America's Inner City Gardeners*. Chelsea Green Publishing, White River Junction, Vermont, USA.

Irazabal, C. and Punja, A. (2009) Cultivating just planning and legal institutions: a critical assessment of the South Central Farm struggle in Los Angeles. *Journal of Urban Affairs* 31(1), 1–23.

Kingsley, J. and Townsend, M. (2006) 'Dig in' to social capital: community gardens as mechanisms for growing urban social connectedness. *Urban Policy and Research* 24, 525–537.

Kortright, R. and Wakefield, S. (2011) Edible backyards: a qualitative study of household food growing and its contributions to food security. *Agriculture and Human Values* 28(1), 39–53.

Kremen, C. and Miles, A. (2012) Ecosystem services in biologically diversified versus conventional farming systems: benefits, externalities, and trade-offs. *Ecology and Society* 17(4), 40.

Lawson, L. (2005) *City Bountiful: A Century of Community Gardening in the United States*. University of California Press, Berkeley, California, USA.

Levkoe, C.Z. (2011) Towards a transformative food politics. *Local Environment* 16(7), 687–705.

Litt, J., Soobader, M.J., Turbin, M.S., Hale, J., Buchenau, M. and Marshall, M. (2011) The influence of social involvement, neighborhood aesthetics, and community garden participation on fruit and vegetable consumption. *American Journal of Public Health* 101(8), 1466.

Louv, R. (2005) *Last Child in the Woods: Saving Our Children from Nature-deficit Disorder*. Algonquin Books, Chapel Hill, North Carolina, USA.

Lynch, K., Binns, T. and Olofin, E. (2001) Urban agriculture under threat: the land security question in Kano, Nigeria. *Cities* 18(3), 159–171.

Mares, T.M. and Peña, D.G. (2011) Environmental and food justice. In: Alkon, A.H. and Agyeman, J. (eds), *Cultivating Food Justice: Race, Class, and Sustainability*. MIT Press, Cambridge, Massachusetts, USA, pp. 197–219.

McClintock, N. (2010) Why farm the city? Theorizing urban agriculture through a lens of metabolic rift. *Cambridge Journal of Regions, Economy and Society* 3(2), 191–207.

McClintock, N. (2014) Radical, reformist, and garden-variety neoliberal: coming to terms with urban agriculture's contradictions. *Local Environment* 19, 147–171.

McLain, R.J., Hurley, P.T., Emery, M.R. and Poe, M.R. (2014) Gathering 'wild' food in the city: rethinking the role of foraging in urban ecosystem planning and management. *Local Environment* 19, 220–240.

Mendes, W., Balmer, K., Kaethler, T. and Rhoads, A. (2008) Using land inventories to plan for urban agriculture: experiences from Portland and Vancouver. *Journal of the American Planning Association* 74, 435–449.

Morales, A. (2011) Growing food and justice: dismantling racism through sustainable food systems. In: Alkon, A.H. and Agyeman, J. (eds), *Cultivating Food Justice: Race, Class, and Sustainability*. MIT Press, Cambridge, Massachusetts, USA, pp. 149–176.

National Gardening Association (2009) *The Impact of Home and Community Gardening in America*. National Gardening Association, South Burlington, Vermont, USA.

Pellow, D.N. and Park, L.S.-H. (2002) *The Silicon Valley of Dreams: Environmental Injustice, Immigrant Workers, and the High-Tech Global Economy*. NYU Press, New York, USA.

Pudup, M.B. (2008) It takes a garden: cultivating citizen-subjects in organized garden projects. *Geoforum* 39, 1228–1240.

Saldivar-Tanaka, L. and Krasny, M.E. (2004) Culturing community development, neighborhood open space, and civic agriculture: the case of Latino community gardens in New York City. *Agriculture and Human Values* 21, 399–412.

Santa Clara County Food System Alliance (2014) Santa Clara County Food System Assessment. Available at: http://www.caff.org/wp-content/uploads/2015/03/Final_VersionASSESS_010814_sm.pdf (accessed 16 March 2016).

Schmelzkopf, K. (1995) Urban community gardens as contested space. *The Geographical Review* 85(3), 364–381.

Smit, J. and Nasr, J. (1992) Urban agriculture for sustainable cities: using wastes and idle land and water bodies as resources. *Environment and Urbanization* 4, 141–152.

Stewart, I.T., Purner, E.K. and Guzmán, P.D. (2013) Socioeconomic disparities in the provision of school gardens in Santa Clara County, California. *Children Youth and Environments* 23(2), 127–153.

Surls, R., Feenstra, G., Golden, S., Galt, R., Hardesty, S., Napawan, C. and Wilen, C. (2015) Gearing up to support urban farming in California: preliminary results of a needs assessment. *Renewable Agriculture and Food Systems* 30, 33–42.

Taylor, J.R. and Lovell, S.T. (2012) Mapping public and private spaces of urban agriculture in Chicago through the analysis of high-resolution aerial images in Google Earth. *Landscape and Urban Planning* 108, 57–70.

Taylor, J.R. and Lovell, S.T. (2014) Urban home food gardens in the Global North: research traditions and future directions. *Agriculture and Human Values* 31(2), 285–305.

Travaline, K. and Hunold, C. (2010) Urban agriculture and ecological citizenship in Philadelphia. *Local Environment* 15(6), 581–590.

Twiss, J., Dickinson, J., Duma, S., Kleinman, T., Paulsen, H. and Rilveria, L. (2003) Community gardens: lessons learned from California Healthy Cities and Communities. *American Journal of Public Health* 93(9), 1435–1438.

Tzoulas, K., Korpela, K., Venn, S., Yli-Pelkonen, V., Kaźmierczak, A., Niemela, J. and James, P. (2007) Promoting ecosystem and human health in urban areas using green infrastructure: a literature review. *Landscape and Urban Planning* 81, 167–178.

Urban Design Lab (2012) The Potential for Urban Agriculture in New York City: Growing Capacity, Food Security and Green Infrastructure. Available at: http://urbandesignlab.columbia.edu/files/2015/04/4_urban_agriculture_nyc.pdf (accessed 16 March 2016).

van Heezik, Y.M., Dickinson, K.J. and Freeman, C. (2012) Closing the gap: communicating to change gardening practices in support of native biodiversity in urban private gardens. *Ecology and Society* 17, 34.

van Veenhuizen, R. and Danso, G. (2007) Profitability and Sustainability of Urban and Periurban Agriculture. Occasional Paper 19. Food and Agriculture Organization, Rome, Italy. Available at: http://www.fao.org/3/a-a1471e.pdf (accessed 16 March 2016).

Vitiello, D. and Nairn, M. (2009) Community Gardening in Philadelphia: 2008 Harvest Report. Penn Planning and Urban Studies, University of Pennsylvania 68. Available at: http://www.farmlandinfo.org/sites/default/files/Philadelphia_Harvest_1.pdf (accessed 16 March 2016).

Vitiello, D. and Wolf-Powers, L. (2014) Growing food to grow cities? The potential of agriculture for economic and community development in the urban United States. *Community Development Journal* 49(4), 508–523.

Wakefield, S., Yeudall, F., Taron, C., Reynolds, J. and Skinner, A. (2007) Growing urban health: community gardening in South-East Toronto. *Health Promotion International* 22, 92–101.

White, M.M. (2011) D-Town farm: African American resistance to food insecurity and the transformation of Detroit. *Environmental Practice* 13, 406–417.

Williams, D. and Brown, J. (2011) *Learning Gardens and Sustainability Education: Bringing Life to Schools and Schools to Life*. Routledge, New York, USA and London, UK.

WinklerPrins, A.M. and de Souza, P.S. (2005) Surviving the city: urban home gardens and the economy of affection in the Brazilian Amazon. *Journal of Latin American Geography* 4, 107–126.

Zezza, A. and Tasciotti, L. (2010) Urban agriculture, poverty, and food security: empirical evidence from a sample of developing countries. *Food Policy* 35(4), 265–273.

4 A Survey of Urban Community Gardeners in the USA

Tammy E. Parece[1]* and James B. Campbell[2]
[1]*Colorado Mesa University, Grand Junction, Colorado, USA;* [2]*Virginia Polytechnic Institute and State University, Blacksburg, Virginia, USA*

4.1 Introduction

Across the world, urban landscapes are a mix of living vegetation and the built environment, with the ratio of green space to human-made structures (e.g. roads, buildings) varying from neighbourhood to neighbourhood. Frequently, these physical variations are correlated with socio-economic factors (e.g. education, income, assets, etc.) because how and where development occurs is controlled by wealth (Wolch *et al.*, 2005; Ernstson, 2013; Anguelovski, 2015). As such, an uneven distribution of power exists and decision-making processes seldom include people with the greatest need for jobs, affordable housing, food and recreational facilities (FAO, 2007; Ernstson, 2013; Cohen and Reynolds, 2015). For example, larger houses, green spaces, and large retail stores are frequently located in more affluent areas, whereas lower-income urban residents live in more densely populated areas, have less access to healthy and nutritious food, and existing green spaces are small or poorly maintained.

Advocates of urban agriculture assert that it provides an opportunity to overcome many of the consequences of this unequal power distribution (McClintock, 2014). Urban agriculture helps alleviate widespread food insecurity in lower-income populations (Koc *et al.*, 1999; FAO, 2010). It combats poverty through income generation and releasing income to be used elsewhere (Bellows *et al.*, 2003; Draper and Freedman, 2010; Dubbeling, 2010; Hampwaye, 2013; Orsini *et al.*, 2013). Urban agriculture also increases local productive green spaces and provides a place for residents to congregate, forming and strengthening social and community relationships (Patel, 1996; Saldivar-Tanaka and Krasny, 2004; Drescher *et al.*, 2006).

However, critics of urban agriculture point out many issues with these assertions, especially because it can only provide a small portion of food needed by urban residents (e.g. Ellis and Sumberg, 1998; Hallsworth and Wong, 2013). Furthermore, it contributes to marginalization of disenfranchised populations because as soon as it improves a neighbourhood, the land becomes more valuable and is reclaimed for development (Voicu and Been, 2008; Draper and Freedman, 2010), or undergoes gentrification and displaces lower-income residents (Wolch *et al.*, 2014). Additionally, financial considerations (e.g. start-up costs, delivery to markets, annual fees and land ownership) are barriers for lower-income populations (Brown *et al.*, 2002; Lee-Smith, 2010; Cabannes, 2012). Risks to human health arise from producing and consuming food grown in degraded environments

*Corresponding author; e-mail: tammyparece@gmail.com

(Bellows, 1999; Furedy et al., 1999; Gallaher et al., 2013; Wortman and Lovell, 2014), and a risk presents for further environmental degradation from chemicals used in agriculture (DeBon et al., 2010).

Does urban agriculture face such a dichotomy? Do answers to this question vary relative to location? Many of the chapters within this book examine these questions, albeit within specific urban areas, e.g. Chapter 6 in Houston, Texas, USA; Chapter 7 in Tamale, North Ghana; and Chapter 13 in Medellín, Colombia. In our chapter, we examine urban agriculture from a broader spatial perspective – across many urban areas – looking for similarities and differences in why people participate in and the benefits they derive from cultivating urban land.

Urban agriculture encompasses many forms, from plants in small containers to farms (Pearson et al., 2010; Table 1.1, Chapter 1, this volume). We specfically look at one form – community gardens – to examine the above questions. We proceed as follows. First we define community gardens and consider whether they only exist in the more affluent areas of the world, i.e. the Global North. Next, we review community garden case studies to see whether any specific benefits have been quantified. Then, we present our methods to examine the aforementioned questions – a survey of urban gardeners located in many different urban areas across the United States. Finally, we discuss how our findings contribute to urban agriculture discourse.

4.2 Community Gardens

Community gardens are not new, dating back at least 100 years (Bassett, 1981; Waliczek et al., 1996; Lawson, 2005). In essence, a community garden is a section of land divided into small plots for use by many individuals (or families), each cultivating their own plot, usually for food production (Patel, 1996; Brown and Jameton, 2000; Brown et al., 2002; Lawson, 2005). In most cases, each gardener keeps produce for personal use, or for family and friends; infrequently, food is sold, and often excess food is given away (Brown et al., 2002). Community garden members share responsibility for common areas, e.g. paths, structures, tools (Smit et al., 2001).

The land upon which a community garden is located is rarely owned by those gardening. A specific community garden may be owned by landlords, local, regional and federal governments, non-profit organizations, or churches – who have granted gardening access (often only temporarily) to a community. The form of agriculture (plants only, plants and small animals, or larger livestock) practised in any particular garden depends on zoning regulations, local, regional and federal laws, and the consent of the property owner (Goldstein et al., 2011; Hodgson et al., 2011).

Community gardens are frequently discussed in conjunction with allotment gardens and guerilla gardens, but while all three types are closely related, fundamental characteristics differ. Allotment gardens are usually sponsored and managed by local governments and were initially developed as cultivation sites for the lower income working class (Petts, 2001). Allotment gardens are common in Europe (Petts, 2001; Andersson et al., 2007; Hardman and Larkham, 2014), and Japan (Matsuo, 2000). Guerila gardens, or informal gardens, are opportunistic – the plots are cultivated without the permission of the land owner, and are found all over the world, but more frequently in the Global South (Hardman and Larkham, 2014).

4.2.1 Community gardens in the Global North and Global South

Community gardens are prevalent throughout the world, but more easily identified in the Global North (see also Chapter 10, this volume). Table 4.1 provides an example of numbers of community gardens in select countries within the Global North.

Community gardens also exist in the Global South, although are more difficult to identify – the term 'community garden' may not be used. Table 4.2 provides information on community gardens located within the Global South.

4.3 Community Garden Research

Urban agriculture research is extensive and a complete review is beyond the scope of this chapter.

Table 4.1. Community gardens in the Global North.

Location	Estimated number	Source
US and Canada	18,000	American Community Garden Association (n.d.)
Australia	579	Australian City Farms & Community Gardens Network (2015)
Japan	3,382	ShiftEast (2010)
United Kingdom	>1,000	Federation of City Farms and Community Gardens (2011)

Table 4.2. Community gardens in the Global South.

Location	Organization (website) or research study author (date)
Africa	The Footprints Network (http://www.footprintsnetwork.org/project/68/Community-food-gardens-KwaZulu-Natal.aspx)
	Women International for a Common Future (http://www.wecf.eu/english/articles/2013/05/ewa_garden_blikkiesdorp.php)
	Slow Food Foundation for Biodiversity (http://www.fondazioneslowfood.com/en)
	Garden Africa (http://www.gardenafrica.org.uk)
	Karaan and Mohamed (1998)
	Bharwani *et al.* (2005)
	Wills *et al.* (2010)
	Shisanya and Hendriks (2011)
	Ruysenaar (2013)
Indonesia	Islamic Relief Worldwide (http://www.islamic-relief.org/wells-latrines-and-community-gardens)
Argentina	Agenda De Ideas: Huertas comunitarias en Rosario Argentina (https://republicavirtual.wordpress.com/2008/05/06/huertas-comunitarias-en-rosario-argentina)
Peru	RUAF (http://www.ruaf.org/publications/community-gardens-villa-maria-del-triunfo-lima-peru-huertas-comunitarias-en-villa-maria)
Brazil	City Farmer (http://www.cityfarmer.info/2008/01/13/cities-without-hunger-community-gardens-sao-paulo-brazil)
Haiti, Malawi, Kenya	Muse D.Territories: Développement local et RSE (http://www.musedt.com/lagriculture-urbaine-un-levier-de-developpement-pour-les-bidonvilles-du-sud)
Philippines, Zambia and Mexico	Wade (1987)

Several literature reviews discuss community gardens in two categories:

1. those generally on urban agriculture which include community gardens (e.g. ETC, 2003; Lesher, 2005; Orsini *et al.*, 2013; Stewart *et al.*, 2013; Hamilton *et al.*, 2014; Mok *et al.*, 2014);
2. those specifically reviewing community garden research (e.g. Draper and Freedman, 2010; Guitart *et al.*, 2012; Chapter 10, this volume).

Publications reporting community garden case studies are also numerous. Such case studies are overwhelmingly surveys, interviews and/or participant observations and identify different benefits derived from participating in community gardens. However, most case studies are limited to one specific urban area or community garden association. Waliczek *et al.* (1996) is a notable exception, distributing their survey across 45 community garden associations throughout the United States.

Benefits identified from community garden case studies can be, generally, divided into four different categories – economic, environmental, health and social. Examples of economic benefits include reducing household expenses and producing income (Karaan and Mohamed, 1998; McCabe, 2014). Examples of environmental benefits include neighbourhood beautification (Waliczek *et al.*, 1996; Ottmann *et al.*, 2012) and improvement of degraded land (Wade, 1987; Patel, 1996; Choo, 2011). Overwhelmingly, health and social benefits are identified most often. Health benefits include eating more fresh foods (Patel, 1996; Ottmann et al., 2012; Pourias et al., 2015) and learning about food and healthy eating (Pudup, 2008; Ottmann et al., 2012). Social benefits identified include strengthening communities, community building and/or community organization (Armstrong, 2000; Quayle, 2008; McCabe, 2014; Chan et al., 2015), social inclusion (Quayle, 2008; Wills et al., 2010; Chan et al., 2015), and promoting community involvement and helping others (Waliczek et al., 1996; Chan et al., 2015). The table within the Appendix provides a more comprehensive listing of benefits identified in community garden case studies.

4.4 Survey of Urban Community Gardeners

Our survey was designed to seek answers to the questions identified in Section 4.1, above. Some specific survey questions explored whether lower-income urban residents are participating in community gardens, the types of benefits being derived, whether participation is a relatively new activity, and whether the participants are cognizant of disparities present within their urban area.

To identify community gardens throughout the USA, we utilized the American Community Garden Association website's search engine – Find a Garden. We chose major cities trying to identify at least one community garden in each state. We then selected community gardens from the resulting list, eliminating any community garden noted as rural, school garden, food pantry garden, or urban farm. In addition, we searched the Garden Clubs of America National Map for garden clubs. In early 2014, we sent out a solicitation email (on four different dates) to each identified contact's email address, asking them to share the survey with their members.

4.5 Results and Discussion

We received responses ($n=177$) from 13 different states. In most states, responses came from one, two or three different cities; for California and Texas, we received responses from ten and nine different cities, respectively. For six states, we only received one response from each. The community gardens named by respondents, numbered from one in Virginia to 11 in California. We received the greatest number of responses from Philadelphia (40) representing five community gardens. Many of our community garden participants responded that they also gardened at home.

The majority of respondents reported two people in their household (51%) and the total number of people in residence ranged from one to six. The majority of respondents reported that they do not have children in their household (76%); only 23% of our respondents reported that they were retired. For the rest of the households, 13% had one child, 10% had two, and one respondent each had three and four children. For households with children, the average number was 1.6 children per household.

The number of family members working in the gardens ranged from only the respondent ($n=89$), up to 11 (reported by one respondent). The average number of people working in the community garden was 1.6. For those households with children, the average number of people working in the garden increased slightly to 2.2. The number of hours worked per week in the garden ranged from 1 hour to 80 hours per week, with an average of 6 hours per week. For those reporting that they were the only ones working in the garden, the average was 6.3 hours per week. One respondent reported working 80 hours per week because he/she also worked at a second community garden. Another respondent for this same community garden reported more than 20 hours per week, also working at a second community garden.

Respondents reported total length of time working in their community garden ranging from 1 month to 34 years, average time: 1 year. For those states with more than one respondent, the average length of gardening time ranged from 2.6 years (Florida) to 15.9 years (Virginia). The greatest number of respondents drive to their gardens (47%), followed by walking (41%). The range of travel distance was 10 feet to 20 miles, with an average of 2.2 miles. The majority of our respondents (92%) reported that they owned a car. About half of our respondents reported that they also garden at other locations (50.5%), including their own backyards, another community garden/public garden/urban farm, their balcony or patio, or another family member's house. Two respondents reported that they have been gardening at home for 9 and 45 years. A majority of respondents (80%) stated they also grew plants other than food (we did not ask for specifics).

The responses to most of our other questions can be generally categorized into four areas – economic, environmental, health or social. For example, the *'Why do you garden?'* question, prompting a free-form answer, provided responses ranging from the short and simple *'Beats watching TV'* to several sentences. Most respondents provided comprehensive answers that fit within multiple categories.

> *Love being outside, watching plants grow, enjoying nature, pleasure of growing my own fresh vegetables organically, sharing with friends and seniors in our community and also visiting with new friends I meet at the garden.* [environmental, social and health]

> *It's therapeutic, keeps my grocery bill down, it benefits people in my community who need help w[ith] supplementing [their food supplies].* [health, economic and social]

> *Fresh, free produce, to be connected to the land and the food, to spend time outdoors, to teach the children where food comes from, to interact with soil and seasons, to learn more about gardening, to have nutrition, to be healthier and happier.* [economic, environmental, social and health]

Furthermore, one of our most comprehensive questions asked participants to identify benefits received from gardening – checking all that apply, including a free-form answer field. The majority of these responses can also be categorized, generally, as either economic, environmental, health or social (discussed in more detail below). *Meeting new people* was the number 1 identified benefit (Table 4.3).

4.5.1 Category profiles of respondents

Economic: To explore the question: *'Are lower-income residents participating in urban agriculture?'*, we examine the answers to our enquiries about employment status, income, and the financial aspect of other responses.

We note that only 82.4% of all respondents answered questions related to employment and income. The response rate by state (i.e. the number

Table 4.3. Benefits identified by respondents.

Benefit	Number of responses (%)
I have met new people	142 (81%)
I am eating more fresh foods	121 (68%)
I have made new friends	102 (58%)
I have more appreciation for being outside	87 (49%)
I feel better	82 (46%)
I am exercising more	79 (45%)
I am now eating healthier	73 (41%)
I have been introduced to new foods and I like them	68 (38%)
I learned how to grow food	65 (37%)
I have learned about cultures from outside the US	35 (20%)
I use other green spaces in my city	28 (16%)
I am eating fewer sweets	26 (15%)
I have more money to spend on other things	22 (12%)
I have lost weight	19 (11%)
I stopped smoking	5 (3%)
My diabetes has got better	3 (2%)
I have earned income from selling food	4 (2%)
Crime has gone down in my neighbourhood	2 (1%)
Other	Relaxing/peaceful/stress-relieving (most frequent), sense of community belonging, lower food costs, educational

agreeing to answer demographic questions divided by the number of respondents to gardening questions) ranged from 70% of Texas and Pennsylvania respondents to 94% of Missouri respondents. All states with one respondent answered these questions. Most respondents are employed in some capacity (73%), 23% are retired, less than 1% identified themselves as unemployed, and in the other category, people identified themselves as a graduate student, stay-at-home mom, homemaker, disabled, and volunteer.

The majority (78%) responded that they earned over US$25,000 (in 2013), and 83% of households with children identified this as their level of income. The percentage of respondents within the highest income category varied by state: Missouri 53%, Texas 59%, Florida 78%, Pennsylvania 79%, Kansas 83%, California 90% and Virginia 93%. Furthermore, the majority of our respondents reported owning a car. From these responses, we conclude that most of our respondents are middle- to higher-income earners (based on USA standards).

Missouri and Texas had the highest percentage of respondents in the lower-income categories. For Missouri, 35% of respondents reported earning between US$15,000 and US$25,000 and 7% earning less than US$10,000. Texas had 14% between US$15,000 and US$25,000, and 22% earning less than US$10,000 (the highest percentage of all states). Texas also had the greatest rate of respondents (25.8%) identifying economic reasons for participating in their community garden, a rate double the average for all respondents (12.6%).

Most respondents reporting earnings in the highest income category did not identify an economic reason for participating in community gardens, nor did they acknowledge earning income from selling their produce, or state that food at retail food outlets was too expensive. But for those respondents who reported income in the lowest income bracket, 71% noted an economic reason (among other reasons) for growing their own food; for example: 'lower food costs', 'it is nice not having to pay grocery store prices for veggitables [sic]', 'reducing family food budgets' and 'supplement my food supply'.

Environmental: Overall, 37.9% of respondents provided an environmental reason for participating in their community garden. Virginia had the greatest rate of respondents identifying environmental reasons (76.5%) followed by Missouri (56.3%); we note that Virginia also had the longest average gardening time (15.9 years). Environmental benefits were rated rather low – using other green spaces in the city (16%) and more appreciation for being outside (49%). Furthermore, when respondents were asked to identify their engagement in other community garden activities, most did not list environmental education as one of the community garden's activities. Lack of environmental benefits can be attributed to either a failure to identify urban agriculture as a beneficial green space, lack of promotion for urban agriculture's ability to alleviate environmental degradation, or perhaps lack of environmental awareness. This dimension of community gardening deserves further investigation.

Health profile: Responses demonstrate that gardeners in all states believe producing their own food provides health benefits, and offers a wide variety of health benefits, from weight loss to mental health. Almost all respondents (96.6%) identified a health reason for participating in their community garden. Respondents stated 'It's happy, healthy, productive, and good for my soul', 'It's relaxing and brings enjoyment to my life', 'great exercise' and 'produce quality, organic food; fun; exercise; peace of mind'. Most respondents included the words 'producing fresh fruits and vegetables' or 'organic foods' when describing their reasons. Healthy benefits identified include eating more fresh foods (68%), feeling better (46%), exercising more (45%), and losing weight (11%), among others (Table 4.3).

These positive health responses are further supported by dissatisfaction with food purchased from retail food outlets, even though some outlets provide healthy choices. We asked about satisfaction with food purchased from farmers' markets, large retail grocery stores and supermarkets, whole food stores, local small corner stores, organic food stores, and cooperative food stores. A majority of respondents felt that the food they grow is better than that sold in retail outlets. Our respondents' lack of satisfaction with their food purchases did not limit itself to the local small corner store, or to large retail groceries and supermarket stores -- it also extended to those retail food stores that would be considered to have healthy options, e.g. farmers' markets, whole food stores, and organic food stores. For those who purchase at farmers' markets, whole foods, small corner and large grocery stores, 90% indicated some type of dissatisfaction with those outlets. For organic and cooperative food stores, 93% expressed some level of dissatisfaction, and for supermarkets 92% noted some level of dissatisfaction.

Social: Responses also show that gardeners value the social aspects of community gardens. Most of our respondents (31.6%) listed many social explanations for their reasons why they participate in urban food gardening. The social reason response rate was almost equal for five states – Texas (41.9%), Virginia (41.2%), Kansas (37.5%), California (37.5%) and Missouri (37.5%). Some specific reasons included 'community involvement, extra food is donated, social interaction of other gardeners, and a catalyst for neighborhood and community development'.

Social aspects also show up in the benefits listed by many, e.g. meeting new people (81%), making new friends (58%) and learning about cultures from outside the US (20%). A variety of additional benefits were provided by respondents – relaxing/peaceful/stress-relieving (most frequent), sense of community belonging, and educational. Furthermore, while half of our respondents stated that they eat all the food they produce, 97% of the other half stated that they gave the food to others, including local food-assistance centres. We equate this last answer to a recognition by our gardeners of the disparities that can exist in access to healthy food options.

The social importance of community gardening was also demonstrated by the provided responses when asked if they participate in activities other than gardening – a majority of respondents (77%) replied in the affirmative. Their activities usually fell into three different categories – educational, community building, or work. Community building activities included luncheons, charity fundraisers, pot-luck meals, and harvest shows/programmes. Work activities included clean-up days, purchasing seeds and plants, plant sales, applying for grants, and other financial duties. Educational activities were the most frequently mentioned, to include programmes on tree pruning, composting, building raised beds, raising chickens, starting a herb garden, organic gardening, preparing fresh vegetables for consumption, and children's educational programmes. In addition, while a majority of our respondents (82%) stated they knew how to grow food prior to joining the community garden, 65% stated that after joining they taught others how to grow food.

4.6 Conclusions

Our survey results are consistent with case studies identified in the broad literature on UA (and summarized in our Appendix below), which list benefits from case studies for individual urban areas or community gardens. Social and health gains are the benefits most widely identified by individual gardeners, and our survey shows that these are highly regarded by urban gardeners. Furthermore, our survey also demonstrates that interest in community gardening is not waning, as it has in past economic recoveries — rather, interest in gardening has increased across all garden sites, from just a few months to decades, confirming its staying power.

We believe our results advance the literature on urban agriculture and community gardens through the number of responses received, and their distribution across the USA. The survey shows consistency in benefits identified, without a difference in spatial distribution. Our survey also demonstrated some consistency with urban agriculture studies in the Global South, e.g. those by Wade (1987), Karaan and Mohamed (1998), Wills *et al.* (2010), Shisanya and Hendriks (2011), and Ruysenaar (2013), all listed within the table in the Appendix.

Furthermore, we feel that the dichotomy discussed in Section 4.1 does not exist. Our survey shows that although participating in community gardening does not alleviate food insecurity, it does provide increased access to healthy foods, and it, undeniably, provides health, social and community benefits. As McClintock so succinctly stated:

> Urban agriculture alone cannot usher in food justice ... Rather than an end unto itself, we should instead view urban agriculture as simply one of many means to an end, one of many tools working in concert towards a unified vision of food justice, and of just sustainability. (2014, p. 166)

We do realize that our survey had some limitations. For instance, we failed to ask what barriers or problems our respondents faced in food gardening within their urban areas, or what additional support they need, or would like. In the future, more income categories should be considered. We failed to ask whether respondents use chemicals, rainwater, or garden organically, and such questions may have added significantly to environmental responses. The community gardens identified by our survey respondents specifically provide opportunities for lower-income populations, but our survey failed to capture responses from this segment – an effect perhaps related to lack of internet access, lack of interest, the length of our survey (21% of people who started the survey did not progress beyond the online consent form), or a focus on important life concerns.

Yes, urban agriculture can only supplement food supplies, but the community gardens and gardeners in our survey recognize food insecurity and are assisting by providing food directly to food-assistance centres. Furthermore, the community gardens represented within our survey also recognize financial barriers for lower-income participants, so provide support through reductions in fees for garden access, which also directly assists in supplementing food supplies for those lower-income residents who take advantage of this assistance (e.g. Arlington Parks, 2015; Manhattan Community Garden, n.d.; Schuylkill River Park Community Garden, n.d.).

Many economic, environmental and human health benefits arise from all forms of urban agriculture. But community gardens provide an additional level of social benefits. Community building, interaction, sharing of knowledge and empowerment are just a few of the social benefits identified in case studies, and in our survey. These benefits represent an integral part of human society, especially in urban settings where people live and work closely together, in a world where more people increasingly live in urban than in rural areas.

Acknowledgements

Our survey, all documents, e-mails and solicitations, was pre-approved by the Virginia Tech Institutional Review Board, IRB No. 14-001, i.e. approved prior to distribution. We acknowledge the contribution of undergraduate research assistant Olivia Jancse, who gathered information on specifically identified community gardens. We would like to thank the Roanoke Community Garden Association for their support and advice in formulating this survey. We would also like to thank the community gardens and garden clubs who forwarded our survey request to their members, and all their members who responded to our survey.

References

Agustina, I. and Beilin, R. (2012) Community gardens: space for interactions and adaptations. *Procedia – Social and Behavioral Sciences* 36(0), 439–448. DOI: 10.1016/j.sbspro.2012.03.048.
Alaimo, K., Packnett, E., Miles, R.A. and Kruger, D.J. (2008) Fruit and vegetable intake among urban community gardeners. *Journal of Nutrition Education and Behavior* 40(2), 94–101.
American Community Garden Association (n.d.) FAQ. Available at: https://communitygarden.org/resources/faq (accessed 12 January 2015).
Andersson, E., Barthel, S. and Ahrné, K. (2007) Measuring social–ecological dynamics behind the generation of ecosystem services. *Ecological Applications* 17(5), 1267–1278. DOI: 10.2307/40062032.
Anguelovski, I. (2015) Alternative food provision conflicts in cities: contesting food privilege, injustice, and whiteness in Jamaica Plain, Boston. *Geoforum* 58(0), 184–194. DOI: 10.1016/j.geoforum.2014.10.014.
Arlington Parks (2015) Community gardens. Available at: http://environment.arlingtonva.us/gardens/community-gardens (accessed 23 January 2015).
Armstrong, D. (2000) A survey of community gardens in upstate New York: implications for health promotion and community development. *Health and Place* 6, 319–327.
Australian City Farms & Community Gardens Network (2015) Directory, data & mapping. Available at: http://directory.communitygarden.org.au/data (accessed 12 January 2015).
Bassett, T.J. (1981) Reaping on the margins: a century of community gardening in America. *Landscape* 25(2), 1–8.

Beilin, R. and Hunter, A. (2011) Co-constructing the sustainable city: how indicators help us grow more than just food in community gardens. *Local Environment* 16(6), 523–538.

Bellows, A.C. (1999) Urban food, health and the environment: the case of Upper Silesia, Poland. In: Koc, M., MacRae, R., Mougeot, L.J.A. and Welsh, J. (eds) *For Hunger-proof Cities: Sustainable Urban Food Systems*. International Development Research Centre, Ottawa, Canada, pp. 131–136.

Bellows, A.C., Brown, K. and Smit, J. (2003) Health Benefits of Urban Agriculture. Community Food Security Coalition's North American Initiative on Urban Agriculture. Available at: http://www.co.fresno.ca.us/uploadedfiles/departments/behavioral_health/mhsa/health%20benefits%20of%20urban%20agriculture%20(1-8).pdf (accessed 6 January 2015).

Bharwani, S., Bithell, M., Downing, T.E., New, M., Washington, R. and Ziervogel, G. (2005) Multi-agent modelling of climate outlooks and food security on a community garden scheme in Limpopo, South Africa. *Philosophical Transactions of the Royal Society B: Biological Sciences* 360(1463), 2183–2194.

Bleasdale, T., Crouch, C. and Harlan, S. (2011) Community gardening in disadvantaged neighborhoods in Phoenix, Arizona: aligning programs with perceptions. *Journal of Agriculture, Food Systems, and Community Development* 1(3), 1–16. DOI: 10.5304/jafscd.2011.013.007.

Brown, K.H. and Jameton, A.L. (2000) Public health implications of urban agriculture. *Journal of Public Health Policy* 21(1), 20–39.

Brown, K.H., Bailkey, M., Meares-Cohen, A., Nasr, J., Smit, J. and Buchanan, T. (2002) Urban Agriculture and Community Food Security in the United States: Farming from the City Center to the Urban Fringe. Urban Agriculture of the Community Food Security Coalition. Available at: http://community-wealth.org/content/urban-agriculture-and-community-food-security-united-states-farming-city-center-urban-fringe (accessed 6 January 2015).

Cabannes, Y. (2012) Financing urban agriculture. *Environment and Urbanization* 24(2), 665–683.

Chan, J., DuBois, B. and Tidball, K.G. (2015) Refuges of local resilience: community gardens in post-Sandy New York City. *Urban Forestry and Urban Greening* 14(3), 625–635. DOI: 10.1016/j.ufug.2015.06.005.

Choo, K. (2011) Plowing over: can urban farming save Detroit and other declining cities? Will the law allow it? *ABA Journal* 97(8), 20.

Cohen, N. and Reynolds, K. (2015) Resource needs for a socially just and sustainable urban agriculture system: lessons from New York City. *Renewable Agriculture and Food Systems* 30(01), 103–114.

Corrigan, M.P. (2011) Growing what you eat: developing community gardens in Baltimore, MD. *Applied Geography* 31, 1232–1241.

DeBon, H., Parrot, L. and Moustier, P. (2010) Sustainable agriculture in developing countries: a review. *Agronomy for Sustainable Development* 30, 21–32.

Draper, C. and Freedman, D. (2010) Review and analysis of the benefits, purposes, and motivations associated with community gardening in the United States. *Journal of Community Practice* 18(4), 458–492.

Drescher, A.W., Holmer, R.J. and Iaquinta, D.L. (2006) Urban homegardens and allotment gardens for sustainable livelihoods: management strategies and institutional environments. In: Kumar, B.M. and Nair, P.K.R. (eds) *Tropical Homegardens: A Time-Tested Example of Sustainable Agroforestry*. Springer, Dordrecht, The Netherlands, pp. 317–338.

Dubbeling, M. (2010) The world urban forum and urban agriculture. *ICT Update* 33, 12.

Ellis, F. and Sumberg, J. (1998) Food production, urban areas and policy responses. *World Development* 26(2), 213–225.

Ernstson, H. (2013) The social production of ecosystem services: a framework for studying environmental justice and ecological complexity in urbanized landscapes. *Landscape and Urban Planning* 109(1), 7–17. DOI: 10.1016/j.landurbplan.2012.10.005.

ETC (2003) *Annotated Bibliography on Urban Agriculture*. ETC, Leusden, The Netherlands. Available at: http://www.ruaf.org/sites/default/files/annotated_bibliography.pdf (accessed 2 November 2016).

FAO (2007) *The Urban Producer's Resource Book*. Food and Agriculture Organization, Rome, Italy.

FAO (2010) *Fighting Poverty and Hunger: What Role for Urban Agriculture?* Economic and Social Perspectives. United Nations, New York City, USA.

Federation of City Farms and Community Gardens (2011) About us. Available at: https://www.farmgarden.org.uk/about-us (accessed 12 January 2015).

Furedy, C., Maclaren, V. and Whitney, J. (1999) Reuse of waste for food production in Asian cities: health and economic perspectives. In: Koc, M., Macrae, R., Mougeot, L.J.A. and Welsh, J. (eds) *For Hunger-proof Cities: Sustainable Urban Food Systems*. International Development Research Centre, Ottawa, Canada, pp. 136–144.

Gallaher, C.M., Mwaniki, D., Njenga, M., Karanja, N.K. and WinklerPrins, A.M.G.A. (2013) Real or perceived: the environmental health risks of urban sack gardening in Kibera slums of Nairobi, Kenya. *EcoHealth* 10(1), 9–20.

Gardiner, M.M., Prajzner, S.P., Burkman, C.E., Albro, S. and Grewal, P.S. (2014) Vacant land conversion to community gardens: influences on generalist arthropod predators and biocontrol services in greenspaces. *Urban Ecosystems* 17, 101–122.

Goldstein, M., Bellis, J., Morse, S., Myers, A. and Ura, E. (2011) *Urban Agriculture: A Sixteen City Survey of Urban Agriculture Practices Across the County*. Emory Law Turner Environmental Law Clinic, Atlanta, Georgia, USA.

Guitart, D., Pickering, C. and Byrne, J. (2012) Past results and future directions in urban community gardens research. *Urban Forestry and Urban Greening* 11(4), 364–373.

Hale, J., Knapp, C., Bardwell, L., Buchenau, M., Marshall, J., Sancar, F. and Litt, J.S. (2011) Connecting food environments and health through the relational nature of aesthetics: gaining insight through the community gardening experience. *Social Science and Medicine* 72(11), 1853–1863.

Hallsworth, A. and Wong, A. (2013) Urban gardening: a valuable activity, but . . . *Journal of Agriculture, Food Systems, and Community Development* 3(2), 11–14.

Hamilton, A.J., Burry, K., Mok, H.F., Barker, S.F., Grove, J.R. and Williamson, V.G. (2014) Give peas a chance? Urban agriculture in developing countries. A review. *Agronomy for Sustainable Development* 34(1), 45–73.

Hampwaye, G. (2013) Benefits of urban agriculture: reality or illusion? *Geoforum* 49, 87–88.

Hardman, M. and Larkham, P.J. (2014) *Informal Urban Agriculture: The Secret Lives of Guerrilla Gardeners*, Vol. 1 of *Urban Agriculture*, ed. C. Aubry, E. Duchemin and J. Nasr. Springer, London, UK.

Hodgson, K., Campbell, M.C. and Bailkey, M. (2011) *Urban Agriculture: Growing Healthy, Sustainable Places*. American Planning Association, Chicago, Illinois, USA.

Karaan, A.S.M. and Mohamed, N. (1998) The performance and support of food gardens in some townships of the Cape Metropolitan Area: an evaluation of Abalimi Bezekhaya. *Development Southern Africa* 15(1), 67.

Koc, M., MacRae, R., Mougeot, L.J.A. and Welsh, J. (1999) Introduction: food security is a global concern. In: Koc, M., MacRae, R., Mougeot, L.J.A. and Welsh, J. (eds) *For Hunger-proof Cities: Sustainable Urban Food Systems*. International Development Research Centre, Ontario, Canada, pp. 1–10.

Lawson, L.J. (2005) *City Bountiful: A Century of Community Gardening in America*. University of California Press, Berkeley, California, USA.

Lee-Smith, D. (2010) Cities feeding people: an update on urban agriculture in equatorial Africa. *Environment and Urbanization* 22(2), 483–499.

Lesher, C.W. (2005) *Urban Agriculture: A Literature Review*. Alternative Farming Systems Information Center, National Agricultural Library, Beltsville, Maryland, USA.

Li, L., Weller, L., Tao, Y. and Yu, Z. (2013) Eco-vegetation construction of the community gardens in US and its implications. *Asian Agricultural Research* 5(7), 92–96.

Manhattan Community Garden (n.d.) Planting Manhattan's garden . . . together. Available at: http://www.tryufm.org/community_garden.htm (accessed 22 January 2015).

Matsuo, E. (2000) Japanese perspectives of allotment and community gardens. XXV International Horticulture Conference Part 13, Brussels, Belgium, 2 August 1998.

McCabe, A. (2014) Community gardens to fight urban youth crime and stabilize neighborhoods. *International Journal of Child Health and Human Development* 7(3), 1–14.

McClintock, N. (2014) Radical, reformist, and garden-variety neoliberal: coming to terms with urban agriculture's contradictions. *Local Environment: The International Journal of Justice and Sustainability* 19(2), 147–171. DOI: 10.1080/13549839.2012.752797.

Mok, H.F., Williamson, V.G., Grove, J.R., Burry, K., Barker, S.F. and Hamilton, A.J. (2014) Strawberry fields forever? Urban agriculture in developed countries: a review. *Agronomy for Sustainable Development* 34, 21–43.

Orsini, F., Kahane, R., Nono-Womdim, R. and Gianquinto, G. (2013) Urban agriculture in the developing world: a review. *Agronomy for Sustainable Development* 33(4), 695–720.

Ottmann, M.M.A., Maantay, J.A., Grady, K. and Fonte, N.N. (2012) Characterization of urban agricultural practices and gardeners perceptions in Bronx community gardens, New York City. *Cities and the Environment* 5(1), Article 13.

Patel, I.C. (1996) Rutgers urban gardening: a case study in urban agriculture. *Journal of Agricultural and Food Information* 3(3), 35–46.

Pearson, L., Pearson, L. and Pearson, C. (2010) Sustainable urban agriculture: stocktake and opportunities. *International Journal of Agricultural Sustainability* 8(1/2), 7–19.

Petts, J. (2001) *Urban Agriculture in London*. WHO Regional Office for Europe, Copenhagen, Denmark.

Pourias, J., Duchemin, E. and Aubry, C. (2015) Products from urban collective gardens: food for thought or for consumption? Insights from Paris and Montreal. *Journal of Agriculture, Food Systems, and Community Development*. DOI: 10.1007/s10460-015-9606-y.

Pudup, M.B. (2008) It takes a garden: cultivating citizen-subjects in organized garden projects. *Geoforum* 39(3), 1228–1240.

Quayle, H. (2008) *The True Value of Community Farms and Gardens: Social, Environmental, Health and Economic*. Federation of City Farms and Community Gardens, Bristol, UK.

Reynolds, K. (2009) Urban Agriculture in Alameda, CA: Characteristics, Challenges and Opportunities for Assistance. University of California Small Farms Program. Available at: http://sfp.ucdavis.edu/files/143920.pdf (accessed 6 January 2015).

Ruysenaar, S. (2013) Reconsidering the 'Letsema Principle' and the role of community gardens in food security: evidence from Gauteng, South Africa. *Urban Forum* 24(2), 219–249.

Saldivar-Tanaka, L. and Krasny, M.E. (2004) Culturing community development, neighborhood open space, and agriculture: the case of Latino community gardens in New York City. *Agriculture and Human Values* 21, 399–412.

Schuylkill River Park Community Garden (n.d.) About us. Available at: http://www.srpcg.org (accessed 22 January 2015).

ShiftEast (2010) City farming blooms with baby boomers. Available at: http://www.shifteast.com/city-farming-blooms-with-baby-boomers (accessed 12 January 2015).

Shisanya, S.O. and Hendriks, S.L. (2011) The contribution of community gardens to food security in the Maphephetheni uplands. *Development Southern Africa* 28(4), 509–526.

Smit, J., Nasr, J. and Ratta, A. (2001) *Urban Agriculture: Food, Jobs and Sustainable Cities*. The Urban Agriculture Network, Inc., Washington, DC, USA.

Stewart, R., Korth, M., Langer, L., Rafferty, S., Da Silva, N.R. and van Rooyen, C. (2013) What are the impacts of urban agriculture programs on food security in low- and middle-income countries? *Environmental Evidence* 2(7), 1–13.

Travaline, K. and Hunold, C. (2010) Urban agriculture and ecological citizenship in Philadelphia. *Local Environment* 15(6), 581–590.

Voicu, I. and Been, V. (2008) The effect of community gardens on neighboring property values. *Real Estate Economics* 36(2), 241–283.

Wade, I. (1987) Community food production in cities of the developing countries. *Food and Nutrition Bulletin* 9(2), 29–36.

Waliczek, T.M., Mattson, R.H. and Zajicek, J.M. (1996) Benefits of community gardening on quality-of-life issues. *Journal of Environmental Horticulture* 14(4), 204–209.

Wills, J., Chinemana, F. and Rudolph, M. (2010) Growing or connecting? An urban food garden in Johannesburg. *Health Promotion International* 25(1), 33–41.

Wolch, J., Wilson, J. and Fehrenbach, J. (2005) Parks and park funding in Los Angeles: an equity-mapping analysis. *Urban Geography* 26(1), 4–35.

Wolch, J., Byrne, J. and Newell, J.P. (2014) Urban greenspace, public health and environmental justice: the challenge of making cities 'just green enough'. *Landscape and Urban Planning* 125, 234–244.

Wortman, S.E. and Lovell, S.T. (2014) Environmental challenges threatening the growth of urban agriculture in the United States. *Journal of Environmental Quality* 42, 1283–1294.

Appendix

Benefits identified from a selection of community garden case studies.

Benefit	Author (date)
Environmental	
Beautifying neighbourhoods	Waliczek et al. (1996); Bleasdale et al. (2011); Hale et al. (2011); Ottmann et al. (2012)
Improvement of degraded land	Wade (1987); Patel (1996); Choo (2011)
Increases diversity of plants or insects	Saldivar-Tanaka and Krasny (2004); Agustina and Beilin (2012); Li et al. (2013); Gardiner et al. (2014)
Environmental education	Pudup (2008); Quayle (2008); Chan et al. (2015)
Economic	
Purchasing local foods	Quayle (2008)
Reduced household expenses	Patel (1996); Karaan and Mohamed (1998); McCabe (2014)
Produces income	Wade (1987); Karaan and Mohamed (1998); Shisanya and Hendriks (2011); McCabe (2014)
Increased property values	Voicu and Been (2008); McCabe (2014)
Health	
Increased exercise	Quayle (2008); Pourias et al. (2015)
Contributes to healthy eating/increased fresh food intake	Patel (1996); Karaan and Mohamed (1998); Armstrong (2000); Alaimo et al. (2008); Quayle (2008); Reynolds (2009); Wills et al. (2010); Bleasdale et al. (2011); Corrigan (2011); Shisanya and Hendriks (2011); Ottmann et al. (2012); Ruysenaar (2013); Pourias et al. (2015)
Improved mental health, i.e. stress relief or relaxation	Armstrong (2000); Bleasdale et al. (2011); Corrigan (2011); Ottmann et al. (2012); McCabe (2014); Chan et al. (2015)
Learning about food and healthy eating	Pudup (2008); Travaline and Hunold (2010); Bleasdale et al. (2011); Corrigan (2011); Hale et al. (2011); Ottmann et al. (2012)
Social	
Strengthening communities, community building and/or community organization	Armstrong (2000); Pudup (2008); Quayle (2008); Travaline and Hunold (2010); Beilin and Hunter (2011); Corrigan (2011); McCabe (2014); Chan et al. (2015)
Social inclusion, sense of belonging	Quayle (2008); Wills et al. (2010); Agustina and Beilin (2012); Chan et al. (2015)
Promoting community involvement, helping others	Waliczek et al. (1996); Corrigan (2011); Chan et al. (2015)
Decreased crime	Wade (1987); McCabe (2014)
Social networking/interaction	Patel (1996); Armstrong (2000); Saldivar-Tanaka and Krasny (2004); Reynolds (2009); Wills et al. (2010); Bleasdale et al. (2011); Hale et al. (2011); Agustina and Beilin (2012); Ottmann et al. (2012); Chan et al. (2015)
Maintaining cultural diversity	Saldivar-Tanaka and Krasny (2004); Li et al. (2013)
Increased self-worth, empowerment	Pudup (2008); Reynolds (2009)
Learning about the community	Agustina and Beilin (2012)

5 Gardens in the City: Community, Politics and Place in San Diego, California

Fernando J. Bosco* and Pascale Joassart-Marcelli
San Diego State University, San Diego, California, USA

5.1 Introduction

In the Global North alternative food practices have become increasingly common as a response to growing dissatisfaction with the industrial and corporate food system. In particular, in the US, scholars and activists have called for a relocalization of food systems, including urban agriculture and community gardening, as a way to foster health, justice and sustainability. Scholars describe these food practices as forms of resistance against the capitalist pressures that strain the food system (Kloppenburg *et al.*, 2000; Hendrickson and Heffernan, 2002; Norberg-Hodge *et al.*, 2002; Heynen, 2009). Similarly, the sites where these activities take place, especially community gardens, have also been conceptualized as 'spaces of resistance' or 'counter-spaces', to the extent that they represent a collective spatial strategy to redistribute value away from the global corporate food system into disenfranchised communities (Schmelzkopf, 2002; Staeheli *et al.*, 2002; Eizenberg, 2013). In addition, and similarly to urban gardens in the Global South (Obosu-Mensah, 1999; Redwood, 2012; Ruysenaar, 2013), many see these sites as increasing the welfare and food security of low-income urban residents, including recent migrants and refugees, by providing an additional source of nutritious food and an environment that facilitates civic participation.

However, other researchers indicate that such practices have become popular among affluent and white urban residents (Slocum, 2007; Guthman, 2008, 2011), with gardening in particular being reimagined as a creative and therapeutic activity (Johnston and Baumann, 2009). As the idea of urban agriculture gains momentum, gardens seem to be sprouting at a disproportionate pace in affluent and gentrifying neighbourhoods, where they symbolize an upscale, educated and ethical lifestyle. Such contradiction puts into question the claim that urban agriculture can foster more democratic and just food systems, at least at the local scale. One also has to wonder whether the local, alternative food spaces that community gardens provide are being appropriated or co-opted by the mainstream, as other successful counter-cultural phenomena and progressive spaces often are.

In this chapter, we address these questions, tensions and contradictions by exploring the growth of community gardens in San Diego. We focus on community gardens because they have recently benefited from changes in local ordinances that facilitated their growth. We explore the type of 'communities' that these gardens

---

*Corresponding author; e-mail: fbosco@mail.sdsu.edu

enable and their roles in resisting, challenging or reproducing neoliberal urban agendas. We do so by paying particular attention to the agendas of community-based organizations, residents, and other actors such as state and non-profit agencies, in supporting urban agriculture. We are interested in understanding how growing support for community gardens plays a role in reshaping urban neighbourhoods along the lines of race and class.

Our analysis builds on recent critical understandings of community, including research in urban political ecology. We argue that most research emphasizing the benefits of community gardens rests on romantic conceptualizations. We draw attention to the fact that community does not emerge naturally from gardens, but that gardens and community shape each other within a broader political and economic context. Therefore, community gardens cannot be viewed as inherently democratic, just or inclusive (Watts, 2004). After reviewing the relevant literature, we investigate the geographic distribution of community gardens in the County of San Diego to identify the types of neighbourhoods where gardens are likely to grow. We then consider three gardens, focusing on the communities they engage and analysing the inclusion and participation of different publics – in particular disenfranchised groups – in the local food agenda. Our case studies illustrate the ambivalent relationships between community-based food initiatives and neoliberal urbanism, a theme consistent with urban political ecology's central premise that urban nature is a social and political project.

5.2 'Community' Gardens: Community as a Neoliberal Strategy?

As the food movement continues to garner attention and arguments for eating local products grow popular, urban agriculture is becoming mainstream in many cities of the Global North. Among the benefits attributed to urban agriculture is the capacity to foster community – a term that appears consistently in association with gardening in presentations by policy consultants and in a wide range of academic publications.

Researchers have highlighted multiple ways in which public gardens can grow community: by bringing together marginalized citizens (Baker, 2004), fostering stronger connections between neighbours, and expanding social capital and collective efficacy (Kingsley and Townsend, 2006; Teig et al., 2009; Alaimo et al., 2010; Firth et al., 2011), creating opportunities for civic participation (Saldivar-Tanaka and Krasny, 2004; Levkoe, 2006), supporting economic development (Colasanti et al., 2012) and overcoming urban blight (Hynes, 1996).

Similarly, policy makers and advocates have embraced public gardens as 'best practice' to foster community development and address many of the problems of late capitalism. For instance, Ron Finley – a self-described guerilla gardener – has received an enormous amount of attention from the media for his 'innovative' idea of turning unused urban space such as parkways and vacant lots into 'community hubs, where people learn about nutrition and join together to plant, work, and unwind' (Finley, 2015). In this sense, urban agriculture has become a cosmopolitan idea that has been appropriated, circulated and emulated by policy elites. It illustrates the sort of neoliberal 'fast policy' (Peck et al., 2012, p. 268) that churns 'silver-bullet' models meant for localities to address the consistent absence of regulatory fixes at broader scales.

In these academic and policy accounts, community always has a positive connotation. Yet, as some literature suggests (Jessop, 2002; Watts, 2004; DeFilippis et al., 2006; Elwood, 2006), these conceptualizations of community are problematic. First, they confuse community and territory. Community is perceived as self-generating: it exists by virtue of people being grouped together in a bounded space. The assumption that volunteers and resources are available and ready to be tapped into is misleading and imposes a heavy burden, especially on residents of low-income areas and women (Joassart-Marcelli and Martin, 2015). Research shows that many neighbourhoods, including those with the greatest needs such as low-income community of colour, do not have active community-based organizations within their boundaries that have the resources and capacity to serve residents and enact lasting change (Joassart-Marcelli and Wolch, 2003). Establishing a public garden requires residents to work together to secure land, obtain permits, pool resources and manage access – a process that is greatly facilitated by the preexistence of formal or informal community organizations.

Second, much of the literature on the social benefits of urban agriculture ignores the power relations underlying communities and the fact that, as organized entities, they are inherently exclusionary and cannot be unequivocally good. Communities are assumed to be inclusive, democratic and homogeneous. Instead, as McCarthy (2006) argues, communities are 'constructed, unstable, and rife with their own internal inequalities'. Community-based organizations are also perceived to be more accountable, transparent and representative than centralized public agencies. However, missions differ widely between non-profit organizations, with some motivated by grassroots change and others acting as providers of state-funded services or agents of developers. Similarly, organizational structures vary from top-down to bottom-up. Given the intense competition associated with the need to obtain funding, through either private donations or government contracts and grants, the goals of community organizations are regularly reinvented and, in many cases, present a great source of internal tension. In that context, moderate strategies, consensus building, partnership development and capacity enhancement tend to be privileged at the expense of more political and radical agendas (DeFilippis et al., 2006).

In the US, the embrace of 'everything local', including urban agriculture, appears to be driven by a growing interest in healthy and safe food, and a desire to live in hip neighbourhoods. These goals stand in sharp contrast to those of activists, who have built community gardens to address the impacts of racism, social exclusion, environmental injustice and economic inequality in their communities (Alkon and Norgaard, 2009; Heynen, 2009; Eizenberg, 2013). The agendas of groups that obtain funding and other assistance to develop community gardens tend to be dominated by an increasingly homogenized and institutionalized vision of urban agriculture, as witnessed in the replication of similar community garden initiatives in very different contexts across the US. Indeed, there is a growing recognition that community gardening has become increasingly popular among affluent and white urban residents. This sort of colour-blind and universal alternative food practice often posits white and middle-class values about 'eating right' as the norm (Slocum, 2007; Guthman, 2008; Alkon and McCullen, 2011), connecting community gardens to an elitist cultural economy, which devalues the practices of low-income and minority residents and contributes to the gentrification of their neighbourhoods (Zukin, 2008; Quastel, 2009).

A third criticism of the romantic visions of community is their failure to acknowledge that communities are situated within larger political and economic structures that influence their capacity and responsibilities. As the urban political ecology literature suggests, urban nature is political. In the context of neoliberal urban governance, community-based organizations are playing a growing role in shaping urban environments and addressing social concerns, including access to green space and food security (Delaney and Leitner, 1997; MacLeod and Goodwin, 1999). While the dismantling of the welfare state has increased the responsibility of non-profits to fill the void, it has also enabled particular organizations to become key actors in urban policy (Wolch, 1990; Jessop, 2002; Joassart-Marcelli, 2012). Among organizations competing for resources, those claiming to strengthen community, prepare people for work, stimulate investment, increase property values and support market-based economic development are more likely to succeed.

Ironically, the alternative food movement, with its emphasis on community, seems to have bought into the sort of market-based neoliberal ideology it often claims to reject (Allen, 1999; Alkon and McCullen, 2011; McClintock, 2014). Not only does the movement increasingly rest on the idea that individuals can change the world through consumption choices (Bryant and Goodman, 2004; Hinrichs and Allen, 2008), it also embraces assumptions regarding the superiority of community-based organizations and non-governmental actors (Feenstra, 1997; Allen, 1999). Narratives of local entrepreneurship, sustainable neighbourhoods and community monitoring (Cortese, 2011; Hewitt, 2013) resonate with urban developers and public officials who seek new ways to generate value by transforming urban landscapes and rebranding places (Harvey, 1996).

As Joseph (2002, p. vii) argues, 'community is deployed by any and everyone pressing any sort of cause'. Yet, if one abandons a romanticized notion of community for a version based on the criticism raised above, it becomes clear that this neo-communitarian approach (Jessop, 2002) is unlikely to generate the type of inclusive place-based solutions its proponents claim to support. As a result, urban agriculture projects are more likely to emerge in places with the greatest capacity. This results in an uneven landscape of food production and access, with disparities and exclusions occurring within the region, as well as at the micro-scale of the garden.

In this chapter, we ask whether the communities involved in urban gardens in San Diego, California, act to resist, challenge and/or reproduce neoliberal urban development policies. Answers to these questions are contingent upon the political, economic and social relations, which partly define these communities and their ability to transform the local food system. Of particular importance are the broader agendas of each community and the various stakeholders who shape it. Our goal is to highlight the contradictory politics of viewing community as inherently good and the dangers of mainstreaming and homogenizing community-based food projects. As McClintock (2014) argues, a better understanding of the internal contradictions of urban agriculture is a necessary step in developing a food politics that challenges the structural inequalities of the dominant industrial food system.

5.3 San Diego's 'Community' Gardens

San Diego has a service-oriented economy highly dependent on tourism and military expenditures. The city government's focus on economic development through public–private partnerships has coalesced downtown, leading many neighbourhoods to an intense competition for resources and creating a fragmented and uneven urban landscape (Chapin, 2002; Davis *et al.*, 2003; Erie *et al.*, 2010). Because the needs of residents at the bottom of the income distribution have often been neglected, urban agriculture and alternative food projects aimed at alleviating community needs have become more common.

Community gardens have been widely adopted in this context, but there is wide variation in their types and aims. On the one hand, some community gardens have been established to increase food security for children and families, while promoting a healthier lifestyle. Others have purposely involved immigrants (many of whom come from agricultural societies) to increase their civic participation. On the other hand, community gardens have also been widely adopted in largely white and affluent or gentrifying neighbourhoods, where they have taken on therapeutic, recreational and educational aims.

Community gardens have grown rapidly in the last 5 years in San Diego County. During an initial round of research in 2010, we accounted for 27 community gardens. Only two of those gardens were actually permitted by the city (Ellsworth and Feenstra, 2010). Today, the San Diego Community Garden Network (SDCGN) – a non-profit umbrella organization that helps create, support and grow community gardens in San Diego County – lists almost 90 active community gardens (Fig. 5.1), with more in the planning stages (SDCGN, 2015). One of the main reasons behind this increase are the changes enacted to the city permitting process for urban agriculture, which was streamlined by a new ordinance adopted by the San Diego City Council in June of 2011. The new ordinance eliminated the need for an expensive permit and removed other requirements, such as mandatory fencing and on-site water meters (Tolin, 2009; Washburn, 2009), therefore reducing community garden start-up costs to only those of materials such as soil, hoses and other garden necessities. The new ordinance also allowed the establishment of community gardens on commercial properties in addition to residential and industrial land.

As Fig. 5.1 shows, there are community gardens established throughout the county; however, the majority of gardens have been established in the City of San Diego, clustering in urban core neighbourhoods, where population density and demographic and socio-economic diversity is high. Fig. 5.1 also shows poverty rates by census track and reveals no obvious spatial connection between community gardens' location and the percentages of people living below the federal poverty line.

The SDCGN's multiple goals of enhancing food security, promoting a sustainable environment and fostering community-building reflect the complex diversity of community gardens in

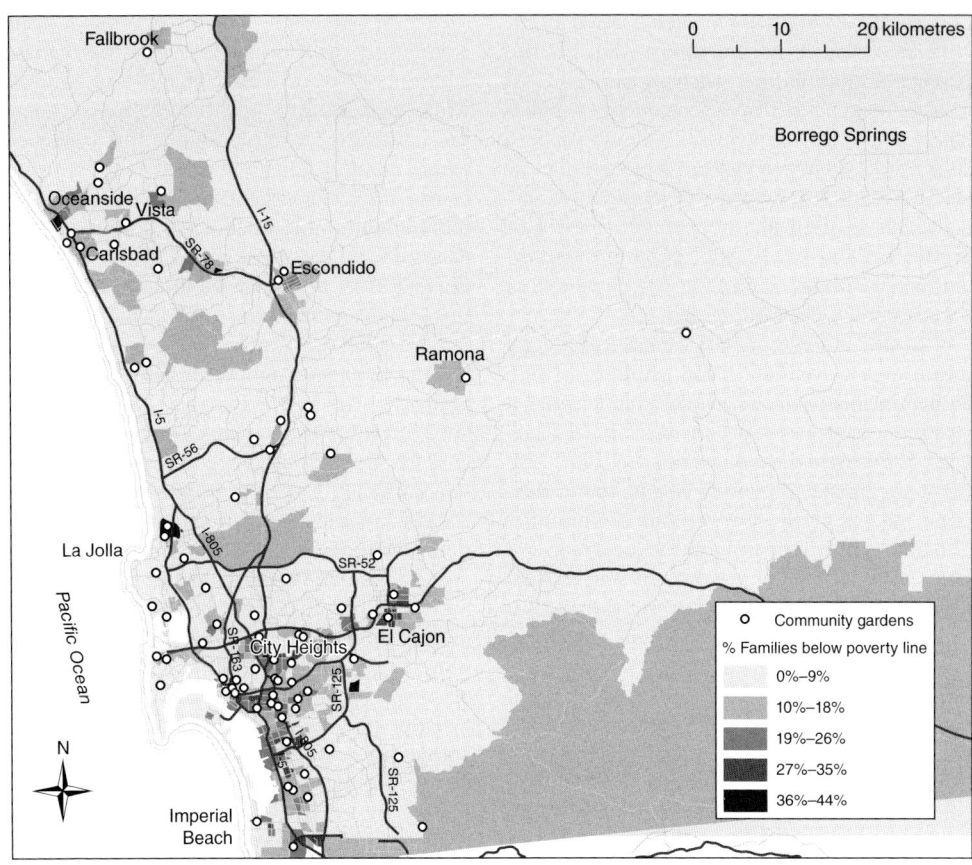

Fig. 5.1. Community gardens and poverty rates (percentage of individuals in households with income below official poverty threshold, by census tract) in San Diego County, California. (Source: 2010 US Census of Population and Housing and San Diego Community Garden Network. Map created by Emanuel Delgado.)

San Diego. The network subscribes to the American Community Garden Association's vision:

> that community gardening is a resource used to build community, foster social and economic justice, eliminate hunger, empower communities, break down racial, ethnic [and generational] barriers, provide adequate health and nutrition, reduce crime, improve housing, promote and enhance education, and otherwise create sustainable communities. (SDCGN, 2015)

The network also envisions community gardens as partners in creating more green spaces in our cities and engaging in environmental stewardship (SDCGN, 2015). Interestingly, among the multiple missions and goals of the network, the attempt at 'building community' receives the most prominent attention, since the organization presents itself to the broader public as a non-profit engaged in 'Building community, one garden at a time.'

SDCGN portrays the diversity of San Diego's community gardens by dividing them into three broad categories: 'community gardens', 'educational/training' and 'school/community'. This categorization, however, is problematic and does not encompass the competing cultural and political meanings of urban agriculture and community. For example, the category of 'community gardens' includes both private gardens with strict membership requirements, as well as those that welcome members of the broader community, thereby masking important distinctions between the communities involved. The 'community' label is uncritically deployed to account for gardens aiming at achieving very different goals. Little is known about the actors, motives and agendas underlying the variety of gardens, even if they are being decidedly portrayed as community spaces. Are San Diego's 'community' gardens places that foster food security, provide leisure activities for an exclusive group of gardeners, or develop the city's image as green and hip? And to what extent is the growth of community gardens being used to revitalize the real estate market through 'green development' (Tornaghi, 2014, p. 553) and to justify continued disinvestments in poorer neighbourhoods as the provision of food for lower-income residents gets passed on to the backs of non-profits, volunteers and citizens? What types of 'community' are being created 'one garden at a time', who do these communities serve and who do they exclude? We begin to answer these questions by turning our attention to three community gardens that illustrate the variety of contexts and competing goals of urban agriculture in the region.

5.4 A Tale of Three Community Gardens

The three community gardens chosen for our analysis are Juniper-Front in Bankers' Hill, New Roots in City Heights, and Mt Hope in Southeastern San Diego (Fig. 5.2). We collected data on these community gardens as part of a broader research project that analyses the changing landscape of alternative food practices. We visited the three gardens often, interacted with stakeholders and observed gardening dynamics. Our methodology is described in detail in Joassart-Marcelli and Bosco (2015).

The Juniper-Front garden was established for senior citizens in 1981 on an empty lot owned by the San Diego Port Authority in Bankers' Hill. The original garden members raised money and enlisted the help of the California Conservation Corps to build the raised plot beds and install water pipelines (Baron, 1993). Today, the garden

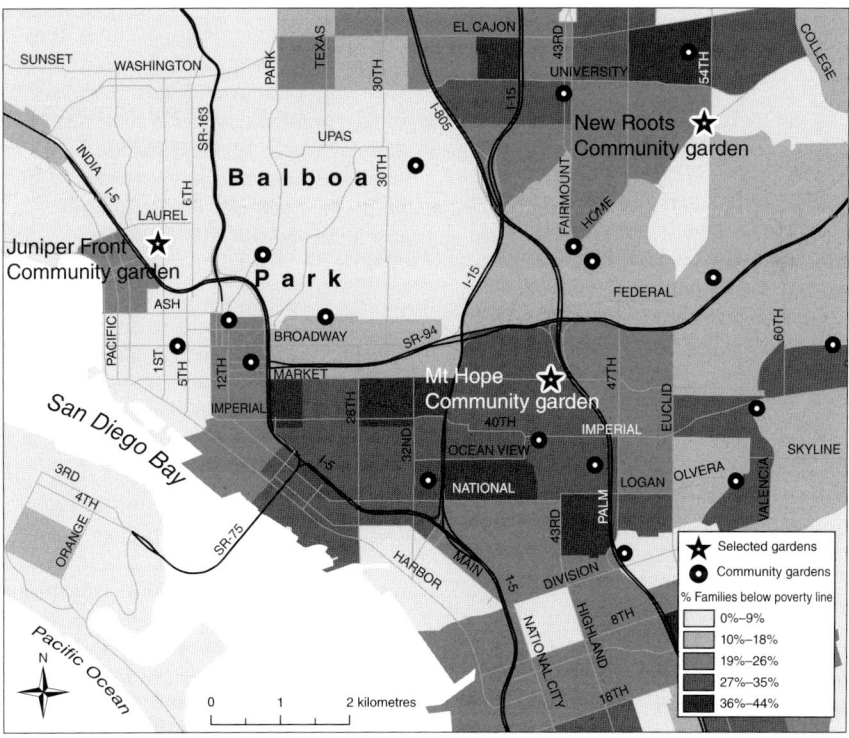

Fig. 5.2. The Juniper-Front, New Roots and Mt Hope community gardens and poverty rates: percentage of individuals in households with income below official poverty threshold, by census tract. (Source: 2010 US Census of Population and Housing. Map created by Emanuel Delgado.)

is no longer restricted to seniors, but has a long waiting list of over 70 people – a more than 2-year wait. The quarter-block where the garden sits is divided into approximately 60 10×12-foot plots leased to members for about US$100 per year.

The International Rescue Committee (IRC) established the New Roots Community Farm in City Heights in 2009 as a way to provide locally grown and culturally appropriate foods for recent Somali Bantu refugees, many of whom were farmers in their home countries. The IRC saw this initiative as a double opportunity for creating positive change: increasing food security for refugees and immigrant families, while helping farmers to reconnect with land and economic opportunities. The New Roots community farm helped the IRC establish itself as a leader in urban agriculture and food justice in San Diego. In 2011, it even drew the attention of First Lady Michelle Obama, who launched her Healthy Eating campaign with a visit to the garden and featured it in her book (Obama, 2012). The garden contains 85 plots on 2.3 acres of public land and serves mostly East African families. The waiting list is also long. The garden's farmers are certified to sell their produce at Farmers' Markets, including the one established by the IRC in City Heights.

The Mt Hope Community Garden began operating in 2012, after the new urban agriculture ordinance. The garden is run by Project New Village, a local non-profit that aims at redressing the deep structural inequalities that have historically disenfranchised Southeastern San Diego. The garden is part of the non-profit's efforts to use food as a way to energize and connect people to their neighbourhood while addressing issues of food security and environmental justice. The Mt Hope community garden fits within this mission by promoting local food production, distribution and healthy eating. The Mt Hope garden is also connected to a farmers' market as a way to provide an additional economic opportunity to growers (SDUT, 2010). However, in comparison with both Juniper-Front and New Roots, Mt Hope struggled to attract and retain farmers, especially in its early stage.

The three community gardens face different opportunities and challenges linked to their location in unique urban neighbourhoods that are themselves distinguished by their relationships to the regional and global economy, demographic composition, histories, and civic institutional environments (Table 5.1). In Bankers' Hill, the location of Juniper-Front, residents are primarily white, upper-middle-class professionals without children. Many are young renters living in converted historical homes or recently built apartments. Based on simple statistics, the majority fit the urban creative class stereotype that is often associated with gentrification (Florida, 2005).

In contrast, City Heights (New Roots) and Southeastern San Diego (Mt Hope) are characterized by high poverty rates and proportions of minorities and immigrants. Both neighbourhoods are within a few miles from downtown, but isolated by limited public transit. City Heights was one of San Diego's first middle-class suburbs, but has suffered greatly from the construction of freeways, municipal neglect and suburbanization in the last 50 years. Following the exodus of businesses and wealthier local residents, City Heights' single-family homes were transformed into apartments for working-class families, and later became one of the largest refugee resettlement areas in the US. Today, City Heights is an ethnically diverse neighbourhood that has received an unusually high level of attention from state and non-profit actors interested in economic revitalization and community health, resulting in over US$400 million being invested there in the last two decades (Kayzar, 2013). Organizations with a track record in the community, including the IRC, have been empowered by these recent developments and have become important regional, state and national players in the local food movement, by supporting refugees' participation in urban agriculture.

Southeastern San Diego is composed of several distinct neighbourhoods that share a similar history of economic decline, violence and political neglect. A century ago, it was the heart of San Diego's black community, which was spatially constricted by discriminatory local covenants (Madyun and Malone, 1981). Mexican immigrants and subsequent generations of Mexican-Americans also settled in the area in large numbers. Today, Blacks and Latinos continue to be the main residents of the neighbourhood (see Table 5.1). In the 1980s, owing in part to failed urban renewal initiatives, multi-family housing expanded, attracting more low-income residents. The area surrounding the community garden is significantly more diverse and poorer

Table 5.1. Study areas[a] and their socio-economic characteristics. (Source: 2007– 2011 American Community Survey 5-Year Estimates.)

	Bankers' Hill	City Heights	Southeastern San Diego	San Diego County
Garden name	Juniper-Front	New Roots	Mount Hope	
Population	2687	3699	4425	3,177,063
Income				
Median household income	$53, 426	$35,223	$41,867	$63,857
People in poverty	5.30%	25%	20%	13%
households with SNAP benefits in past 12 months	0%	13%	20%	4%
Public transit user	8.3%	19.5%	16.30%	3.20%
Race/ethnicity				
Latino	15.50%	38%	62%	33%
White	71.70%	7.1%	3%	48%
Black	5.20%	18.5%	18%	5%
Asian	5.8.%	36.4%	14%	11%
Other	1.80%	0%	3%	3%
Housing				
Owner-occupied units	18.80%	33.6%	40%	55%
Structure with two or more units	90%	52.8%	20%	34%
Household characteristics				
Average size	1.57	3.73	3.97	2.79
Households with children	7%	51.2%	47%	32%
Married with children	N.A.	66.9%	48%	72%

[a]Census tracts of community garden location.

than the rest of the county and it is often perceived as one of the most dangerous neighbourhoods of San Diego. Many adults are employed in low-wage service jobs that do not pay enough to escape poverty (Center on Policy Initiatives, 2014). Southeastern San Diego has undergone limited population growth and its urban development initiatives have received much less attention from the media, academics and large foundations.

The three neighbourhoods are homes to a variety of communities with overlapping and conflicting interests shaped by their historical and geographic contexts. The groups involved in urban agriculture represent a subset of actors that have come to dominate this agenda in their respective neighbourhoods. As we show, the construction and empowerment of these different communities in turn play a role in how urban agriculture is practised and how community gardens themselves support different communities by becoming parts of networks that may represent parochial – though still laudable – interests. But they could also move beyond the scale of the territorial community in which the garden is embedded, and reach out to have effects at the scale of the regional food system and beyond.

The dynamics at play in the three community gardens challenge assumptions regarding the inclusive, democratic and just nature of community by drawing attention to the absence of representation of certain groups, the increasingly homogenized vision of urban agriculture that is emerging from the tensions between various goals, and the resources disparities linked to a broader political-economic context.

5.4.1 Exclusionary practices underlying community

Because the terms neighbourhoods and communities are often used interchangeably, there is an assumption that community gardens are public spaces open to all neighbourhood residents. This, however, is rarely true. All three gardens are subdivided into individual plots assigned to local residents and have extensive waiting lists. Given the limited number of plots (46 at Mount

Hope, 60 at Juniper-Front, and 85 at New Roots), it is not surprising that growers in each garden represent relatively homogeneous groups that do not necessarily reflect the general demographic characteristics of the neighbourhood, suggesting some degree of exclusion.

In the Juniper-Front garden, the majority of the gardeners are white, and approximately one-third of them are senior citizens – reflecting the original legacy of the garden. As younger individuals have joined the garden, conflicts have arisen, with younger individuals wishing to use the garden as a platform for increased neighbourhood sociality and connect to the larger sustainability discourse embraced by the city and other groups. On the other hand, older members see the garden as a source of leisure. Regardless, all gardeners see the garden as a cultural amenity that increases the attractiveness of the neighbourhood. Despite the 'community' label, access to the garden is limited to members only. This restriction is visible in the fences that surround the garden and the signs warning passers-by. During interviews, gardeners equated a sense of community and belonging to the garden to the experience of belonging to an HOA (a Homeowners Association), where membership is restricted and members are required to obey certain rules. One gardener explained that, despite the exclusivity of the garden and the limited interactions gardeners have with people outside the garden, neighbouring residents are 'happy that the garden is here and not a noise-making tenement house [or] apartment[s]' (Porcella, 2012).

We observed similar processes of segregation and exclusion in the other two gardens, despite the fact that they very consciously aimed at serving diverse and disenfranchised populations. Even though the New Roots community farm and the Mt Hope community garden are more open and welcoming to the general public, they tend to appeal to specific populations and, by default, marginalize or discourage participation of others. In the case of Mt Hope, the organizing efforts of garden leaders have targeted primarily African-American residents of Southeastern San Diego, which represent about one-fifth of the neighbourhood's population (Table 5.1). Although other residents are invited to participate and are certainly not intentionally excluded, the message put out by Project New Village may be less appealing to them. In the case of New Roots in City Heights, a similar process can be observed in relation to the privileging of refugees from East Africa, to the detriment of longstanding residents and other groups of refugees and immigrants who are also under-served and disenfranchised.

The exclusion of some residents is also linked to the organizational structure of the gardens. Establishing the gardens required considerable efforts and leadership, including fundraising and political organizing to reform the city's zoning ordinances. This meant that a single group often came to represent the interests of gardeners, especially in negotiations with landowners, city agencies and neighbours. In the case of Mt Hope and New Roots, a single non-profit organization with a history of social service in the neighbourhood, took on those tasks. For example, Project New Village, the organization in charge of the Mt Hope garden, is led by African-American activists who are motivated by a social justice agenda that aims at bringing young African-Americans into farming and helping people of colour grow their own food in the community (Project New Village, 2012). Their urban agriculture initiatives are driven by a desire to offer 'new opportunities for job growth, skill development and prevention in high risk communities [of colour]' (Florido, 2011). The organization notes that, even though San Diego County is home to the largest number of small farms in the nation, only 27 of these farm operators are African-American – less than 1%. The worthy goal of providing a corrective for this situation has the unintended effect of excluding other groups and agendas.

Similarly, the New Roots community farm is the project of the IRC, whose funding and activities are directly tied to refugee assistance. Although other organizations in City Heights have been involved in promoting urban agriculture and have contributed to the IRC's success in establishing the garden, the lease and management of the space is ultimately the responsibility of the IRC. The Juniper-Front garden was due to the labour of one determined resident, Ethel Baron, who persuaded the Senior Citizen Center of San Diego to guarantee the lease from the Port Authority. About 15 years later, the garden was incorporated into a non-profit organization for the purpose of managing the garden. Regardless of the differences between them, it is clear that, in all three cases considered here, the

Table 5.2. Community dynamics in three gardens.

Community garden	Management	Goals	Community dynamics
Juniper-Front (Bankers' Hill)	Non-profit: neighbourhood residents	Leisure	Exclusionary: white, upper-middle-class, older participants
New Roots (City Heights)	Non-profit: International Rescue Committee	Refugee integration Food security	Mostly exclusionary: refugees and low-income participants
Mt Hope (Southeastern San Diego)	Non-profit: Project New Village	Civic empowerment Community pride Food security Racial justice	Moderately exclusionary: limited range of community participants

combination of a strong organizational structure, a well-defined agenda, and a targeting of certain groups at the expense of others contributed to the creation of territorially exclusionary communities through the practices of urban agriculture. Table 5.2 summarizes the gardens' community dynamics discussed so far and sets the stage for the remainder of our discussion.

5.4.2 Urban best practices and the increased homogenization of community

Recent literature suggests that, under the competitive model of neoliberal urban governance, communities' ability to raise support for their activities rests on the adoption of projects that build on local assets, generate consensus, strengthen partnerships and are inspired by evidence-based research (DeFilippis *et al.*, 2006; Peck *et al.*, 2012). Radical activism and conflict-based politics that seek to alter structural inequities are unlikely to generate much political or financial support. Instead, solutions must come from the mobilization of existing community resources including volunteers, workers, land, buildings, and so on. This ideology has permeated the process of grant-writing, resulting in increasingly homogenized policy solutions based on common language, themes and best practices. In the realm of urban agriculture in the US, images of communities pulling together to address their most pressing needs abound, often calling upon nostalgic and gendered visions of gardening, such as the wartime victory gardens featured on the Juniper-Front website in 2015. In the narratives of the three leading organizations studied here, the provision of food appears secondary to the goal of building social capital and strengthening community. Despite differences between gardens and disagreements among members, urban agriculture and local food-movement stakeholders in San Diego share a common language that makes disproportionate use of terms like 'togetherness', 'social bonds', 'connecting with heritage', 'community engagement', and analogies such as 'cultivating community', 'growing roots' and 'harvesting community'. This universalizing narrative suppresses differences, masks disagreements and, often inadvertently, reproduces exclusion by erasing other motives or bringing them to the background.

Even members of the Mt Hope Garden, who are motivated by the radical potential of urban agriculture to address racial inequality and other structural injustices, tend to adopt that language. According to one gardener: 'we know the resources our community has . . . we know the resources our community needs and are taking steps day by day to build stronger bonds' (Project New Village, 2012). As the organization puts it, this is about 'neighbors growing, selling and cooking with their neighbors' (Project New Village, 2015) – a seemingly innocuous activity. The garden is described as a 'huge step in bringing community together' (Project New Village, 2012). To many, building community entails reclaiming something that has been taken away:

> We are trying to take control of that [our food] back and put it in the hands of our communities because who knows better what we need than ourselves . . . [Gardening] was a way of life here, long before it became popular.

Nowhere is this narrative about community more prevalent than in the New Roots community farm. The IRC's success in City Heights is linked to its ability to brand its approach by calling upon values of community, assimilation, self-sufficiency and entrepreneurship. As Michelle Obama claimed during her visit (see Fig. 5.3):

> The World is watching what is going on on this plot of land. You are truly doing phenomenal work. It's a model for the nation, for the world . . . This is what we need to be doing in communities all across the country. (IRC, 2015a)

According to the IRC (2015b), 'the New Roots program enables refugees to reestablish their ties to the land, celebrate their heritage and nourish themselves and their neighbors in their new communities'.

The similarity of discourses used across organizations in justifying their activities and claiming resources is striking. These statements emphasize the prevalent place that community, constructed as a colour-blind and conflict-free entity, takes in contemporary urban agriculture in the Global North cities as both a resource and a desirable outcome. Calling upon this universally good vision of community may exclude or devalue other forms of urban agriculture with more radical goals and strategies. It also supports a policy approach that ignores the specific circumstances of different people in different places and naturalizes the community as a desirable entry point for social change.

5.4.3 'We can do this ourselves'

The growing popularity of community gardens needs to be understood in the context of urban neoliberal governance that shifts the responsibility of social reproduction and neighbourhood revitalization on the very communities in need of assistance. City governments throughout the Global North, and increasingly in the Global South (Thornton, 2009; Ruysenaar, 2013), have

Fig. 5.3. A welcome sign commemorates First Lady Michelle Obama's visit to the New Roots Community Farm. (Photograph by Fernando J. Bosco.)

adopted policies allowing neighbourhoods to meet their own food needs through urban agriculture and farmers' markets. The 2011 San Diego ordinance that relaxed requirements for growing and selling food on vacant lots can be seen as a victory for community groups, but it also reflects a desire by the city to capitalize on an increasingly popular food movement, resulting in an institutionalization and homogenization of so-called alternative food practices. This is what Joassart-Marcelli and Bosco (2015) call 'an uncritical embrace of localism' that hides and leaves untouched the power inequalities that have led activists to advocate and develop locally based alternative food networks in the first place. In addition, this homogenization of community garden practices in the Global North also stands in contrast with practices in cities of the Global South, where urban agriculture is increasingly tied to a *solidarity economy*, where participants engage in alternative economic practices that try to subvert dominant discourses of what is possible and most effective (Massicotte, 2014).

The consequences of an uncritical embrace of localism in San Diego are particularly obvious in Mt Hope, where claims to 'gain control of the local food system' are refuted by the recognition that local resources and support for the garden are limited. Larger scale socio-economic disparities continue to shape the garden and challenge the capacity of its members to meet their goals. The suggestion, made by one gardener: 'we can do this ourselves ... take ownership of [our] food and create [our] own food system' seems misplaced given the challenges faced by the garden to secure land, raise funds and attract volunteers.

A similar process is at play at New Roots community farm, where the responsibility for the garden's success falls primarily on the shoulders of refugee farmers, with very little assistance from the state. Yet, unlike Project New Village, the International Rescue Committee is part of a global network that provides access to resources including funding, knowledge and institutional infrastructure. The IRC's general website, for instance, features a number of colourful stories about the New Roots programme in San Diego and elsewhere, highlighting the success of farmers in becoming self-sufficient entrepreneurs (IRC, 2015b). These narratives make invisible the federal refugee resettlement contracts and grants awarded to the IRC and the large private donations enabled by professional public-relation efforts.

The dynamics surrounding the Juniper-Front garden are different, given the admitted lack of interest in addressing social disparities in access to food. Conflating what takes place in Bankers' Hill with what is happening in City Heights and Southeastern San Diego under the label of 'community garden' masks the disparities of socio-economic processes shaping those places. Juniper-Front is a leisure space, which implies less pressing needs. Yet, Juniper-Front and other similar gardens are used in the public imagination and policy discourse as examples of urban agriculture that low-income communities of colour could emulate, without any recognition of the differences in circumstances and minimizing the struggles that many community groups face.

5.5 Conclusion

The three case studies that we present contribute to recent research that highlight the ambivalent relationships between urban agriculture and neoliberal urbanism in cities of the Global North (McClintock, 2014; Joassart-Marcelli and Bosco, 2015) by drawing attention to the material and discursive role of 'community'. Our findings can be summarized in three key points. First, we highlight how, in the context of urban agriculture, commonly adopted definitions of community conflate 'community' with 'territory'. These confused definitions mask the diversity of actors that are embedded in what are, in reality, several overlapping communi*ties* with different interests, often – but not always – occupying a shared territory or place.

Second, our discussion of the experiences of three community gardens in San Diego shows that leaders and organizations in charge have had very specific agendas and have targeted certain groups at the expense of others. This is done to ensure the success, or at least the continuity, of urban agricultural practices under a specific set of 'best practices' that can also help cement organizations in a competitive environment of scarce resources. The community gardening practices we discuss here have led to the creation of territorially and socially exclusionary communities, albeit with some degree of variation. Even in the two community gardens (Juniper-Front and Mt Hope) that aim at serving diverse and disenfranchised populations, exclusionary practices exist to the extent that only a limited range of community participants are involved.

Third, we show how the three gardens openly embrace a community framework to guide their activities, relegating the provision of food to a lower priority level. Even though the three gardens have different goals (ranging from growing food for leisure to food security) they all privilege 'cultivating', 'harvesting' or 'growing community' rather than actually growing food. We show that this heavy emphasis on 'growing community' has been shaped by an uncritical embrace of localism and a discourse of resilience that is characteristic of contemporary neoliberal urban governance. Many of those involved in the community gardens in our study celebrate the more flexible municipal regulatory environment that allows for the expansion of urban agriculture in San Diego. In doing so, they become less attentive to the remaining power inequalities in the region beyond their own circumscribed community. To us, the danger here is that accepting responsibilities for functions of social reproduction, such as feeding a community, can blind urban agricultural supporters and participants from the reasons why they advocate for an increase in local and alternative food networks in the first place.

Altogether, our findings suggest that an urban agriculture policy – including support for community gardens – that relies on (exclusionary) communities to create spaces to grow food will *not* generate the sort of democratic, diverse and just communities envisioned by proponents. Such an agricultural policy will also fall short of creating a truly fair food system that is supported by a variety of locally sourced, nutritious food that is accessible to all – even to those who are not personally involved in urban agriculture.

Acknowledgements

Research for this chapter was supported by a grant from the National Science Foundation, Geography and Spatial Sciences Program and Anthropology Program, Division of Behavioral and Cognitive Sciences, award number 1155844. The authors would like to thank Emanuel Delgado and Audrey Porcella for their research assistance in this project.

References

Alaimo, K., Reischl, T.M. and Allen, J.O. (2010) Community gardening, neighborhood meetings, and social capital. *Journal of Community Psychology* 38(4), 497–514.
Alkon, A.H. and McCullen, C.G. (2011) Whiteness and farmers markets: performances, perpetuations . . . contestations? *Antipode* 43(4), 937–959.
Alkon, A.H. and Norgaard, K.M. (2009) Breaking the food chains: an investigation of food justice activism. *Sociological Inquiry* 79(3), 289–305.
Allen, P. (1999) Reweaving the food security safety net: mediating entitlement and entrepreneurship. *Agriculture and Human Values* 16(2), 117–129.
Baker, L.E. (2004) Tending cultural landscapes and food citizenship in Toronto's community gardens. *Geographical Review* 94(3), 305–325.
Baron, E. (1993) Interview with Ethel Baron. 4 June. Transcript. San Diego Historical Society Oral History Program, San Diego, California, USA.
Bryant, R.L. and Goodman, M.K. (2004) Consuming narratives: the political ecology of 'alternative' consumption. *Transactions of the Institute of British Geographers* 29(3), 344–366.
Center on Policy Initiatives (2014) Making ends meet: when wages fail to meet basic costs of living in San Diego County. Available at: http://docs.sandiego.gov/councilcomm_agendas_attach/2014/edir_140324_2c.pdf (accessed 11 November 2016).
Chapin, T. (2002) Beyond the entrepreneurial city: municipal capitalism in San Diego. *Journal of Urban Affairs* 24(5), 565–581.

Colasanti, K., Hamm, M. and Litjens, C. (2012) The city as an 'agricultural powerhouse'? Perspectives on expanding urban agriculture from Detroit, Michigan. *Urban Geography* 33(3), 348–369.

Cortese, A. (2011) *Locavesting: The Revolution in Local Investing and How to Profit from it*. Wiley, Hoboken, New Jersey, USA.

Davis, M., Mayhew, K. and Miller, J. (2003) *Under the Perfect Sun: The San Diego Tourists Never See*. New Press, New York, USA.

DeFilippis, J., Fischer, R. and Shragge, E. (2006) Neither romance nor regulation: re-evaluating community. *International Journal of Urban and Regional Research* 30(3), 673–689.

Delaney, D. and Leitner, H. (1997) The political construction of scale. *Political Geography* 16(2), 93–97.

Eizenberg, E. (2013) *From the Ground Up: Community Gardens in New York City and the Politics of Spatial Transformation*. Ashgate, Burlington, Vermont, USA.

Ellsworth, S. and Feenstra, G. (2010) *Assessing the San Diego Food System: Indicators for a More Food Secure Future*. UC Davis Agricultural Sustainability Institute, Davis, California, USA.

Elwood, S. (2006) Beyond cooptation or resistance: urban spatial politics, community organizations and GIS-based spatial narratives. *Annals of the Association of American Geographers* 96(2), 323–341.

Erie, S.P., Kogan, V. and MacKenzie, S.A. (2010) Redevelopment, San Diego style: the limits of public–private partnerships. *Urban Affairs Review* 45(5), 644–678.

Feenstra, G. (1997) Local food systems and sustainable communities. *American Journal of Alternative Agriculture* 12(1), 28–36.

Finley, R. (2015) The gangsta gardener: Ron Finley. Available at: http://ronfinley.com (accessed 11 November 2016).

Firth, C., Maye, D. and Pearson, D. (2011) Developing 'community' in community gardens. *Local Environment: The International Journal of Justice and Sustainability* 16(6), 555–568.

Florida, R. (2005) *Cities and the Creative Class*. Routledge, New York, USA.

Florido, A. (2011) She cultivates community in Southeastern San Diego. Available at: http://www.voiceofsandiego.org/all-narratives/q-and-a/she-cultivates-community-in-southeastern-san-diego (accessed 21 March 2015).

Guthman, J. (2008) 'If they only knew': color blindness and universalism in California alternative food institutions. *The Professional Geographer* 60(3), 387–397.

Guthman, J. (2011) *Weighing In: Obesity, Food Justice, and the Limits of Capitalism*. University of California Press, Berkeley, California, USA.

Harvey, D. (1996) *Nature, Justice and the Geography of Difference*. Blackwell, Malden, Massachusetts, USA.

Hendrickson, M.K. and Heffernan, W.D. (2002) Opening spaces through relocalization: locating potential resistance in the weaknesses of the global food system. *Sociologia Ruralis* 42(4), 347–369.

Hewitt, C.P. (2013) *Financing Our Foodshed: Growing Local Food with Slow Money*. New Society Publishers, Gabriola, British Columbia, Canada.

Heynen, N. (2009) Bending the bars of empire from every ghetto for survival: the Black Panther party's radical antihunger politics of social reproduction and scale. *Annals of the Association of American Geographers* 99(2), 406–422.

Hinrichs, C.C. and Allen, P. (2008) Selective patronage and social justice: local food consumer campaigns in historical context. *Journal of Agricultural and Environmental Ethics* 21(4), 329–352.

Hynes, H.P. (1996) *A Patch of Eden: America's Inner-City Gardeners*. Chelsea Green Publishing Company, White River Junction, Vermont.

IRC (2015a) The IRC in San Diego, CA. Available at: https://www.rescue.org/united-states/san-diego-ca (accessed 11 November 2016).

IRC (2015b) IRC website. Available at: http://www.rescue.org (accessed 11 November 2016).

Jessop, B. (2002) Liberalism, neoliberalism and urban governance: a state-theoretical perspective. *Antipode* 34(2), 452–472.

Joassart-Marcelli, P. (2012) For whom and for what? An investigation of the roles of nonprofits as providers to the neediest. In: Salamon, L. (ed) *The State of Nonprofit America*. The Brookings Institution Press, Washington, DC, USA, pp. 657–681.

Joassart-Marcelli, P. and Bosco, F.J. (2015) Alternative food projects, localization and neoliberal urban development: farmers' markets in southern California. *Métropoles* 15(1), 2–23.

Joassart-Marcelli, P. and Martin, N. (2015) Migrant civil society: shaping community and citizenship in a time of neoliberal reforms. In: Haymes, V., Vidal de Haymes, M. and Miller, R. (eds) *The Routledge Handbook on Poverty in the United States*. Routledge, New York, USA, pp. 547–554.

Joassart-Marcelli, P. and Wolch, J. (2003) The intrametropolitan geography of poverty and the nonprofit sector in southern California. *Nonprofit and Voluntary Sector Quarterly* 32(1), 70–96.

Johnston, J. and Baumann, S. (2009) *Foodies: Democracy and Distinction in the Gourmet Foodscape*. Routledge, New York, USA.

Joseph, M. (2002) *Against the Romance of Community*. University of Minnesota Press, Minneapolis, Minnesota, USA.

Kayzar, B. (2013) Interpreting philanthropic interventions: media representations of a community savior. In: Curti, G., Craine, J. and Aitken, S. (eds) *The Fight to Stay Put: Social Lessons Through Media Imaginings of Urban Transformation and Change*. Steiner Verlag, Stuttgart, Germany, pp. 209–228.

Kingsley, J.Y. and Townsend, M. (2006) 'Dig in' to social capital: community gardens as mechanisms for growing urban social connectedness. *Urban Policy and Research* 24(4), 525–537.

Kloppenburg, J. Jr, Lezberg, S., DeMaster, K., Stevenson, G.W. and Hendrickson, J. (2000) Tasting food, tasting sustainability: defining the attributes of an alternative food system with competent, ordinary people. *Human Organization* 59(2), 177–186.

Levkoe, C.Z. (2006) Learning democracy through food justice movements. *Agriculture and Human Values* 23, 89–98.

MacLeod, G. and Goodwin, M. (1999) Space, scale and state strategy: rethinking urban and regional governance. *Progress in Human Geography* 23(4), 503–527.

Madyun, G. and Malone, L. (1981) Black pioneers in San Diego 1880–1920. *The Journal of San Diego History* 27(2). Available at: https://www.sandiegohistory.org/journal/1981/april/blacks (accessed 11 November 2016).

Massicotte, M.J. (2014) Solidarity economy and agricultural cooperatives: the experience of the Brazilian landless rural workers movement. *Journal of Agriculture, Food Systems, and Community Development* 4(3), 155–176.

McCarthy, J. (2006) Neoliberalism and the politics of alternatives: community forestry in British Columbia and the United States. *Annals of the Association of American Geographers* 96(1), 84–104.

McClintock, N. (2014) Radical, reformist, and garden-variety neoliberal: coming to terms with urban agriculture's contradictions. Urban Studies and Planning Faculty Publications and Presentations. Paper 93. Available at: http://pdxscholar.library.pdx.edu/usp_fac/93 (accessed 20 March 2015).

Norberg-Hodge, H., Merrifield, T. and Gorelick, S. (2002) *Bringing the Food Economy Home: Local Alternatives to Global Agribusiness*. Kumarian Press, Bloomfield, Connecticut, USA.

Obama, M. (2012) *American Grown: The Story of the White House Kitchen Garden and Gardens Across America*. Crown Publishers, New York, USA.

Obosu-Mensah, K. (1999) *Food Production in Urban Areas: A Study of Urban Agriculture in Accra, Ghana*. Ashgate Publishing, Aldershot, UK.

Peck, J., Theodore, N. and Brenner, N. (2012) Neoliberalism resurgent? Market rule after the great recession. *The South Atlantic Quarterly* 111(2), 265–286.

Porcella, A.L.A. (2012) Reap what you sow: social capital in community gardens. MA thesis, San Diego State University, San Diego, California, USA.

Project New Village (2012) People's produce project good food legacies of southeastern San Diego (video). Available at: https://www.youtube.com/watch?v=23gg6xG4o2c&t=596 (accessed 20 March 2015).

Project New Village (2015) Project New Village website. Available at: http://www.projectnewvillage.org (accessed 12 March 2015).

Quastel, N. (2009) Political ecologies of gentrification. *Urban Geography* 30(7), 694–725.

Redwood, M. (ed.) (2012) *Agriculture in Urban Planning: Generating Livelihoods and Food Security*. Routledge, New York, USA.

Ruysenaar, S. (2013) Reconsidering the 'Letsema Principle' and the role of community gardens in food security: evidence from Gauteng, South Africa. *Urban Forum* 24(2), 219–249.

Saldivar-Tanaka, L. and Krasny, M. (2004) Culturing community development, neighborhood open space, and civic agriculture: the case of Latino community gardens in New York City. *Agriculture and Human Value* 21, 399–412.

Schmelzkopf, K. (2002) Incommensurability, land use, and the right to space: community gardens in New York City. *Urban Geography* 23(4), 323–343.

SDCGN (2015) San Diego Community Garden Network website. Available at: http://www.sdcgn.org (accessed 20 March 2015).

SDUT (2010) For Southeastern San Diego, a farmers market of its own. *San Diego Union-Tribune*, 29 November. Available at: http://www.sandiegouniontribune.com/sdut-southeastern-san-diego-farmers-market-call-its-own-2010nov29-story.html (accessed 11 November 2016).

Slocum, R. (2007) Whiteness, space and alternative food practice. *Geoforum* 38(3), 520–533.
Staeheli, L., Mitchell, D. and Gibson, K. (2002) Conflicting rights to the city in New York's community gardens. *GeoJournal* 58, 197–205.
Teig, E., Amulya, J., Bardwell, L., Buchenau, M., Marshall, J.A. and Litt, J.S. (2009) Collective efficacy in Denver, Colorado: strengthening neighborhoods and health through community gardens. *Health and Place* 15, 1115–1122.
Thornton, A. (2009) Pastures of plenty? Land rights and community-based agriculture in Peddie, a former homeland town in South Africa. *Applied Geography* 29(1), 12–20.
Tolin, R. (2009) The $46,000 question. Available at: http://sdcitybeat.com/news-and-opinion/news/46-000-question/ (accessed 20 March 2015).
Tornaghi, C. (2014) Critical geography of urban agriculture. *Progress in Human Geography* 38(4), 551–567.
Washburn, D. (2009) Building a farm and saving a culture. Voice of San Diego, 4 June. Available at: http://www.voiceofsandiego.org/topics/news/building-a-farm-and-saving-a-culture/ (accessed 11 November 2016).
Watts, M. (2004) Antinomies of community: some thoughts on geography, resources and empire. *Transactions of the Institute of British Geographers* 29(2), 195–216.
Wolch, J. (1990) *The Shadow State: Government and Voluntary Sector in Transition*. Foundation Center, New York, USA.
Zukin, S. (2008) Consuming authenticity: from outposts of difference to means of exclusion. *Cultural Studies* 22(5), 724–748.

6 'Growing food is work': The Labour Challenges of Urban Agriculture in Houston, Texas

Sasha Broadstone[1] and Christian Brannstrom[2]*
[1]*Stokes Nature Center, Logan, Utah, USA;* [2]*Texas A&M University, College Station, Texas, USA*

6.1 Introduction

Urban agriculture (UA) contributes to improved food access, eating habits and community interaction. Although UA offers modest contribution to food supply in cities, many scholars report the importance of UA in improving food sovereignty and producing alternative food systems. Details on the management, production and distribution strategies of UA in the Global North (GN) are poorly documented, but relatively better knowledge exists in the Global South (GS) on these topics. Knowledge and understanding of these aspects of UA are important for practitioners and policy makers in the GN and GS. Here we ask: how, and under what conditions, do UA sites in the GN produce food? We answer this question through an agricultural systems approach applied to UA organizers in Houston, Texas, regarding management strategies and food production practices. UA site objective and site-access regimes were closely related and influential in determining decision-making strategies, division of labour, and destination of the harvest. As our title suggests, labour was a major concern among respondents.

6.2 Background

Scholars have associated urban agriculture with healthier eating patterns, increased environmental education, improved food security and stronger community ties (Gottlieb and Joshi, 2010; Kortright and Wakefield, 2011; Smith and Miller, 2011; Block *et al.*, 2012). UA may also promote democratic decision making, urban land reclamation, and community empowerment through efforts toward healthy food access (Pudup, 2008; Guitart *et al.*, 2012; Ghose and Pettygrove, 2014). 'Psychologically satisfying' UA food harvests shared with friends and family (Page, 2002, p. 49; Kortright and Wakefield, 2011) may compensate for the tiring and tedious drudgery of farm work.

However, an emerging gap exists between what UA *could* be, and what it *is* in practice. Relatively little is known about labour, management, production and distribution strategies of UA at the scale of farming plots, especially in GN sites of UA. Guitart *et al.* (2012) argued that there is a substantial lack of research regarding the cultivation methods of UA practitioners, i.e. *how* people grow food. Scholars have reported that UA participants in the GN do not trust the

*Corresponding author; e-mail: cbrannst@geos.tamu.edu

conventional food system and want to know and control the conditions in which the food is grown (Kortright and Wakefield, 2011; Smith and Miller, 2011).

Social scientists have analysed labour relations in GS and GN agricultural systems, especially in cases of incomplete mechanization, for decades. Detailed consideration of labour is missing in studies of UA in the GN, but labour in the GS is often approached as a 'double burden' for women. As Hovorka (2006) argues, UA could be a 'double-edged sword for women' because it may increase their labour burden. If development agencies promote UA as a livelihood-improvement strategy in the GS, then women will do even more work. Other studies report the high participation rate of women in UA in the GS (WinklerPrins and de Souza, 2005, p. 112; Gallaher et al., 2013, p. 395).

Labour and work – in their varied, nuanced, symbolic, and material meanings and forms – present a promising analytical lens for the desired crossover between studies of GN and GS studies of UA. This requires changing the passive construction 'food is grown' (Taylor and Lovell, 2014, p. 294) to the question 'who grows food in UA?' Recently, Gray et al. (2014, p. 200) indicated that mainly women and children cultivated in a home gardening project near San Jose, California, the start of a promising trend. Labour, therefore, presents an important comparative category of analysis between the GN and GS for studies of UA, even if rationale, materiality and meaning of work may differ internationally.

We understand UA as a farming system (Turner and Brush, 1987) that produces nature in cities, following the objective of urban political ecology (UPE) to 'disentangle the interwoven knots of social process, material metabolism, and spatial form' in the urban environment (Heynen et al., 2006, p. 8). We take material metabolism to mean food-production practices and harvest destination, while social process involves knowledge, land tenure and labour. Here we describe how the social process is contingent upon mobilizing workers in a multi-ethnic and highly segregated city. Labour relations strongly influence how material metabolism occurs in terms of access to UA plots, decision making, and harvest destination. Spatial form includes the geospatial relations among UA sites, associations between UA sites and socio-demographic variables, and the spatial extent of UA sites, summarized elsewhere (Broadstone, 2013).

Our definition of UA follows Table 1.1 (Chapter 1, this volume) in that we study community gardens (allotments, in which organizations lease plots of land to individuals who are responsible for farming) and non-profit urban farms (all-access, in which labour, harvest and decision making are shared), which include 'institutional gardens' such as church land (managed as all-access urban farms) and 'interstitial' spaces such as rights-of-way managed as either allotments or all-access.

We offer three claims, based on a case study of Houston's UA, with relevance in GN and GS. First, we emphasize how UA organizers offered numerous reasons why 'growing food is work' and appear to balance the meaning of farming, mainly social interactions and education, with the drudgery of work. This claim adds to literature that stresses community development and social capital formation in UA in the GN (Gray et al., 2014) and the importance of UA in social relations in the GS (WinklerPrins and de Souza, 2005). Second, we note how UA site goals may be a means to overcome the drudgery of farm work and may be used to secure land, especially on church properties. The importance of church land as a location for UA in Houston may be replicated elsewhere in the GN. Third, we describe a local knowledge network that has disseminated best practices suited to Houston's agricultural climatology.

Houston, a global energy capital and fourth-largest city in the US, with 6.5 million people living in the metropolitan statistical area (Melosi and Pratt, 2014), is a multi-ethnic metropolis (25% white, 24% African-American; 44% Hispanic; 6% Asian; 29% foreign-born). It is characterized by immigration, historical segregation of African Americans, hydrocarbon-driven economic growth, meagre social welfare and considerable corporate welfare in the form of urban revitalization and public art (Feagin, 1985; Podagrosi et al., 2011). Houston's pro-growth political forces directed the city's outward spatial expansion, while simultaneously preserving a market-friendly approach that utilized outside resources to increase private-sector profit. This growth strategy stimulated urban expansion through annexation, partially aided by national programmes that funded projects such as Houston's freeway system (Thomas and Murray, 1991).

Houston's UA activity is primarily in the cool season. For example, January mean daily minimum temperature is 4–7°C. Urban farming declines in July, when the daily maximum temperature is around 35°C. Annual precipitation, around 1300 mm (Nielsen-Gammon, 2011), normally requires supplemental irrigation for UA. In addition to hot and humid conditions in the summer months, another obstacle to outdoor work is Houston's surface ozone problem, caused by emissions from vehicles, a major petrochemical complex, and meteorological conditions, which has become severe enough for Houston to exceed the National Ambient Air Quality Standard frequently (Liu et al., 2015).

6.3 Data and Methods

To develop our sample frame we defined UA as socially organized food production visible on Google Earth imagery, mainly through the presence of raised beds. This definition excludes backyard gardens that are individual or family enterprises. We developed a list of UA sites by mapping published data (UA affiliate status, US$40 per year fee) from Urban Harvest, a Houston-based charitable organization that promotes and supports UA ($n=67$ as of November 2012). We augmented the Urban Harvest database with 12 additional sites identified through interviews with UA organizers. We verified the final list of 79 sites using Google Earth, following procedures described in Taylor and Lovell (2012). Of the 79 potential UA sites, 9 (7 schools, 2 private land) showed no visible evidence of UA, yielding a total sample of 70 UA sites. After eliminating schools from the sample (children did not occupy an analytical category in this research), we made email or telephone contact with 45 site organizers and determined that 26 UA sites were producing food during January to April 2013, our study period, and they accepted participation in the study. Through interviews with UA organizers, 12 more unaffiliated gardens were identified and contact was attempted. Of this group, 5 additional UA organizers agreed to participate. Some UA sites were near downtown Houston (Fig. 6.1), while others were located in Houston's many low-density neighbourhoods.

In total, we visited 31 UA sites and interviewed organizers following survey and semi-structured interview protocols during the main period of Houston's vegetable production calendar. (Survey and semi-structured interview protocols were approved by Texas A&M University, IRB protocol 2013-0059.) The sample (organizers responsible for the UA site) was evenly balanced between men and women (15 males, 16 females), but included more whites ($n=20$) than blacks ($n=6$) or Hispanics ($n=3$). (Two organizers declined to state their ethnicity.) Nearly three-quarters of organizers ($n=22$) reported extensive cultivation experience. Half the sample reported working 4 hours or more per week on managing the UA site, while half reported working fewer than 4 hours per week. Sixteen per cent of organizers reported that they provided the primary source of funding for their UA site. Indeed, managing a UA site is a multifaceted and challenging job. Organizers, who receive no salary, appear to be passionate about the site's objective and internally motivated to take the time and effort required to manage a site. We used a survey aimed at obtaining qualitative data on UA site management strategies regarding the production of food, division of labour, and the distribution of harvest. The first author did not participate in farming during site visits; rather, she scheduled interviews to coincide with participant work days to observe the garden as site of production and to observe how organizers interacted with volunteers, participants and passers-by. Interviews lasted between 30 and 45 minutes. After transcribing the interviews, we coded responses in qualitative software using a code list created from the interview prompts and from emerging categories we had not anticipated.

6.4 Findings

Table 6.1 presents a summary of our main findings organized in terms of the number of UA sites reporting site objective, tenure, cultivation practices, labour, harvest destination and development goals. Below we discuss the key characteristics, emphasizing material metabolism and social processes, evidenced in our qualitative data.

Fig. 6.1. View of downtown Houston from a UA site. (Source: Sasha Broadstone.)

6.4.1 Site objective

Improving food access was a priority objective for 68% of UA organizers. As one organizer stated: 'We focus on getting food to people . . . we try [to] grow food that services . . . the Hispanic community.' Another organizer mentioned how for one person working at the site, UA was 'a main source of her food supply', so the organizer did not bother pressing for the US$25 allotment fee. However, organizers are also aware of the limitations of UA to feed large numbers of people: 'our involvement is a very small drop in the bucket . . . the amount of food we contribute is a small amount, but it's appreciated'.

Community outreach, the primary objective for 19% of UA organizers surveyed, is a fundamental component for UA sites affiliated with religious organizations. 'Our main purpose', said one organizer, 'is to connect with our community by meeting both their physical and spiritual needs'. One organizer explained that the UA site aimed to 'promote community harmony', and went on to note that the UA site was located in 'a very, very diverse community'. For this organizer, UA 'might be something to maybe promote some interaction because in this community . . . different ethnic groups tend to stay in their own communities, and they don't necessarily interact with the other communities'. Most organizers who emphasized community outreach favoured an inclusive approach, to which UA was well suited, according to one organizer, whose UA was linked to a church. One organizer told us that the church 'had all of this land and it wasn't being used, so we included this garden as is a part of outreach', without restricting access on basis of religious beliefs: 'People are welcome to come, they don't have to come to church, we don't preach to them, none of that, they can just garden here.'

Education was the primary objective for 12% of UA organizers, who reported that they enjoyed teaching others the fundamentals of food production. For them, the process offered a natural bridge to the discussion of

Table 6.1. Summary of survey and semi-structured interview responses of UA sites (n=31). (Source: Field work, January–April 2013.)

Category	Response or characteristic	No. UA sites	% UA sites
Primary objective	Food access	21	67.7
	Community outreach	6	19.4
	Education	4	12.9
Access type	All-access	21	67.7
	Allotment	10	32.3
Land access	Owner	19	61.3
	Free-use	8	25.8
	Lessee	4	12.9
Land ownership	Religious organization	13	41.9
	City of Houston	11	35.4
	Private non-religious	4	12.9
	School	3	9.7
Plot size (square metres)	> 465	16	51.6
	465–929	6	19.4
	930–1,858	4	12.9
	>1,858	5	16.1
Irrigation	Hose	20	64.5
	Low-flow drip	5	16.1
	Combination	5	16.1
	Sprinkler	1	3.2
Water source	Municipal water	27	87.1
	Well water	3	9.7
	Rain catchments + city water	1	3.2
Cultivation	Strictly all organic	25	80.6
	Individual choice	6	19.4
Harvest destination	Individual choice	21	67.7
	Donation only	10	32.3
Number of participants	1–5	8	25.8
	6–10	10	32.3
	11–25	8	25.8
	> 30	5	16.1
Challenges (number of mentions exceeds n=31)	Recruiting and retaining committed participants	26	83.9
	Secure land tenure	12	38.7
	Reliable funding source	7	22.6
	Physical ability	9	29.0

other food-related topics, such as healthy-eating practices and nutrition. As one organizer put it, 'we talk about what to eat and what better way to do that than to show people how to grow their own food?' Site organizers recognized that UA cannot produce enough to fully address city-wide hunger. Instead, they aimed to teach people how to grow their own food to provide them with additional, healthy food access options. One organizer admitted that 'you can't feed everyone in that community' with the UA site, 'but with the beds you have, you can show them how to do it, and give them the option to be a part of it'. Some organizers teach participants how to grow their own food to reinforce ideas such as sustainability and locally accessed food. One organizer phrased it in simple terms:

[There is] a more sustainable way to access food, and that's growing [your] own. It doesn't have to be in a fancy garden like this, it can be in containers, or in one little plot, but we want people learning that you can in fact grow your own food.

For this organizer, UA was 'not about donating to people in the neighborhood,' but rather about getting volunteers to 'understand that this is really about getting out and learning how to grow food and having fun'.

6.4.2 Plot access regime

A second aspect of the social process of producing nature in cities relates to the rules for obtaining access to land to grow food. Most UA sites (68% of sample) use the 'all-access' model, which organizers say includes advantages such as production flexibility and efficiency resulting from centralized decision making (Table 6.1). In all-access sites, labour, harvest and decision making are shared among organizers and farmers (volunteers). All-access sites normally have a plan for crop production, rotation and bed uniformity. In two-thirds of all-access sites, organizers make most management decisions, while one-third rely on input from organizers and participants. UA organizers claim that this means of site access is flexible and allows for quick adjustment to changing circumstances, especially labour. By contrast, allotment sites (32% of sample) lease plots of land to individuals (farmers) who are responsible for labour, planting decisions, crop management and harvest. Allotment sites are thought to provide the advantage of more sustainable funding, because farmers pay a monthly fee to site organizers for plot access, but decision making is normally decentralized to farmers to cultivate allotments.

UA organizers provided varied rationales for all-access tenure, but they cited flexibility in terms of time and labour management as a major benefit of all-access sites. For one UA organizer, 'everything is shared . . . I'm not worried about how much stuff I'm going to have to eat and stuff I have to take care of'. Organizers also claimed that working together at the all-access site fostered feelings of community ownership and lowered the barriers of participation for anyone interested: 'We wanted everybody to get involved', according to one organizer.

All-access UA may also be a good choice for sites where committed, long-term participation is harder to obtain. According to this rationale:

> This is all community, no set anything, really loose, really fluid . . . allotment beds were kind of a failed experiment [at our old UA site] . . . we ended up with these beds that we didn't really feel comfortable maintaining ourselves because they weren't really our responsibility, so [the growing space] was kind of wasted.

Other organizers mentioned labour as a key issue in their preference for all-access tenure. At a UA site, organizers had planned for allotments leased at US$5 per year, 'but once we did that nobody signed up for the plots, so we just went ahead and planted the garden and kept it going, and now we harvest the food and give it out in our food pantry'.

6.4.3 Land tenure

Land tenure is a major concern among UA organizers and an important material basis for UA as a social process that produces nature in cities. Surveyed UA sites were mainly operated by the land owner (61%), while 8 sites were used freely without formal lease (Table 6.1). Four were leased through contract with the landowner. Religious organizations owned 13 UA sites and the City of Houston owned 11 sites. Schools owned three UA sites and private entities owned four. Nearly all sites ($n=27$) were located on land exempt from property tax.

Nearly two-thirds of sites surveyed owned the land of the UA production area. Ownership creates a stable environment and may encourage long-term development goals and management practices; moreover, 87% of these sites were exempt from property tax. The siting of UA sites on church land is an especially important phenomenon. Most UA sites (68%) on non-public property were located on land owned by a religious organization. UA sites located on private residential property made up less than 10% of sites surveyed, and even fewer sites were owned by universities. As one organizer noted: 'A lot of times churches got land, and they don't do anything with it. It's sitting there, and they could help their parishioners tremendously, you'd be surprised, we get tons of stuff [food] out of here.'

A major theme observed in discussion of land tenure was the notion of putting apparently unused land to productive use. One organizer explained how their church had purchased an easement to use as a parking lot during a building expansion. After the construction was complete, the area was not utilized and could not be developed because of an overhead electrical utility easement for transmission. In addition to making use of vacant land, the site was selected because 'It's sunny, well exposed, and we had water that we could get out here relatively easy, and it's visible from the church.'

One-quarter of UA sites are located on municipal land, which confers tax exemption from property tax and also may generate other benefits such as a free water supply. The example of one organizer shows the tangible benefits:

> City land, city water, and no water bill. It's part of the park so it's like a service of the city. We're just responsible for everything that goes inside. They come out and inspect it every once in a while to make sure we're doing the right thing.

The remaining 13% lease their land, usually for a nominal fee, making it 'essentially free', according to one organizer. However, the duration of leases for UA is variable, increasing the risks of investment in land, and contracts come with a 6-month vacate notice. One organizer reported on leases from the Houston Housing Authority, from which the UA site was leased for 'a very small fee, just enough to make it legal', while another organizer reported paying a lease of US$1 per year to a school district for UA.

One-fifth of UA organizers mentioned land tenure as a challenge to production. Interviews revealed that UA organizers must navigate considerable transaction costs (e.g. the costs of seeking land and negotiating contracts) to obtain urban farmland. UA organizers must find out who owns unused land suitable for UA, how a UA site can get permission to use the land, and how to obtain access to water. After organizers identify and resolve these issues, the remaining problem is risk. For example, without secure tenure, the landowner could force the UA site to leave with little or no notice, meaning that substantial investments of labour and capital in soil preparation are lost.

Gaining lawful access to some of these unused sites is a major challenge. One organizer reported that access to land is the highest hurdle to creating a new UA site, especially in low-income areas:

> It's more than just access to food, it's access to land that you can build a garden on, and have a long enough tenure, you know, the period of time which you can use the land, that's important. In underserved neighborhoods access to land is a critical item. Ultimately the real gain that you have is community, but you got to have the garden first, and access to land is the barrier.

Another organizer discussed the challenges of turning land into UA in a neighbourhood with abandoned houses and vacant lots. The narrative merits quotation at length. In early 2013, this organizer was operating on a privately owned site of ~1951 square metres in a residential area. The organizer's account of the site's recent history recounts the numerous challenges of UA.

> When I first came to the garden, it was so grassy and growed up, it was almost taller than the house! Wasn't nobody keepin' that yard clean, and these people were afraid to live down here in they own house, cause they didn't know if there was somebody down in here [the overgrown site]. All them other little houses were fallin' down, and it took me a whole year, fightin' with the city to get them to tear 'em down, to get 'em down from around those poor peoples livin' down here alone by themselves. We threatened [the city council] that we'd bring the cameras out here, had to keep remindin' them, because it's been years, the houses, they've been asking them to be torn down for years. I made threats that I was gonna bring news cameras out here and let them see what was goin' on out here. And the city don't like to be bothered with those news cameras.

After city workers demolished the dilapidated houses and the site was producing food, this organizer wanted to expand production into the neighbouring vacant lots. However, identifying land tenure and obtaining permission to use the land was a struggle. This organizer related how he called a city councilman 'several times', but his calls were never returned:

> I wanted to know, does that city own that land? Because [the city] had red-tagged that land. I told [city councillor] I want to discuss with him that lot next door to the garden because I would like to expand to that lot there. We wanted to do some more fruit trees on that lot there. I keep callin', and nobody never returns my call.

6.4.4 Production practices

The production practices or material metabolism of UA are non-mechanized and adjusted to the agricultural climatology of Houston as interpreted through the main organization, Urban Harvest. More than half ($n=16$) of sites were less than 465 square metres of farmland, while only five reported more than 1858 square metres. The 70 sites we surveyed through Google Earth have a total production area of 40,515 square metres, roughly comparable to 160 UA sites in Chicago that had a production area of 54,518 square metres (Taylor and Lovell, 2012). UA sites in Houston grow an array of different crops, depending on season and preference of participants. In early spring, when this research was conducted, the cold crops of different greens, beets, broccoli, onions and leeks were nearing their last harvests. Participants were clearing beds to make way for warm-season crops such as tomatoes, peppers, corn, beans and squash. In hot Houston summers, only okra and eggplant can survive for long without constant water and attention, and labour is difficult to arrange because of heat, humidity and surface ozone. Twenty of the surveyed sites have incorporated fruit production, and all but three sites grew culinary herbs (Fig. 6.2).

Municipal water, priced at normal household rates, is the main source for irrigation at nearly all UA sites, which mainly use a hose for irrigation. Despite the labour, time and water-saving qualities of drip irrigation, only five sites reported using drip irrigation because of high installation cost and maintenance requirements. Eighty-one per cent of sites reported that they are strictly organic, probably the result of the fact that nearly all surveyed sites are affiliated with Urban Harvest, which is a strong supporter of organic principles. The other 19% of UA sites surveyed explained that organic principles are encouraged but not enforced. For example, one organizer 'recommends' mostly organic crop production while not making it a site rule, saying that herbicides 'should be regulated and seldom used' and another indicated that the UA site was 'mostly' organic, 'but if we need to get rid of certain kind of bug without hurting anyone or causing cancer then so be it'. Some organizers admitted to using pesticide to deter fire ants because organic products were 'totally useless'.

All surveyed sites employed raised-bed cultivation, following methods espoused by Urban Harvest because of Houston's often waterlogged clay soils. They all used a layer of newspaper as a foundation to deter weeds, then filled in the rest with organically enriched soil. Compost, either

Fig. 6.2. UA production site in Houston showing cold-season crops and herbs. (Source: Sasha Broadstone.)

bought or made (or both), is used by 39% of all sites, and some type of organic fertilizer is utilized by 48% of sites. One-fifth of organizers stated that they used Nature's Way and Micro-Life at their site. Nature's Way Resources is a Houston business that provides organic supplies for gardeners and landscapers; the firm is also an Urban Harvest partner, which provides special rates to affiliated sites on soil, mulch, compost, fertilizers, and the MicroLife brand, which offers a range of organic and biological fertilizers. Only 10% of organizers used mulch and 13% used manure for soil enrichment. Three sites have a member or an organizer who is capable of performing periodic soil analysis, which aids in the decision-making process for soil amendments. Regardless of what formula they use, all organizers understand how essential soil amendments are to vegetable production.

Cultivation varies according to season and UA site objective, usually following guidelines in what is considered the essential book for urban farming in Houston, Bob Randall's *Year Round Vegetables, Fruits and Flowers for Metro-Houston*, which started as a dozen stapled pages mimeographed to urban farmers as a booklet in the mid-1980s. Randall's publication has evolved into a book of more than 300 pages in its 12th edition. Randall, one of Urban Harvest's founders, developed specific guidelines for different regions of metropolitan Houston. Organizers and participants emphasized that Randall's book is essential to UA site management. In 2013, the first author of this chapter observed a meeting at a UA site during which members and organizers discussed the crops they would plant the following week. Every person had a copy of Randall's book, which they consulted frequently throughout the discussion.

6.4.5 Harvest destination

We queried UA organizers about harvest destination (Table 6.1) and found that most sites (68%, *n*=21) left destination of crops to individual choice. Organizers reported that 65% of urban farmers reserve the harvest for their own use, while 25% donate either all or a portion of their harvest to food banks, food pantries and soup kitchens. Only two surveyed sites sell their produce at farmer's markets in addition to other harvest destinations.

One-third of sites surveyed, all of which are all-access, reported donating the entire harvest to food banks, food pantries and soup kitchens. These sites often allow 'client preferences' to influence what they cultivate. As one organizer stated:

> I think the most important thing, as we're a donation garden, is to know and to learn what food the people we're giving the food to want, and it doesn't make a lot of sense to grow something they won't eat.

6.4.6 Capital

Capital flows for UA are essential to sustain production. Almost half (45%) of sites rely primarily on financial support from a charitable or non-profit organization, such as a church, civic council, or some other benefactor. When asked about additional or secondary sources of funding, 81% reported donations and grants, while 16% received additional funding from non-profits. Churches play a critical role in capital. For example, one organizer explained that church-member donations provided critical assistance in starting UA and in paying the cost of water for irrigation. Of the sites that are funded primarily by members, 39% are supported either through allotment plot fees or through member contributions.

Securing funding to cover production costs is another concern of 23% of UA organizers. One-quarter of UA sites rely on outside funds procured by their organizer. Forty per cent of allotment organizers found member access fees to be insufficient to cover all costs. For example, one organizer said that member fees did not 'even cover the watering'. This organizer claimed to be 'pretty good at finding the money pots', such as donations and grants. Maintaining a UA site once it is already established is one thing, but, according to one organizer, financing the creation of a site can be the biggest challenge:

> Money is always a problem if they don't have the financial resources to build a garden. We spent almost $30,000 building this garden, and the land was free. But when you have to clear the trees, and when you have to level the land and

build a fence, because that's what [sites] want. Then you set the blocks in, and get the soil, and even though you have volunteers doing all of that, it's big dollars. And it doesn't have to be that way, but a lot of times it can be very expensive. And in Houston we have to use the raised bed concept, because the soil has way too much clay in it, so that's just the way that works.

6.4.7 Labour

When we queried UA organizers about their main challenges, 84% indicated that obtaining committed participants was their biggest (and sometimes only) challenge (Table 6.1). The most commonly cited reason for the lack of interested volunteers was summarized by one organizer, who said:

> growing food is work. Everybody likes the sound of growing their own food, or joining a community site, but after a week or two in the dirt and the heat and the bugs, a lot of people just give up.

In fact, 55% of UA organizers similarly referred to agriculture as hard work. For example, one organizer commented:

> One of the challenges I've found is that people really like the convenience of being able to go to a store, if they have one, or to a fast food restaurant, which is most likely what they're going to do. Growing food is work. It is not easy, and it requires time, and it requires resources, and requires a lot of blood, sweat, and tears.

The theme of hard work in UA was repeated in numerous (and colourful) ways, showing the importance of obtaining labour for successful UA. We cite three organizers below to convey the spirit and content of this issue.

> It's the labor, that's the problem, the labor. Everybody loves to reap that benefit, it's the labor. Because the labor is not easy, you come out here and you sweat, it's not easy. People don't want to do the labor, but like if you say come we got to pick something, you get a whole slew of people. But if you say we got do this, plant that, everybody busy. But if it's time to pick the fruit, now, OK you got a lot of volunteers.

> A lot of people build a garden and think people will come, but it ain't like that, you can build a community garden, but people just don't like to do work, they want the harvest, but they don't want to come out here and do the Monday, Wednesday, Friday. You reap what you sow; you got to do the labor.

> I've had people come, and they'll maybe come for a week, and you know it sounds like such a wonderful thing to do, growing food for the poor, and they find out that it's not all fun . . . it's hard work. And so they don't stay long, unfortunately.

Other organizers mentioned the problem of lack of commitment among UA farmers. One organizer told us that

> we'll get people that'll come and they'll try for a while then give up. People who will come on a fairly frequent basis, that's hard to find. We'll get some people who come for 2–3 weeks, and then they're done.

Obtaining volunteers was

> the hardest thing – to get people committed on a regular basis. I've invested in the materials to have a year-round garden, if I could have volunteers I would personally fund the operating cost myself, if I just had people doing it.

For this organizer, group coordination was not a problem; rather, 'It's the day-to-day, the week-to-week, the ongoing, the weeding, the things that people don't like to do'.

The lack of commitment was related to the expressed concern regarding ensuring an even distribution of labour that 23% of UA organizers indicated. For example, one organizer said 'a lot of times in a community garden you have a few people doing most of the work', and compared running an urban agriculture site to running a business, a field in which she gained ample experience before retiring. Even for her, coordinating people with differing ideas and personalities can still be a challenge:

> A lot of times in a community garden you have a few people doing most of the work, and then people just want to keep their own garden. So I'd like to see 20 people over here working on this compost, because everyone uses it, but nobody wants to work it. Everyone does a great job of taking care of their own bed, and we have a great group who makes sure everything gets plenty of water, but [not] when it comes to cleaning out the shed or turning the compost.

Nearly one-third of organizers cited physical ability required to complete the manual labour as a challenge for their UA site. One organizer noted, 'we're all older and we can't do the physical work now'. For another organizer, 'To keep a community garden going you need to bring new people in to take over, like we're all senior citizens now, we're not going to be able to do this for ever.'

Some organizers argued that Houston's summer climate made it especially difficult to find volunteers. For example, one organizer noted that 'it is hard, especially in the summers, for people to be motivated, you know to come out, because there's not a lot to harvest, but there is still a lot to do, like mow'. Another organizer made a similar argument:

> I used to bring people out, convinced them to come out and garden, but they quit, they quit because they don't like the bugs. 'I'm itching; you cannot make me go out there again!' [Spring] is nice, but in the summer time, it's horrible, bugs everywhere, mosquitoes, ants, it's combat.

Besides some personal discomfort from the work involved, or the hot, bug-infested conditions, there are other barriers to participation, such as available time and physical ability. One organizer claimed that:

> In underserved neighborhoods you see two other types of personal barriers: older folks, who can only do so much, in many cases. The bottom line, however, is that they are available, and they are typically retired and they have time. Time is not the issue – it's physical capability. And the second thing is you've got people who have the physical ability but they're working two and three jobs a week or even day and they just don't have the time.

6.5 Discussion

This study of UA in a GN site focused on site objective, production challenges (especially labour), site access and land tenure. We found that allotment, with its decentralized decision making, is the preferred tenure arrangement, but all-access is most common. Church land is highly important to the persistence of UA, in spite of labour difficulties. The UPE literature encouraged us to consider the social process and material metabolism of UA. An agricultural systems approach helped us to define these terms empirically for data collection. UA organizers give meaning to their sites, create rules for volunteers to access land, respond to land-tenure institutions, and report recruiting labour as their main challenge. Attention to labour has begun to emerge in studies of UA in the Global North (Gray et al., 2014) and has long been an important focus in the Global South, especially as regards women's work (WinklerPrins and de Souza, 2005; Hovorka, 2006; Gallaher et al., 2013).

Our findings support three generalizations with possible relevance in the Global North and South. First, UA inspires meanings about self, food and place that outweigh the negative aspects of work in humid and 'buggy' conditions. Certainly, UA organizers offered numerous reasons why 'growing food is work', a perennial concern among farmers everywhere and famously interpreted by Chayanov (1986) in terms of drudgery and self-exploitation. Complaining about farm work is not new, yet Houston's UA manages to produce food in spite of the drudgery. The establishment and persistence of UA is likely to be a result of the meaning and importance that organizers and farmers ascribe to their own work, to the food they grow, and to the site in which they produce food. Chayanov considered tradeoffs between family demand and drudgery; UA organizers and volunteers appear to balance meaning and social interactions with drudgery.

Second, UA is a means to accomplish objectives, such as food access, that cannot otherwise be accomplished in particular neighbourhoods. The ways in which UA organizers state the goals of their site indicate that the mission or objective may be a way to overcome the drudgery of farm work and to secure apparently abundant, low-cost and irrigation-ready land on Houston's churches. The considerable presence of UA on church land indicates the importance of engagement with religious leaders and staff for further development of UA.

Third, a network of UA organizers has disseminated best practices suited to Houston's agricultural climate and site organizers have implemented these practices for their site's particular social and economic characteristics. It is likely that this local knowledge network appears in other UA sites, as others have reported (WinklerPrins and de Souza, 2005); similarly, public policies could support this network, which has obvious positive benefits to UA production.

6.6 Conclusion

Our survey, based on UPE and farming systems approaches applicable to UA in the GN and GS, found that the all-access regime is considerably more prevalent than allotments; churches are desirable locations for UA; and obtaining a consistent workforce is the most commonly cited challenge for UA organizers. UPE approaches are highly useful for conceptualizing the potential role of UA in cities, but scholars also should follow the insights from farming systems approaches, which help determine how UA sites produce food, obtain access to land and water, and the destination of UA crops, regardless of whether the UA site is located in the GN or GS. UA sites, potentially, produce many important social and political outcomes, but knowledge of their production challenges and site characteristics is necessary for informing policies that may support the development of UA. Shifting from a normative view of UA – what it should be – to empirical and practical approaches, whether in the GN or GS, helps illuminate the many challenges that UA organizers and volunteers face while producing food in urban settings.

Acknowledgements

We thank Urban Harvest for their assistance and all the urban agriculture organizers who participated in our study for sharing their agricultural insights, successes and harvests.

References

Block, D.R., Chávez, N., Allen, E. and Ramirez, D. (2012) Food sovereignty, urban food access, and food activism: contemplating the connections through examples from Chicago. *Agriculture and Human Values* 29(2), 203–215.

Broadstone, S.B. (2013) Growing food is work: a spatial and social analysis of urban agriculture in Houston. MSc thesis, Texas A&M University, College Station, Texas, USA.

Chayanov, A.V. (1986) *The Theory of Peasant Economy*, trans. Christel Lane. University of Wisconsin Press, Madison, Wisconsin, USA.

Feagin, J.R. (1985) The global context of metropolitan growth: Houston and the oil industry. *American Journal of Sociology* 90(6), 1204–1230.

Gallaher, C.M., Kerr, J.M., Njenga, M., Karanja, N.K. and WinklerPrins, A.M.G.A. (2013) Urban agriculture, social capital, and food security in the Kibera slums of Nairobi, Kenya. *Agriculture and Human Values* 30, 389–404.

Ghose, R. and Pettygrove, M. (2014) Urban community gardens as spaces of citizenship. *Antipode* 46(4), 1092–1112.

Gottlieb, R. and Joshi, A. (2010) *Food Justice*. MIT Press, Cambridge, Massachussetts, USA.

Gray, L., Guzman, P., Glowa, K.M. and Drevno, A.G. (2014) Can home gardens scale up into movements for social change? The role of home gardens in providing food security and community change in San Jose, California. *Local Environment* 19(2), 187–203.

Guitart, D., Pickering, C. and Byrne, J. (2012) Past results and future directions in urban community gardens research. *Urban Forestry and Urban Greening* 11, 364–363.

Heynen, N., Kaika, M. and Swyngedouw, E. (2006) Urban political ecology: politicizing the production of urban natures. In: Heynen, N., Kaika, M. and Swyngedouw, E. (eds) *In the Nature of Cities: Urban Political Ecology and the Politics of Urban Metabolism*. Routledge, New York, USA, pp. 1–20.

Hovorka, A. (2006) Urban agriculture: addressing practical and strategic gender needs. *Development in Practice* 16(1), 51–61.

Kortright, R. and Wakefield, S. (2011) Edible backyards: a qualitative study of household food growing and its contributions to food security. *Agriculture and Human Values* 28, 39–53.

Liu, L., Talbot, R. and Lan, X. (2015) Influence of climate change and meteorological factors on Houston's air pollution: an ozone case study. *Atmospheres* 6, 623–640.

Melosi, M.V. and Pratt, J.A. (2014) The energy capital of the world? Oil-led development in twentieth-century Houston. In: Pratt, J.A., Melosi, M.V. and Brosnan, K.A. (eds) *Energy Capitals: Local Impact, Global Influence*. University of Pittsburgh Press, Pittsburgh, Pennsylvania, USA, pp. 30–57.

Nielsen-Gammon, J.W. (2011) The changing climate of Texas. In: Schmandt, J., North, G.R. and Clarkson, J. (eds) *The Impact of Global Warming on Texas*, 2nd edn. University of Texas Press, Austin, Texas, USA, pp. 39–68.

Page, B. (2002) Urban agriculture in Cameroon: an anti-politics machine in the making? *Geoforum* 33(1), 41–54.

Podagrosi, A., Vojnovic, I. and Pigozzi, B. (2011) The diversity of gentrification in Houston's urban renaissance: from cleansing the poor to supergentrification. *Environment and Planning A* 43, 1910–1929.

Pudup, M.B. (2008) It takes a garden: cultivating citizen-subjects in organized garden projects. *Geoforum* 39, 1228–1240.

Smith, C. and Miller, H. (2011) Accessing the food systems in urban and rural Minnesotan communities. *Journal of Nutrition Education and Behavior* 43, 492–502.

Taylor, J.R. and Lovell, S.T. (2012) Mapping public and private spaces of urban agriculture in Chicago through the analysis of high-resolution aerial images in Google Earth. *Landscape and Urban Planning* 108, 57–70.

Taylor, J.R. and Lovell, S.T. (2014) Urban home food gardens in the Global North: research traditions and future directions. *Agriculture and Human Values* 31, 285–305.

Thomas, R.D. and Murray, R.W. (1991) *Progrowth Politics: Change and Governance in Houston*. University of California Press, Berkeley, California, USA.

Turner, B.L. and Brush, S.L. (1987) Purpose, classification, and organization. In: Turner, B.L. and Brush, S.L. (eds) *Comparative Farming Systems*. Guilford Press, New York, USA, pp. 3–48.

WinklerPrins, A.M.G.A. and de Souza, P.S. (2005) Surviving the city: urban home gardens and the economy of affection in the Brazilian Amazon. *Journal of Latin American Geography* 4(1), 107–126.

7 The Marketing of Vegetables in a Northern Ghanaian City: Implications and Trajectories

Imogen Bellwood-Howard* and Eileen Bogweh Nchanji
Institute of Social and Cultural Anthropology, Göttingen University, Göttingen, Germany

7.1 Introduction

The proximity of urban production sites to markets is one factor that has let urban agriculture (UA) flourish in both the Global North (GN) and the Global South (GS) (Drechsel and Dongus, 2010; Danso *et al.*, 2014). Studying markets for urban produce provides an opportunity to consider consumption alongside production and income generation, and economic alongside social and ecological concerns (Chagomoka *et al.*, 2014; Yusuf *et al.*, 2014). This theme thus acts as a lens through which to consider the multifunctionality of UA (Atukunda and Maxwell, 1996; Mougeot, 2000).

In this chapter, we argue that the market function of UA, alongside specific characteristics of the urban zone, allows urban farmers and marketers to reconnect the ecological to the social and economic within their livelihood strategies. Referring to urban political ecology and livelihoods frameworks, we show this happening to varying extents across the GN and GS, due to different extents of politicization and connection between producers and consumers. We draw on primary data about vegetable marketing in Tamale, northern Ghana, and compare this with case studies from the Global North. The chapter concludes by considering the implications of these similarities and differences for the future of market-oriented UA in the Global North and South (Amoah *et al.*, 2007; van Veenhuizen and Danso, 2007; Cobbina *et al.*, 2013).

7.2 Urban Political Ecology and Livelihoods Perspectives

Urban political ecology (UPE) starts by rejecting perceptions of a divide between natural countryside and urban society, as society and 'nature' co-exist in the urban ecosystem (Wachsmuth, 2012). It examines the struggles between different actors making a living in the city by turning natural resources into economic commodities (Keil, 2005; Heynen *et al.*, 2006). UPE writers have been intrigued by Marx's concept of 'metabolic rift', specifically the proposal of commercialized agriculture and urbanization as the locus of industrialized people's divorce from nature, the historical source of their ecological and socio-economic metabolism. Metabolic rift occurs as production and consumption are separated through commodification (Wittman, 2009), ecologically, as nutrients are transferred in food from production to consumption sites, and socio-economically, with division of agricultural wage labour between rural and urban areas (Foster, 1999; Moore, 2000). McClintock (2010)

---

*Corresponding author; e-mail: ibellwoodh@gmail.com

proposes UA as a way to 'heal' metabolic rift through cycling of organic matter in urban areas and livelihood provision. This chapter explores how commodification, considered responsible for metabolic rift, facilitates urbanites' metabolism of nature into livelihood through labour. This happens in a specific fashion in the urban zone, with its dense physical and social infrastructure.

The livelihoods framework identifies the capabilities individuals have to mobilize and transform fungible capitals (Scoones, 1997; Bebbington, 2000). Following criticism of the livelihoods framework as apolitical, 'political capital' was added to the original list of natural, economic, human, physical and social capitals (Ashley and Carney, 1999), facilitating its use in a more politically aware analysis. Our Ghanaian case study material focuses on micro-political negotiations between individual actors, making the livelihoods framework's idea of capital fungibility an appropriate analytical tool with which to describe urban agricultural metabolism. The next section identifies five unique characteristics of UA that are well illustrated by the following Northern Ghanaian case studies.

7.3 Characteristics of Market-oriented Urban Agriculture

Our research pointed to five relevant characteristics of market-oriented urban agriculture.

Quality considerations are particularly interesting since they relate to consumers' ideas about health. Weinberger and Msuya (2004) describe how health-conscious Tanzanian consumers prefer indigenous leaf vegetables because of their reputation for being easy to grow organically. Richer, educated, middle-class customers converging in cities have often been the driving force behind these concerns. Farmers seeking price premiums for quality goods have therefore targeted high-end outlets, such as the Umuchi supermarket chain in Kiambo district, Kenya, that sells vegetables at a higher price than local markets (Ngugi *et al.*, 2007). Quality perceptions that facilitate the metabolism of natural into economic capital thus have a particular role to play in situations of social change, such as the emergence of a middle class and an internationally inspired health discourse.

Dense social and physical urban infrastructure facilitates a particularly effective flow of resources. Information travels easily through loose and open networks of mobile colleagues, including in leisure situations, blending the distinction between the economic and the social and relating to Polanyi (1944), Polanyi (1971) and Granovetter's (1985) ideas about socially embedded economies. Flexible transport networks and accessible input and output markets enable people to act upon information. In urbanizing zones, the open morphology of vacant building sites and low-rise development makes farm sites visible to marketers, lowering entry barriers and transaction costs. This theme of tight infrastructure supporting trade resounds with Weinberger and Pichop's (2009) description of how better urban transport infrastructure encourages local urban vegetable sourcing in urban Benin, Cote d'Ivoire, Senegal, Kenya, Tanzania, Uganda and South Africa.

Another characteristic is the existence of **numerous opportunistic economic transactions**. Farmers and marketers who are neighbours may develop long-term trading relationships, yet more particular to the urban zone are the fleeting spontaneous and anonymous transactions facilitated by the proliferation of many farms and markets in contiguous areas. The flexibility of not always being bound to a particular trading partner is especially useful for marketers.

We also observed that **long-term social liaisons minimized economic risk**. Urban actors use these more temporary relationships as a way in to more long-term liaisons. Long-term, faithful trading relationships are often encountered in African agricultural and trading situations, and are a way to deal with ecological risks such as seasonal gluts and shortages. Lyon (2000) describes the relationships of trust and reciprocity between agricultural traders in Brong Ahafo, central Ghana, that also appear in Clark's account of Ghana's Kumasi market (2010). Weinberger and Pichop (2009) similarly describe the verbal contracts that the majority of farmers and marketers they surveyed were party to across East, West and Southern Africa. Traders in their survey also retained customers by giving them gifts when goods were scarce. This element is particularly interesting because it demonstrates the relationship between flexibility and consistency, as opportunism acts as a route into more fixed reciprocal relationships.

Community and fidelity narratives often prevail among farmers, marketers and consumers as economic activities are embedded in social and kin networks. There are, of course, pragmatic motivations behind these actions when they act to reinforce long-term trading affiliations. WinklerPrins and de Souza (2005) describe how exchange of urban produce is rooted in kin relations in a Brazilian 'economy of affection', with similar seasonal reciprocal motivations, and how spaces analogous to what we term 'backyard farms' in this paper are important loci for the reproduction of kin relations (WinklerPrins and de Souza, 2009). Pietila (2007) describes how such discourses move into a more recognizable commercial arena. In her work, vegetable marketers in Kilimanjaro mobilize family discourses to cultivate reciprocal relationships with customers, illustrating an intentional re-embedding of economic activity in the social fabric of urban life. In the following pages, we demonstrate how these processes play out in the setting of Tamale in northern Ghana.

7.4 Study Site and Methods

7.4.1 Tamale, Northern Ghana

Tamale, a rapidly growing city of approximately 370,000 inhabitants (Ghana Statistical Service, 2012) is the capital of Ghana's Northern Region (Fig. 7.1). Located in the Guinea Savanna, it has a monomodal climate with a single rainfall season between June and October (Danso *et al.*, 2014).

Contiguous urban fields farmed by different farmers are grouped in larger open space sites around water sources such as reservoirs, open gutters or standpipes. The mean size of these fields is 0.31 ha. Many farmers also crop on smaller, isolated plots between buildings, with a mean area of 0.11 ha, using domestic water, reservoirs or wastewater (Bellwood-Howard *et al.*, 2015a). In the rainy season, subsistence maize may be combined with market crops, including a high proportion of leaf vegetables, which dominate irrigated farms in the dry season. Most of Tamale's local leaf vegetables are grown locally rather than imported (Giweta, 2011). Up to 80% of Tamale's residents are from the Dagomba ethnic group (Ghanadistricts, 2006). Labour roles are gender-specific: male farmers sell to female marketers (Danso *et al.*, 2014): there is a long tradition of research on West African market women (Lyon, 2000; Clark, 2010). Wholesale marketers bring vegetables from the farms to the markets, where they are distributed for retail (Fig. 7.2).

7.4.2 Methods

We used ethnographic methods in our research, with primary data collected by the two authors. One concentrated on in-depth observation of three groups of marketers, participating in harvesting and marketing roughly once a week with each group for 2 years. These groups were purposively selected on the basis of accessibility and regular market activity. Each case study comprised one or two main marketers and up to seven peripheral figures, and was constructed through the use of participant and non-participant observation, interviews and informal conversations and production of an ethnographic film. Case study data was triangulated using interviews with 30 other marketers, 15 farmers and 15 consumers. A portion of these were snowball-sampled from randomly located GPS points and the rest purposively selected on the basis of their relation to the sites used in the case studies. The second researcher made an exhaustive survey of farmers and marketers working at all open space farming sites in Tamale, collecting interview, observational and interactional data from approximately 400 farmers and 20 marketers. Notes were taken after each day of participant observation, and interviews and conversations were usually voice-recorded before being transcribed. The characters in the case studies reported here were mainly harvesting indigenous vegetables. However, the socio-economic patterns elicited from their narratives equally apply to marketers who harvest vegetables of temperate origin, such as lettuce and cabbage.

Fig. 7.1. Map of case study cultivation and sale sites. (Sources: OpenStreetMap.org, 2015 and Ghanamap (http://ghanamap.facts.co/Ghana-Map.jpg), 2016.)

7.5 Case Studies: Vegetable Marketing Systems in Tamale

Our research breaks into three cases, illustrating three loose, interlinked systems. They are described here in the order that permits the analysis to move from a more to a less socially embedded situation.

7.5.1 Zagyuri: Community trade relationships and health concerns

The first case study describes Zagyuri, formerly a peri-urban village 7 km to the north of Tamale, now absorbed into the suburbs of the rapidly expanding city. This example shows the importance of strong relationships between farmers, wholesalers and retailers.

In the dry season, Zagyuri farmers cultivate jute mallow (*Corchorus olitorius*), a popular leaf vegetable used in soup. The leaves of the plant can be harvested up to six times. Zagyuri's farmers crop on an area of about 2 ha of land customarily owned by five community members, who use it to grow maize in the rainy season. The land is on the edge of a site used by the army, so has not been developed residentially, yet the army has not evicted the farmers due to the site's peripheral location. In fact, in one way, the soldiers' presence has been beneficial to the farmers, who

Fig. 7.2. Unofficial market stalls in Aboabo market, Tamale, Ghana. (Photo by Imogen Bellwood-Howard.)

irrigate their crops with sewage from the barracks. The vegetable farmers here cultivate vegetables to supplement other income sources or provide pocket money. They are all male, mostly in their teens or twenties, and are related to the long-term land users, the maize farmers. Most of the female marketers who harvest the jute mallow are also relatives of the farmers, and sell their vegetables in Tamale's central market. Marketers often provide seeds to the farmers, who are then obliged to sell the leaves from that particular crop only to that specific woman.

The Zagyuri marketers also retail other goods, and several also harvest vegetables from sites further afield. They tend to make more profit from these other trades than from selling the jute mallow from their hometown. This is because, despite purchasing the vegetables on credit, they often give the gross proceeds of their sales to the farmers they harvested from. They also refrain from charging those farmers for the seeds they supply, for reasons to be explained shortly. They thus make very little profit. Nevertheless, they justify continuing to harvest from Zagyuri in three ways. First, many consider that in facilitating the sales of their male relatives they fulfil their duty as wives and mothers and enhance household income. Second, jute mallow is so popular that many claim it attracts retail customers, who then buy the other more lucrative goods they are selling. The final reason marketers continue harvesting from Zagyuri is to maintain long-term relationships with farmers, in order to smooth out seasonal variation in supply. As a tropical crop, jute mallow performs poorly in the cooler, drier windy season called the Harmattan that falls between November and February. In this season of scarcity, jute mallow prices are high, and marketers want access to competent farmers' plots in order to sell large volumes in the more lucrative marketplace. When such a competent farmer uses a marketer's seeds during the Harmattan, she has a good chance of making a profit. This explains why many marketers do not collect the seed money back from their supplier farmers – this apparently 'irrational' economic behaviour is a tactic to prolong the trading relationship. It can be noted that these seasonal fluctuations in supply and demand happen to some extent for all crops, depending on the season in which they flourish. Crops of temperate origin, such as cabbage, are abundant in the cooler Harmattan.

The health concerns associated with wastewater irrigation have been widely discussed (Amoah et al., 2007; Gallaher et al., 2013; Drechsel and Keraita, 2014). Jute mallow from Zagyuri is notorious within Tamale for this reason, and fears have been exacerbated by popular radio call-in shows. Zagyuri's marketers are aware of consumers' hygiene and health concerns, so conceal the provenance of their products. Many consumers develop loyal relationships with certain retailers, using these to justify their confidence in the hygienic nature of their supplier's vegetables, despite never having seen the source of the goods. Ironically, one consumer who often unwittingly purchased Zagyuri jute mallow from her regular retailer made this type of association. This case study therefore shows how farmers and marketers manipulate discourses of community and quality to continue making a living throughout the year, as they develop social capital in the form of long-term relationships at the sites of both production and sale.

7.5.2 Choggu: dense infrastructure and temporary trade relations

The second case study similarly illustrates the maintenance of long-term relationships to manage risk, this time in the neighbourhood of Choggu, closer to central Tamale than Zagyuri. Here, however, it is also possible to observe the importance of the more fleeting associations that are possible in the urban zone (see also Chapter 11, this volume).

In Choggu, many farmers grow vegetables in their backyards or interstitial spaces close to their houses. Between May and October, vegetables are rainfed, and some use piped water to irrigate dry season gardens. Amaranthus (*Amaranthus* spp.) is one popular vegetable, which, like jute mallow, can be harvested weekly over its 2–4-month lifetime (see Fig. 7.3). Amaranthus in backyards is often harvested by groups of women, led by a senior marketer. Many senior marketers buy from multiple farmers in the same area, and attract new suppliers by approaching farmers they see cultivating in their backyards around the neighbourhood. When they see that the amaranthus is ready, they choose a fixed day each week to come and cut the leaves, tie them into bunches and pay a set price for each, very rarely on credit. They wholesale most of the goods. The leader is accompanied to each field by a group of up to 20 other women, the composition of which changes weekly. Once the bunches are tied, the leader takes the majority and shares the rest between these other marketers, who pay immediately. The women in this loose group learn which day of the week each backyard field is generally harvested as they interact socially and, commonly, through mobile phones. For several years now, the majority of market women have had a personal mobile phone. Marketers make sure they attend on the harvesting day, relying on hearing about changes in the schedule from each other, the leader or the farmer. Information about new farms may reach them in these same ways, or they may simply observe farms being harvested as they move around the open landscape of the neighbourhood.

These women are not always related to each other, yet they have autonomously formed a loose, informal, obligation for the leader to distribute vegetables to each of the other peripheral marketers. However, the crops do not always do well enough to permit all to get a share. This is particularly the case in the cool, dry, Harmattan season, when amaranthus, a tropical crop like jute mallow, performs poorly. The leader then benefits from the system: she can take almost all the vegetables if she wants. Conversely, in the hot, wet season when crops flourish and vegetables are cheap, farmers struggle to persuade marketers to harvest their crops. The system then works to their advantage: the leader is still obliged to harvest from her supplier farmer during this period, even if cheaper vegetables are available elsewhere. On these occasions, if the peripheral marketers learn about better deals, they can abandon the leader and the farmer for those locations. They act on information updates by using taxis to transport themselves and their goods from farm to market. Some call reliable drivers on their mobile phones to arrange collection. However, it is also entirely possible for them to use any vehicle they encounter on the roadside. As each field is harvested only once a week, each woman may be a leader on some days and on other days a peripheral marketer following these more flexible strategies.

Although these farmers and marketers all live in adjacent communities, they rarely draw on community or kinship discourses to leverage favourable prices and long-term partnerships. Inner-city neighbourhoods such as Choggu are not as socially homogeneous as Zagyuri, so faithful trading relationships are based more on partners' mutual perception of each other's reliability than kinship. Nevertheless, there are some situations where family connections facilitate trade; for example, when one party is an elderly relative of the other. So, people in a more

Fig. 7.3. A visible backyard farm in Nobisco, Tamale, Ghana. (Photo by Imogen Bellwood-Howard.)

central urban setting still pursue long-term relationships to smooth out fluctuations in demand and supply. However, they frame these more around maintaining trade than social, community or familial bonds. Simultaneously, they use good travel and information infrastructure to take advantage of fleeting, flexible associations within a large pool of individuals.

7.5.3 Nobisco: opportunistic transactions between acquaintances

The final case study, in another suburb of Tamale, Nobisco, illustrates best the opportunities afforded by having access to multiple, unconnected production sites in the urban zone, and the imperative that this places upon developing and maintaining lasting partnerships at the retail site. It also highlights the occasional disjuncture between maintaining strong personal relationships and providing good-quality vegetables.

Not all marketers have established a farm where they are the leader. After purchasing goods from a leader's farm in the early morning, peripheral marketers roam around their area opportunistically searching for vegetables. Nobisco is a promising area for such activity as the good piped-water infrastructure supports many backyard farms, and the open morphology of this suburb makes gardens visible and accessible, as shown in Fig. 7.3.

Many new, small-scale farmers who have not yet established a strong bilateral relationship with a senior marketer are happy to sell to any itinerant trader passing their farm. More established farmers may also make such spontaneous sales if their regular leading marketer fails to purchase all their goods. This may happen if she fails to attract enough peripheral marketers to join her in a season when goods are abundant. Instantaneous sales are opportunities to embark upon longer-term relationships. At the start of such an association, the season and the relative abundance of goods become important as bargaining tools, because there is no chance of negotiating the price of the goods downwards on the basis of a prior relationship. The farmer thus has an advantage when goods are scarce, but when goods are plentiful, marketers can manipulate this to their benefit.

It is also important to note that in this situation, quality almost never reduces unit cost, it only determines the quantity of goods that are not sold, as marketers may reject leaves that are pale in colour or damaged by insects. Quality becomes more important at the consumption end of the market chain. As marketers perambulate around Nobisco's streets, they make single retail sales to consumers. One of these stated that she was buying these vegetables because, being from an area with such a good piped-water supply, they were likely to be irrigated with treated water rather than wastewater. Radio discussions have been instrumental in bringing these quality issues into the public consciousness – in particular, Tamale's 2011 cholera outbreak was speculatively related to wastewater irrigation on the station Filla FM.

Traders' marketplace strategies mobilize informal relationships of obligation during extremes of availability, and these interact with quality in another way. When goods are abundant, wholesalers deliver large quantities on credit to their faithful, regular retailer customers in the market, expecting them to sell to consumers and pay back later. As recompense, they reserve goods for these customers when vegetables are in short supply. The expectation that these loyal customers will try to sell even very poor-quality vegetables pits the discourses of health and community against each other. This reinforces the erroneous nature of the assumption made by many consumers in the Zagyuri case study, that longstanding relationships with sellers are a quality guarantee.

This case study illustrates further how fleeting market transactions with acquaintances and strangers can give traders access to farmers' surpluses, making profit for both and serving as a potential basis for more long-term affiliations. It also shows how, when developing such affiliations to facilitate trade, quality becomes a secondary issue.

These three case studies show how farmers and marketers mobilize social capital to make a living from urban agriculture. It is important to remember that the relationships they devise are entirely unregulated, and transactions are made and honoured only on the basis of verbal agreements. Through these relationships, actors engage with community and quality discourses and take advantage of particularly dense urban physical and social infrastructures, effectively metabolizing natural into financial capital. The next section will examine how those same livelihood, quality and community concerns play out in the Global North and the implications of the similarities and differences across locations.

7.6 Discussion: Marketing UA Produce in the South and the North

The propensity of the northern Ghanaian actors to manipulate social and ecological dimensions to economic ends invites comparison to the situation McClintock (2010) describes as 'healing the metabolic rift'. There are nuanced differences between locations in terms of how this relationship between nature and capital functions. This section considers how these differences are related to different extents of politicization of UA, reflecting on the multifunctionality of urban agriculture.

7.6.1 Achieving income and food security

Local actors in Tamale certainly perceive UA primarily as a way to generate income and supplement food supply, generating a livelihood from natural resources. In the Global North, the livelihood role of UA is associated more with self-provisioning than income generation. This has historically been the case, for example in the UK's wartime 'dig for victory' campaign and the USA's 'victory gardens', which had political overtones related to subsistence. That political emphasis is carried into current concerns with food sovereignty and justice.

Toronto is an example of a concerted, politicized effort to improve northern urban livelihoods through UA. Commodity price shifts in the 1970s prompted the People's Food Commission's 1980 report on food insecurity in Canada, followed by Toronto Health Department's 1983 *Access to Nutritious Food Report*. In 1985, Toronto's mayor founded the non-profit organization Foodshare (Foodshare, 2016). Then, as today, Foodshare laid emphasis on UA as a sustainable alternative to emergency food banks. Other initiatives primarily aiming to feed the disadvantaged included direct marketing from farmers to local consumers through a 'good food box' scheme and farmers' markets. Foodshare's activities aim to bring municipal and community attention to the structural inadequacies in the current local and global food system (Johnstone and Baker, 2005). Since the Toronto Food Policy Council formed in 1991, businesses and organizations performing various activities connected to UA have proliferated, including commercial city farms with vegetable-box delivery services, seed suppliers, backyard garden construction services, lifestyle food producers, a network of community gardens, and backyard Community Supported Agriculture schemes. Food security and income-generation imperatives are thus now complemented by lifestyle and leisure interests. UA has also been linked to food security on a macro-scale by northern national organizations: the American Community Food Security Coalition emphasized as long ago as 2003 that UA was a 'major instrument against hunger and poverty' (Brown and Carter, 2003).

As in Tamale, actors in Toronto use UA as a way to reconnect their livelihoods to nature. This metabolism is structured at the community scale in Toronto, with the political involvement of consumer groups and local authorities alongside producers. In Tamale, UA is more related to income generation within individual's livelihood strategies. The most contentious political issue relating to UA in Tamale concerns production: there are tensions over land access in the context of an accelerating property market (Naab *et al.*, 2013; Nchanji and Bellwood-Howard, 2014). In the north the politicization of metabolism is more holistic, as production and livelihood concerns are directly linked to consumption in a way that Schneider and McMichael (2010) refer to as 'democratic'. This results in individual actors taking explicit action to tighten the linkages between ecological and economic elements of the UA system.

7.6.2 Quality, community and environmental values

There is discernible interest in health and community in Tamale, alongside the more ubiquitous price imperatives. Some perceived health risks are connected to environmental issues, specifically wastewater, fertilizer and pesticide use. However, producers act on these only to the extent that they affect income, concealing the origin of their goods, for example. Organic nutrient cycling, the basis of the claim for UA's ecological 'healing' potential, is unpopular, and pesticide use is common.

Conversely, in Portland, Oregon, the city authorities' food security motivations are mediated through projects with community and health focuses. The sustainable food programme's objectives are 'community resiliency, equity, and environmental, economic, and personal health' (City of Portland, 2015b). The community gardens programme connects individual plot owners in a system of work parties, functioning through intern managers, including a Spanish language outreach officer. The 'produce for people' programme emphasizes that the food used in community soup kitchens is 'organic and locally grown', supporting local producers (City of Portland, 2015a). The local authorities' orientation is reflected in local business and NGO activities. Zenger farms' environmental stewardship programme conserves green areas for community use and inculcates environmental management skills in future generations of urban farmers through internship schemes. Revitalizing the Lents International Farmers Market in 2005, they emphasized its important role in providing culturally appropriate foods to diverse ethnic populations (Zenger Farms, 2015). In the Portland Urban Farm Collective, formal guidelines structure how community groups barter food grown in backyards, donating the excess to charity. Education and community-building objectives come before food security, and the group forbids non-organic techniques (Urban Farm Collective, 2015). Organic backyard farming Community Supported Agriculture scheme 'the Sellwood Garden Club' persists, due to the significance its customers attach to community identities (Coleman, 2009). The organizer of a similar model, 'your backyard farmer', emphasizes the importance of 'fresh, safe and local' food alongside 'a sense of community' (Your Backyard Farmer, 2015).

Such discussions are evident elsewhere: Montreal's municipal community gardening programme emphasizes community, equal opportunities and quality of life, as well as food production (Reid and Pedneault, 2006). Community gardens in Flint, Michigan, emerged as attempts to combat gang violence in the late 1990s, and are now associated with community development and improving urban aesthetics (Alaimo *et al.*, 2008). Community narratives are thus connected to environmental concerns through a broader definition of public and environmental health, comprising recognition of consumers' concerns and their conscious connection to producers.

The relationship of price and quality is intriguing. In Tamale, mutual support between wholesalers and retailers allows the sale of poor-quality goods for low prices. In the north, similar community discourses are associated with expensive, high-quality, organic goods sold at farmers' markets. When unsupervised, such market segmentation leads to inequality (Meenar and Hoover, 2012; Hoover, 2013). McClintock (2010), drawing on Polanyi, claims that all commodification of food has this effect. Local political bodies in Portland have addressed this, making food stamps available to local markets (Grace *et al.*, 2005). Political involvement is thus necessary to reconnect consumers' and producers' concerns. This has hitherto been lacking in Tamale, hence the persistence of the ecological rift.

In the south, explicit attention to such concerns is commonly reported in Latin America; for example, in the international peasant organization Via Campesina (Clausen, 2007; Wittman, 2009; Altieri and Toledo, 2011). Such holistic politicism is the domain of Toronto's Foodshare (Johnstone and Baker, 2005). Wekerle (2004), examining the same case study, thus reflects that, rather than distracting from structural inadequacies by providing extra-state solutions, the advocacy role of UA activists encourages policy change. This intersection between radical and neoliberal agendas is interrogated in depth by McClintock (2014). The interacting roles of local government, business and NGOs in the Toronto and Portland case studies raise questions about governance, a theme integral to food networks literature (Raynolds, 2004; Qazi and Selfa, 2005; Lockie, 2006). Similar connection of consumer and producer concerns may happen in Tamale through a series of multi-stakeholder policy workshops that began in 2013, facilitated by RUAF (Bellwood-Howard *et al.*, 2015b).

7.6.3 Multifunctional agriculture

The convergence of multiple rationales for UA is termed 'multifunctional agriculture'. Vancouver's food action plan, for example, links UA to the city's sustainability agenda (Mendes *et al.*, 2008). In the Netherlands, a 'multifunctional' farmer collaborated with Delft municipality and 'Delft Initiatives for Nature' to dedicate some land to recreation, simultaneously acting as a natural water filter. This helped the water board achieve their targets, for which the farmer received subsidies (Deelstra *et al.*, 2001). UK allotments illustrate how spaces traditionally allocated to food security objectives have acquired recreational, environmental and social meaning (van Leeuwen *et al.*, 2010; Crouch and Wiltshire, 2012).

Migrant farming communities are particularly demonstrative of multifunctional agriculture (Baker, 2004). Vietnamese migrants in New Orleans produce produced food for local ethnic markets and their own consumption, but also to remind themselves of home and because they considered gardening improved their health (Airriess and Clawson, 1994). Baker (2004), writing about Toronto, considers that immigrants, through gardening, are drawn into the political arenas of the food security movement.

In Tamale, local actors are just beginning to draw the various functions of UA together. In West Africa's less overtly politicized UA scene, aligning the interests of geographically proximate consumers and producers to heal the ecological rift is an ongoing process. Simultaneously, workers control their own labour to support their own livelihood by commodifying food.

7.6.4 Urban infrastructure within politicized livelihood strategies

In both north and south, actors use the specific characteristics of the urban environment in their livelihood strategies. Tamale's dense infrastructure lowers the entry requirements for UA as people can act on information quickly and make multiple, diverse trading links to insulate themselves from risk. Northern farmers also use such elements to give themselves an advantage in the marketplace, but do this in a more overtly holistic fashion, positioning themselves in relation to quality and community discourses.

Green City Acres in Kelowna, British Columbia, is an example. The founder cultivates on less than an acre using Small Plot Intensive (SPIN) farming (Green City Acres, 2015). Developed in Saskatoon, Canada, SPIN farming minimizes input costs by relying on voluntary community-based labour. It maximizes income through sales to nearby, high-value markets, making it especially appropriate for urban farms (Christensen, 2007). Indeed, Institute for Innovations in Local Farming concluded that SPIN farming was a viable model for commercial agriculture in Philadelphia (Urban Partners, 2007). SPIN emphasizes that it is not a political movement but a business concept (SPIN Farming, 2015). Accordingly, Green City Acres' customers include high-end restaurants committed to local, organic procurement. The farm has been lauded as a model business on the 'Radical Personal Finance' podcast, noting the low overheads it incurs through its urban location (Sheats, 2012). Nevertheless, alongside its entrepreneurialism, the farm espouses the ideologies of the food movement with its mission to 'foster social and environmental change' and 'empower people to start growing their own'. This convergence of political ideological objectives with practical income generation and food security imperatives again demonstrates the holistic nature of northern UA.

Here, commodification, held responsible for the separation of ecology and society, engenders their reconnection through use of the urban infrastructure, as it does in Tamale, Brong Ahafo and Kilmanjaro. Northern actors, however, achieve this by manipulating political and ideological elements in a holistic corporate identity.

When consumers' and producers' concerns converge in a holistic and politicized food movement, UA also begins to heal what has been termed the ecological rift by incorporating ecological functions.

What are the implications of this for the future of UA? Food systems may change in cities like Tamale as some producers act on consumer preferences for healthy and organic products, and produce produced with irrigated water, and not sewage. A premium sector would benefit from building on existing dense, flexible trading networks. Simultaneously, majority preoccupations with price over quality means that the standard market segment will continue to dominate. The resilience of its accessible distribution networks means that it will likely persist as an important livelihood source for many southern men and women.

The start of mobilization around UA in Tamale raises questions regarding future southern UA trajectories. To what extent will consumers' and marketers' voices inform advocacy movements? How differentiated will the market become, and what will the implications be? How holistic will future visions of UA be?

To conclude, elements of Tamale's market vegetable system bear similarity to UA in the Global North, considering how dense, urban, market-focused infrastructures and discourses of livelihood security, health and community prevail in both. These facilitate farmers' and marketers' metabolism of natural capital, through social and physical structures, into social and economic benefits. These concerns currently cohere in a more holistically and consciously political movement in the north than in our case studies. Nevertheless, the seed of an urban food movement that is germinating in Tamale prompts questions about how far global UA futures may converge.

7.7 Conclusion: Implications for the Trajectory of Market UA

Farmers and marketers in Tamale encounter similar themes to their northern counterparts. In both settings, producers use their own labour to metabolize natural capital into a livelihood through a dense urban market infrastructure.

Acknowledgement

This work was carried out as part of the UrbanFoodPlus Project, funded by the German Federal Ministry for Education and Research (BMBF) under the initiative GlobE – Research for the Global Food Supply, grant number 031A242-C.

References

Airriess, C.A. and Clawson, D.L. (1994) Vietnamese market gardens in New Orleans. *Geographical Review* 84(1), 16–31. DOI: 10.2307/215778.

Alaimo, K., Packnett, E., Miles, R.A. and Kruger, D.J. (2008) Fruit and vegetable intake among urban community gardeners. *Journal of Nutrition Education and Behavior* 40(2), 94–101. DOI: 10.1016/j.jneb.2006.12.003.

Altieri, M. and Toledo, V. (2011) The agroecological revolution in Latin America: rescuing nature, ensuring food sovereignty and empowering peasants. *The Journal of Peasant Studies* 38(3), 587–612.

Amoah, P., Drechsel, P., Abaidoo, R. and Henseler, M. (2007) Irrigated urban vegetable production in Ghana: microbiological contamination in farms and markets and associated consumer risk groups. *Journal for Water and Health* 5(3), 455–466. DOI: 10.2166/wh.2007.041.

Ashley, C. and Carney, D. (1999) *Sustainable Livelihoods: Lessons from Early Experience*. DfID, London, UK.

Atukunda, G. and Maxwell, D. (1996) Farming in the city of Kampala: issues for urban management. *African Urban Quarterly* 11(2–3), 264–275.

Baker, L.E. (2004) Tending cultural landscapes and food citizenship in Toronto's community gardens. *Geographical Review* 94(3), 305–325. DOI: 10.2307/30034276.

Bebbington, A. (2000) Capitals and capabilities: a framework for analysing peasant viability, rural livelihoods and poverty. *World Development* 27(12), 2021–2044.

Bellwood-Howard, I., Haring, V., Karg, H., Roessler, R., Schlesinger, J. and Shakya, M. (2015a) Characteristics of Urban and Peri-urban Agriculture in West Africa: Results of an Exploratory Survey Conducted in Tamale, Ghana, and Ouagadougou, Burkina Faso. IWMI Working Paper 163. International Water Management Institute (IWMI), Colombo, Sri Lanka. DOI: 10.5337/2015.214.

Bellwood-Howard, I., Chimsi, E., Ganiyu, S. and van Veenhuisen, R. (2015b) *Urban and Periurban Agriculture in Tamale: A Policy Narrative*. RUAF-URBANET, Tamale, Ghana.

Brown, K. and Carter, A. (2003) *Urban Agriculture and Community Food Security in the United States: Farming from the City Center to the Urban Fringe: A Primer prepared by the Community Food Security Coalition's North American Urban Agriculture Committee*. Community Food Security Coalition's North American Urban Agriculture Committee, Venice, California, USA.

Chagomoka, T., Afari-Sefa, V. and Pitoro, R. (2014) Value chain analysis of traditional vegetables from Malawi and Mozambique. *International Food and Agribusiness Management Review* 17(4), 59–86.

Christensen, R. (2007) SPIN farming: improving revenues on sub-acre plots. *Urban Agriculture Magazine*, pp. 25–26.

City of Portland (2015a) Parks & Recreation. Available at: http://www.portlandoregon.gov/parks (accessed 16 October 2015).

City of Portland (2015b) Planning and Sustainability. Available at: http://www.portlandoregon.gov/bps (accessed 16 October 2015).

Clark, G. (2010) *Onions Are My Husband: Survival and Accumulation by West African Market Women*. University of Chicago Press, Chicago, Illinois, USA.

Clausen, R. (2007) Healing the rift: metabolic restoration in Cuban agriculture. *Monthly Review* 59(1), 40–52.

Cobbina, S.J., Kotochi, M.C., Korese, J.K. and Akrong, M.O. (2013) Microbial contamination in vegetables at the farm gate due to irrigation with wastewater in the Tamale Metropolis of Northern Ghana. *Journal of Environmental Protection* 4, 676–682.

Coleman, P. (2009) The front lawn farm: the Sellwood Garden Club farms people's lawns – and everybody wins. *The Portland Mercury*. Available at: http://www.portlandmercury.com/portland/the-front-lawn-farm/Content?oid=1572199 (accessed 20 June 2015).

Crouch, D. and Wiltshire, R. (2012) Designs on the plot: the future for allotments in urban landscapes. In: Viljoen, A. and Howe, J. (eds) *Continuous Productive Urban Landscapes*. Taylor and Francis, London, UK, pp. 124–132.

Danso, G., Drechsel, P., Obuobei, E., Forkuo, G. and Kranjac-Berisavljevic, G. (2014) Urban vegetable farming sites, crops and cropping practices. In: Drechsel, P. and Keraita, B. (eds) *Irrigated Urban Vegetable Production in Ghana: Characteristics, Benefits and Risk Mitigation*, 2nd edn. International Water Institute (IWMI), Colombo, Sri Lanka, pp. 7–27.

Deelstra, T., Boyd, D. and van den Biggelaar, M. (2001) *Multifunctional Land Use: An Opportunity for Promoting Urban Agriculture in Europe*. The International Institute for the Urban Environment, Delft, The Netherlands.

Drechsel, P. and Dongus, S. (2010) Dynamics and sustainability of urban agriculture: examples from sub-Saharan Africa. *Sustainability Science* 5(1), 69–78. DOI: 10.1007/s11625-009-0097-x.

Drechsel, P. and Keraita, B. (eds) (2014) *Irrigated Urban Vegetable Production in Ghana: Characteristics, Benefits and Risk Mitigation*, 2nd edn. International Water Institute (IWMI), Colombo, Sri Lanka.

Foodshare (2016) The land of milk and honey. Available at: http://foodshare.net/timeline/foodshare-is-born (accessed 9 November 2016).

Foster, J. (1999) Marx's theory of the metabolic rift: classical foundations for environmental sociology. *American Journal of Sociology* 105(2), 366–405.

Gallaher, C.M., Mwaniki, D., Njenga, M., Karanja, N. and WinklerPrins, A.M.G.A. (2013) Real or perceived: the environmental health risks of urban sack gardening in Kibera slums of Nairobi, Kenya. *EcoHealth* 10, 9–20.

Ghana Statistical Service (2012) *2010 Population and Housing Census. Summary Report of Final Results*. Ghana Statistical Service, Accra, Ghana.

Ghanadistricts (2006) Northern: Tamale Metropolis. Available at: http://www.ghanadistricts.com/DistrictSublinks.aspx?s=1153&distID=139 (accessed 24 November 2016).

Giweta, M. (2011) Mainstreaming wastewater management in urban planning: a case study of Tamale Metropolis, Ghana. Master's thesis, Wageningen University and Montpellier Sup Agro/IRC, The Netherlands and France.

Grace, C., Grace, T., Becker, N. and Lyden, J. (2005) *Barriers to Using Urban Farmers' Market: An Investigation of Food Stamp Clients' Perception*. Oregon Food Bank, Portland, Oregon, USA.

Granovetter, M. (1985) Economic action and social structure: the problem of embeddedness. *American Journal of Sociology* 91(3), 481–510.

Green City Acres (2015) The Farmers. Available at: http://www.greencityacres.com/about/the-farmers (accessed 16 October 2015).

Heynen, N., Kaika, M. and Swyngedouw, E. (2006) Urban political ecology: politicizing the production of urban natures. In: Heynen, N., Kaika, M. and Swyngedouw, E. (eds) *In the Nature of Cities: Urban Political Ecology and the Politics of Urban Metabolism*. Routledge, London, UK, pp. 1–20.

Hoover, B.M. (2013) White spaces in Black and Latino places: urban agriculture and food sovereignty. *Journal of Agriculture, Food Systems, and Community Development* 4(3), 109–115.

Johnstone, J. and Baker, L.E. (2005) Eating outside the box: Foodshare's good food box and the challenge of scale. *Agriculture and Human Values* 22, 313–325.

Keil, R. (2005) Progress report – urban political ecology. *Urban Geography* 26(7), 640–651. DOI: 10.2747/0272-3638.26.7.640.

Lockie, S. (2006) Networks of agri-environmental action: temporality, spatiality and identity in agricultural environments. *Sociologia Ruralis* 46(1), 22–39.

Lyon, F. (2000) Trust and power in farmer–trader relations: a study of small scale vegetable production and marketing systems in Ghana. PhD thesis, Durham University, Durham, UK.

McClintock, N. (2010) Why farm the city? Theorizing urban agriculture through a lens of metabolic rift. *Cambridge Journal of Regions, Economy and Society* 2, 191–207.

McClintock, N. (2014) Radical, reformist, and garden-variety neoliberal: coming to terms with urban agriculture's contradictions. *Local Environment* 19(2), 147–171.

Meenar, M.R. and Hoover, B.M. (2012) Community food security via urban agriculture: understanding people, place, economy, and accessibility from a food justice perspective. *Journal of Agriculture, Food Systems, and Community Development* 3(1), 143–160.

Mendes, W., Balmer, K., Kaethler, T. and Rhoads, A. (2008) Using land inventories to plan for urban agriculture: experiences from Portland and Vancouver. *Journal of the American Planning Association* 74(4), 435–449. DOI: 10.1080/01944360802354923.

Moore, J. (2000) Environmental crises and the metabolic rift in world-historical perspective. *Organization and Environment* 13(2), 123–157.

Mougeot, L. (2000) Urban agriculture: definition, presence, potential and risks. In: Bakker, N., Dubbeling, M., Gundel, S., Sabel-Koschella, U. and de Zeeuw, H. (eds) *Growing Cities, Growing Food: Urban Agriculture on the Policy Agenda*. Zentralstelle für Ernährung und Landwirtschaft der Deutschen Stiftung für internationale Entwicklung in Feldafing, Feldafing, Germany, pp. 1–42.

Naab, F.Z., Dinye, D.R. and Kasanga, R.K. (2013) Urbanisation and its impact on agricultural lands in growing cities in developing countries: a case study of Tamale, Ghana. *Modern Social Science Journal* 2(2), 256–287.

Nchanji, E. and Bellwood-Howard, I. (2014) *Governance for Development in Urban Agriculture*. In: Harmattan School proceedings, University for Development Studies, Tamale, Ghana.

Ngugi, I.K., Gitau, R. and Nyoro, J. (2007) *Access to High Value Markets by Smallholder Farmers of African Indigenous Vegetables in Kenya*. International Institute for Environment and Development (IIED), London, UK.

People's Food Commission (1980) *Land of Milk and Money: The National Report of the People's Food Commission*. Between the Lines, Ontario, Canada.

Pietila, T. (2007) *Gossip, Markets, and Gender: How Dialogue Constructs Moral Value in Post-Socialist Kilimanjaro*. University of Wisconsin Press, Madison, Wisconsin, USA.

Polanyi, K. (1944) *The Great Transformation: The Political and Economic Origins of Our Time*. Beacon Press, Boston, Massachusetts, USA.

Polanyi, K. (1971) The place of economies in societies. In: Polyani, C., Arensberg, C. and Pearson, H. (eds) *Trade and Market in the Early Empires. Economies in History and Theory*. Henry Regnery Company, Chicago, Illinois, USA, pp. 239–270.

Qazi, J. and Selfa, T. (2005) The politics of building alternative agro-food networks in the belly of agro-industry. *Food, Culture and Society* 8(1), 45–72.

Raynolds, L. (2004) The globalization of organic agro-food networks. *World Development* 32(5), 725–743.

Reid, D. and Pedneault, A. (2006) *Montreal's Community Gardening Program*. Ville de Montreal, Montreal, Canada.

Schneider, M. and McMichael, P. (2010) Deepening, and repairing, the metabolic rift. *The Journal of Peasant Studies* 37(4), 461–484.

Scoones, I. (1997) *Sustainable Rural Livelihoods: A Framework for Analysis*. IDS working paper 72. Institute of Development Studies, Brighton, UK.

Sheats, J. (2012) Making $80k on 1/3 acre with an urban farm without owning land? Yes, please! Interview with Curtis Stone. Radical Personal Finance. Available at: http://radicalpersonalfinance.com/making-80k-on-13-acre-with-an-urban-farm-without-owning-land-yes-please-interview-with-curtis-stone-rpf0040 (accessed 17 June 2015).

SPIN Farming (2015) SPIN – A new way to learn to farm. Available at: http://spinfarming.com/whatsSpin (accessed 16 October 2015).

Urban Farm Collective (2015) Frequently asked questions. Available at: http://urbanfarmcollective.com/people/frequently-asked-questions (accessed 16 October 2015).

Urban Partners (2007) *Farming in Philadelphia: Feasibility Analysis and Next Steps*. Institute for Innovations in Local Farming, Philadelphia, Pennsylvania, USA.

van Leeuwen, E., Nijkamp, P. and de Noronha Vaz, T. (2010) The multifunctional use of urban greenspace. *International Journal of Agricultural Sustainability* 8(1–2), 20–25. DOI: 10.3763/ijas.2009.0466.

van Veenhuizen, R. and Danso, G. (2007) *Profitability and Sustainability of Urban and Peri-urban Agriculture*. Agricultural Management, Marketing and Finance Occasional Paper 19. Food and Agriculture Organization, Rome, Italy.

Wachsmuth, D. (2012) Three ecologies: urban metabolism and the society–nature opposition. *The Sociological Quarterly* 53(4), 506–523. DOI: 10.1111/j.1533-8525.2012.01247.x.

Weinberger, K. and Msuya, J. (2004) *Indigenous Vegetables in Tanzania – Significance and Prospects*. AVRDC Technical Bulletin 31. AVRDC, Shanhua, Taiwan.

Weinberger, K. and Pichop, G. (2009) Marketing of African indigenous vegetables along urban and peri-urban supply chains in sub-Saharan Africa. In: Shackleton, C., Pasquini, M. and Drescher, A. (eds) *African Indigenous Vegetables in Urban Agriculture*. Earthscan, London, UK, pp. 225–244.

Wekerle, G. (2004) Food justice movements: policy, planning, and networks. *Journal of Planning Education and Research* 23, 378–386. DOI: 10.1177/0739456X04264886.

WinklerPrins, A.M.G.A. and de Souza, P.S. (2005) Surviving the city: urban home gardens and the economy of affection in the Brazilian Amazon. *Journal of Latin American Geography* 4, 107–126.

WinklerPrins, A.M.G.A. and de Souza, P.S. (2009) House-lot gardens as living space in the Brazilian Amazon. *FOCUS on Geography* 52, 31–38.

Wittman, H. (2009) Reworking the metabolic rift: La Via Campesina, agrarian citizenship, and food sovereignty. *The Journal of Peasant Studies* 36(4), 805–826.

Your Backyard Farmer (2015) About the farmers. Available at: http://www.yourbackyardfarmer.com/about.html (accessed 16 October 2015).

Yusuf, S.A., Ashagidigbi, W.M. and Mustapha, T. (2014) Choice and level of dry season vegetable market participation under tropical conditions. *International Journal of Vegetable Science* 1–22. DOI: 10.1080/19315260.2013.876567.

Zenger Farms (2015) Lents International Farmers Market joins the Portland Farmers Market family. Available at: http://www.zengerfarm.org/farmers-market (accessed 16 October 2015).

8 Hunger for Justice: Building Sustainable and Equitable Communities in Massachusetts

Timothy F. LeDoux* and Brian W. Conz
Westfield State University, Westfield, Massachusetts, USA

8.1 Introduction

Over the past few decades, the role of urban agriculture in ameliorating social and environmental inequities in the global agri-food system has received growing attention from academics, food activists, practitioners, local government officials, non-governmental organizations and planners. Urban agriculture has been celebrated for its abilities to alleviate food insecurity, hunger and malnutrition, improve food accessibility and sovereignty, strengthen communities and promote economic development, and fashion greener, healthier and more resilient cities. More importantly, it has been seen as an important step in recombining food production and consumption with social relationships that have been eroded by the global agri-food system. Ironically, despite its global reach, research on urban agriculture often has been partitioned into initiatives occurring in the Global South and practices arising in the Global North.

The artificial divide between urban agriculture in the Global South and North serves to hide, more than illuminate, the underlying synergies between urban agriculture around the globe. At a fundamental level, people farm urban environments in order to exercise their right to the city and to reclaim their metabolism from an industrial agri-food system that has systematically attempted to disconnect people from their food in an attempt to obfuscate larger socio-spatial inequities and maintain exploitative socio-ecological relationships. By reclaiming the metabolic right to the city, it becomes possible to envision non-capitalist and more equitable and just, autonomous foodscapes while addressing the fundamental structural inequities shaping the landscape. Yet such efforts are fraught with tensions as alternative food systems are constantly being reshaped and co-opted by the dominant neoliberal agri-food system (see also Chapter 5, this volume).

This chapter examines these tensions in the Global North by documenting the attempts of the food justice movement in the lower Pioneer Valley of Massachusetts to challenge inequities in the agri-food system, while addressing broader socio-economic and racial disparities in the region. It documents how marginalized communities and their allies have begun to use urban agriculture to alleviate food insecurity and improve food accessibility, while reimagining new ecological and socio-spatial relationships in a region devastated by decades of de-industrialization and neoliberal economic reforms. To that end, this chapter leverages insights from urban political ecology to examine how Holyoke-based Nuestras Raíces and Springfield-based Gardening the Community attempt to exert their metabolic

*Corresponding author; e-mail: tledoux@westfield.ma.edu

right to the city and re-vision the food system. The reassertion of their right to the city underlies their attempts to dismantle the institutional racism and socio-economic stratification embedded, not only in the current food system, but also in the region.

8.2 Urban Political Ecology, Urban Agriculture and Food Justice

Food justice advocates remain at a crossroads with respect to the emancipatory powers of urban agriculture initiatives. Various assemblages of neoliberalism have begun to appropriate such initiatives and the broader alternative food movements in which they are housed. The rise of public–private partnerships, the reliance on community and non-profit organizations to fill the holes in the social safety net and the dependence on market-based mechanisms to address the deep inequities embedded in the food system have underwritten the rolling back of the state. Subsequently, a growing body of literature not only has highlighted the failure of urban agriculture initiatives and alternative food movements to address socio-spatial inequities in the agri-food system and within their own movements, but also their struggles to create truly equitable and sustainable alternatives (Allen and Guthman, 2006; Born and Purcell, 2006; Slocum, 2007; Allen, 2010; Engel-Di Mauro, 2012; Alkon, 2013; Sbicca, 2014; Chapter 5, this volume).

Nevertheless, as McClintock (2010, 2014) has noted, tensions, contradictions and setbacks are to be expected from initiatives attempting to circumvent existing power geometries in an effort to create a truly transformative, equitable and just food system. In many ways, the struggle to come to grips with them have reinvigorated the food justice movement's commitment to addressing the socio-economic, racial and patriarchal inequities embedded in the agri-food system and society (Alkon and Agyeman, 2011; Gottlieb and Joshi, 2013). Subsequently, food justice movements have sought insights from urban political ecology to address better the systemic disparities and unequal outcomes imbued in the agri-food system (Agyeman and McEntee, 2014).

Urban political ecology (UPE) provides a reflexive and relational theoretical framework that examines how socio-ecological processes create uneven and power-laden urban environments. By scrutinizing the production of nature and how it is embedded in socio-economic relationships, UPE breaks apart powerful discourses that treat nature as separate from society and the city. In so doing, it reveals how socio-spatial relationships operating at different scales come together to create and reconfigure urban environments in ways that privilege certain groups over others often along lines of race, class and gender (Keil, 2003, 2005).

In this context, an urban political ecology framework views socio-ecological processes as co-produced hybrids that are mediated through socio-spatial relationships and rationalized by culturally specific discourses and discursive practices (Swyngedouw, 1996). Consequently, it becomes possible to examine how urban built environments – whether urban parks, community gardens or children's playscapes – aid capitalist consumption, reflect unequal social relationships and promote the circulation of capital (Harvey, 1996). Central to this project is the creation of a nature–city dichotomy that attempts to obfuscate the inequities and socio-ecological processes that exploit people and nature (Swyngedouw and Heynen, 2003; Heynen *et al.*, 2006). Such a dichotomy also serves to disconnect people from the ecological processes (metabolic rift) occurring in cities and to normalize the uneven outcomes and socio-spatial relationships associated with food production (Kaika and Swyngedouw, 2000; Domene and Saurí, 2007; Wachsmuth, 2012).

Accordingly, it has not only become unnatural for processes such as urban agriculture to occur in the urban environments of the Global North, but it also has allowed for privileged conceptions of urban agriculture and socio-ecological relationships to emerge that disproportionately harm minority and low-income communities, while dissuading non-capitalist alternatives. This can be seen clearly in economic development officers' and community planners' relatively long neglect (or late arrival) to incorporate the food system into their planning initiatives (Pothukuchi and Kaufman, 2000).

At a broader level, this dichotomy has severed the bonds uniting urban agriculture initiatives in the Global South with those in the Global North. In the Global South, urban agriculture is viewed through the lens of food security, and in the Global North, it is viewed through an alternative food movement lens that attempts to ameliorate inequities in the agri-food system while promoting community empowerment and sustainable economic development. Rather than seeing these efforts as distinct, an urban political ecology framework allows one to see that fundamental to urban agriculture initiatives across the globe is the attempt to claim one's metabolic right to the city.

As Shillington (2013) has noted, forms of urban agriculture have become one way in which people claim their right to the city and autonomy over their metabolism. There is nothing more essential and reflective of broader socio-ecological processes than the production and consumption of food. Subsequently, how individuals appropriate urban space is a critical component to resisting the social upheaval and contradictions embedded in capitalism and a vital step in creating sustainable and equitable autonomous spaces (Heynen, 2006; Wilson, 2013). This everyday struggle to feed oneself leads to urban spaces that prioritize the inhabitants' well-being over the accumulation of capital and development initiatives that seek to sacrifice such well-being.

By recognizing this common thread, cities and the nature within them become contested spaces and the hunger, food insecurity and marginalization found within them are reflections of the inherent contradictions within capitalism and the global agri-food system. The construction of the nature–city dichotomy is an attempt to prevent marginalized groups from asserting their rights to the city while devaluing non-capitalist socio-ecological relationships and obfuscating the socio-ecological inequities (Gibson-Graham, 2006; Gibson-Graham, 2008; Sbicca, 2012). Subsequently, by challenging such universal conceptualizations of capitalism in the agri-food system, it opens up the possibility for relational urban agricultural initiatives that are fundamentally non-capitalist to evolve and co-exist. Such initiatives not only challenge a hegemonic economic monism and representation of space that constrains alternative political and economic possibilities, but also enable co-constituted, open relationships situated in place to evolve as alternatives (Gibson-Graham, 1996; Massey, 2005).

It is against this backdrop that this chapter turns to highlighting how two food justice movements in the lower Pioneer Valley of Massachusetts have attempted to assert their metabolic right to the city, while addressing the deeper socio-spatial inequities in the agri-food system and the broader structural inequities at the local level. At the same time, it highlights how attempts to navigate existing power structures has led to protean movements that must balance the inherent tensions in their efforts to create more equitable and sustainable alternatives and to open the possibilities of non-capitalist alternatives devoid of market-logic solutions.

8.3 Springfield and Holyoke

The postindustrial cities of Springfield and Holyoke lie nestled on the meandering banks of the Connecticut River in the lower Pioneer Valley of Massachusetts. The landscapes of the cities are dotted with vestiges of the golden era of manufacturing which brought economic prosperity to tens of thousands of working class families. Historically, Springfield leveraged its position as the nation's first national armoury to forge cutting-edge precision manufacturing methods that not only ushered in an era of industrialization in the region, but helped place it at the very heart of industrialization in the US and globally. Over time, the region became a leading metalworking and fabrication centre (Forrant, 2003). While Springfield was rising to prominence as a bastion of precision manufacturing, another industrial revolution was occurring 13 km across the river to the northwest. Holyoke, one of the first planned industrial cities in the US, leveraged its location on the western banks of the Connecticut River along Hadley Falls to develop an elaborate series of canals and dams to fuel hundreds of textile and paper mills that ushered in an era of economic growth (Hartford, 1990).

From the mid-1800s into the 1960s, manufacturing in Springfield and Holyoke provided well-paying jobs for generations of Irish, French Canadian, German, Polish and Italian immigrant workers with little formal education. Eventually, these workers and their families would establish small businesses in their communities and lay the cultural foundations that would continue to be built upon with successive African-American and Puerto Rican migration movements to the region. However, by the mid-20th century, the industrial foundations of Springfield and Holyoke began to wither away.

Large-scale restructuring of the American economy, shifting capital and investment patterns, the de-skilling of workers, the disinvestment in older factories and the resulting de-industrialization hit Springfield and Holyoke, as well as its sister cities in the northeast, hard (Bluestone and Harrison, 1982). The closing of the historic Springfield armoury in the late 1960s, followed by the closing of many of the region's major metalworking and fabrication factories and paper mills during the 1970s and 1980s, saw the gradual erosion of high-paying manufacturing jobs that had sustained the economic backbone of the cities (Jacobson-Hardy and Weir, 1992; Forrant, 2005).

By the late 1980s, waves of mergers and consolidations, combined with increased capital flight and changing social dynamics, shut many of the remaining factories. As the manufacturing jobs evaporated, increases in the education level of the workforce did not keep pace with the rest of the state and nation, nor the smaller high-tech industries emerging in its wake (Federal Reserve Bank of Boston, 2009). Consequently, many older workers found themselves too young to retire but too old to reinvent themselves. During this era of de-industrialization, the social fabric of the cities also was undergoing a profound transformation.

Migrant farm workers from Puerto Rico were settling in the region to take advantage of the high number of affordable housing units and low-paying jobs in the declining manufacturing centres and on large corporate farms spread across the valley. As manufacturing jobs waned and migration increased, tensions emerged among the older White communities and the newer Puerto Rican communities (Morales, 1986). Left in its wake were extreme socio-economic and racial divides that created a spatially divided region, where low income Whites, Puerto Ricans and African Americans are segregated in the city centres and middle-class and affluent Whites live in the surrounding suburban centres.

This vociferous racial and economic divide has created uneven neighbourhoods in which Whites have better life opportunities than Puerto Ricans and African Americans. Today, both groups in each city experience high unemployment levels, a spatial job mismatch, enduring job and housing discrimination and exclusionary zoning regulations that prohibit multi-family housing in many of the suburban communities (Marzan, 2009; Federal Reserve Bank of Boston, 2010, 2011; PVPC, 2013). The accumulation of these processes has led to a spatial concentration of poverty around the old industrial downtown areas and a predominantly impoverished minority community that lags behind the rest of the region.

In 2013, the predominantly minority cities of Springfield (19.2% African-American, 40.5% Hispanic, and 35.4% White) and Holyoke (3.4% African-American, 48.3% Hispanic and 46.1% White) were home to an estimated 153,428 and 40,029 residents respectively. Both cities are characterized with 2013 unemployment rates over 8% (Springfield 8.6% and Holyoke 8.2%) and low participation in the workforce (Springfield 42.4% and Holyoke 44.2%). In both cities, less than half the population owns a home. The 2013 median household income in Springfield was US$34,311 and US$31,628 in Holyoke, which is well below the state average of US$66,866. Consequently, widespread food insecurity and poverty exists in both cities. Over one-third of households in Springfield (36.07%) and Holyoke (36.38%) receive food assistance from the supplemental nutrition assistance programme, and 29.4% and 31.5% of the population in Springfield and Holyoke respectively live below the poverty line (US Census Bureau, 2013).

Against this economic backdrop, a retail deconcentration that has further exacerbated food security in the region has occurred. Consequently, the inner-city residents face tremendous food access and security issues, while also dealing with housing discrimination and a spatial mismatch that imposes an undue burden on them to maintain well-paying jobs in the suburbs. It is within this climate of urban disinvestment that the food justice movements, spearheaded by Nuestras Raíces and Gardening the Community have emerged. The remainder of this chapter turns to examining how these organizations have attempted to reconceive the local food system in a broader effort to tackle the systematic vestiges of socio-spatial and socio-ecological inequity in the region.

8.4 Nuestras Raíces and Gardening the Community

In this section, we provide brief case histories of the two organizations with attention to their development, goals, community context, and their food production and distribution strategies. We also examine the networks that have produced these organizations and the larger food justice movement of which they are a part. These are diverse and multi-scalar networks, which include grassroots community organizers, elder custodians of ethnic food traditions, university students, volunteers and researchers, food- and social-justice activists, food policy councils, concerned community members, state agricultural agencies, organic farmers and neighbourhood youth. Documenting these connections reveals the extent to which community and urban food systems in the Global North share much with their counterparts in the Global South.

Holyoke-based Nuestras Raíces and Springfield's Gardening the Community (GTC) are grassroots efforts to address contemporary issues of food insecurity and to achieve participatory community development and youth empowerment. Evaluating these projects from the perspective of the actual contribution to food security in terms of the volume of food reveals a variable contribution relative to the urban populations they seek to serve. Currently, the food production capacity of each organization continues to grow in real terms. There also remains considerable capacity to increase production on additional vacant lots which could add hundreds of thousands of pounds of food to the supply chain (Berg et al., 2014). However, in looking at the history and accomplishments of each organization and mapping the intersecting web of networks that culminates in each, it is clear that the contribution of urban agriculture projects like Nuestras Raíces and GTC addresses not only food security, but also other aspects of urban metabolism associated with the food system. These include control over the waste stream, the built environment, the water cycle, microclimate and wildlife habitat, as well as the social spaces where food is processed, prepared, shared and consumed. By asserting this control and leveraging it for the purposes of education, empowerment and community health, both organizations strive to tackle broader injustices in the region in attempts to create more just and sustainable socio-ecological relationships. By laying claim to their right to the city and to their metabolism, they lay the framework for a more equitable and just food system and region.

8.4.1 Nuestras Raíces

> Community gardens are the heart and soul of Nuestras Raíces but many important projects grow out of the gardens.
>
> Daniel Ross, former Executive Director, Nuestras Raíces

Founded in 1992 by a handful of community members with a rich agricultural heritage, Nuestras Raíces ('Our Roots') is a grassroots urban agriculture organization based in Holyoke, Massachusetts. With a mission dedicated to creating healthy, just and sustainable environments, while celebrating and harnessing the strong agricultural ties embedded in the city's Puerto Rican community, Nuestras Raíces has become a national model for advancing social and environmental justice and promoting community lead 'agri-cultural' development. Over the past two decades, the organization has created a wide range of programmes and regional movements centred on the connection between food and the environment, which have fundamentally transformed the city of Holyoke.

Today, Nuestras Raíces operates a network of 11 community gardens, which include a youth garden, greenhouses, a 30-acre inner-city farm and a farmers' market. It actively implements nutrition and gardening workshops, educational agricultural tours, a farmers' incubation and small business development programme designed to support minority agri-business, youth development programmes, a women's empowerment group (Raíces Latinas), job training programme, and it is a major stakeholder in a socially responsible energy services company, Energia, LLC (Nuestras Raíces, 2013a, b). Overall, these programmes are designed to increase the health and emotional well-being of the community by reconnecting people to the environment and their food. These programmes break apart the nature–city dichotomy to cultivate culturally appropriate and affordable locally grown nutritious foods, support green enterprise and public education while creating intergenerational and multiracial spaces to support grassroots initiatives that seek to create an equitable, just and sustainable city.

At the heart of the organization are the intergenerational and multiracial spaces created by the community gardens, which support roughly 100 families, and the urban farm. At one level, the community gardens located in the low-income and predominantly Puerto Rican sections of Holyoke provide access to affordable, culturally appropriate produce to residents and allow them to maintain and pass along rich cultural traditions that do not know of any human/nature dichotomy. The community gardens create shared spaces for social interactions and cultural expression. However, at a broader level, these community gardens have served as space of resistance to the disinvestment and racial barriers in the region. They have provided a radical space in which community members can make their voices heard and organize in ways to hold accountable local government and businesses leaders. Moreover, the community gardens serve as a co-evolving space in which non-exploitative intimate relationships that build upon and celebrate local knowledge and Puerto Rican heritage can thrive. Nuestras Raíces has taken advantage of these spaces to organize and communicate about the different issues affecting the community.

The community gardens represent spaces that do not preclude non-capitalist possibilities that move beyond mere opposition to the dominant capitalist economic model. Such spaces place the nature–city connection and human well-being at the centre of activities and eschew the mantra that one must make sacrifices in order to promote economic development. It provides a way forward that denies the subjectivity of a historically racist city bureaucracy, which views the Puerto Rican residents of Holyoke as social and economic deviants who need economic development brought to them (Williams, 1992; Graham, 2001). Consequently, for residents involved with Nuestras Raíces, reclaiming abandoned and blighted spaces has become a catalyst for organizing and executing their right to the city and a reassertion of their right to metabolism. It defies the system and denies traditional capitalist opportunities in favour of activities that fundamentally serve to increase the health and emotional well-being of the community rather than generate economic profit.

These spaces of non-capitalist activities not only include volunteer labour, self-provisioning, the sharing of food with others, the establishment of childcare networks and safe spaces for children, but also an open space in which residents share their knowledge and organize against broader injustices in the city. At an institutional level, these spaces reflect how Nuestras Raíces has striven to advance its own causes and infused its own social fabric with a transparent structure for progressive change. Moreover, it is a reflection of how the organization itself has utilized non-capitalist mechanisms to grow.

The construction and design of Nuestras Raíces' headquarters (Centro Agricola), which houses a greenhouse, plaza, community kitchen and a library, were all achieved through unpaid non-market mechanisms. The equipment in the community kitchen, such as the walk-in freezers, were salvaged from an abandoned restaurant, while the boiler for the building, the frame and glass for the greenhouse along with the HVAC services, plumbing supplies and concrete, were all donated by local business. Countless hours of volunteer labour helped transform the vacant lot and abandoned building in which their offices are located into a majestic plaza that serves as meeting place for the community (Toensmeier, 2013).

The construction of the Centro Agricola also was utilized as an opportunity to train youth in carpentry skills as well as provide apprenticeship carpenters opportunities to earn experiences towards becoming journeymen (Constantine, 1999; Lauer, 1999). These alternative economic practices were aided by grants that allowed Nuestras Raíces to occupy a space that did not accommodate, nor completely resist, the dominant market-based economy. In many ways, it demonstrates the different representations of the economy that is grounded in the community. Furthermore, it serves to remind us that there is no single logic of sustainable economic development and that such efforts can be grounded in informal non-market exchanges.

As Healy and Graham (2008) have noted, the cooperation, volunteer labour, gift giving and other acts of generosity central to Nuestras Raíces are an example of non-capitalist opportunities that has a multiplicative effect, leading to employment, economic development and a greater well-being among the community. This multiplicative effect has been estimated to contribute millions of dollars to the local economy and provided Holyoke residents with additional food (Klindienst, 2006; Oehler et al., 2007). In essence, through the community gardens, the community leverages their cultural heritage to create an alternative path that broadens social well-being rather than relying on some outside economic development strategy. At the core of these activities has been the reclaiming and decommodification of the environment. The development of community gardens is centred on volunteered labour and cooperation among residents and the spaces are governed in a participatory manner that allows for a diverse community economy to exist (Graham and Cornwell, 2009). Moreover, it creates a radical space that allows residents to organize to tackle broader systematic social and environmental inequities. This can be seen in Nuestras Raíces and the community efforts to mobilize against broader social and environmental injustices afflicting the community.

In 2006, Nuestras Raíces was vital in bringing together a multi-stakeholder community environmental health coalition to tackle environmental racism in the community. The South Holyoke neighbourhoods in which the organization was most active were dealing with some of the highest asthma rates in Massachusetts, especially among children, as well as other respiratory illnesses (Ross, 2008a, b). Many of these air quality issues were tied to dozens of brownfield sites from old abandoned factories, gas stations and auto repair shops. Nuestras Raíces was able to mobilize its members and youth, the majority of residents who lived in the low-income neighbourhoods most adversely affected by the poor air quality and brownfield sites, to implement community health assessment, mapping and monitoring (Burke, 2006). Over the course of several months, community members collected environmental data with the help from coalition partners, local higher education institutions and the Environmental Protection Agency to demonstrate that the network of highways and auto repair businesses in the community resulted in substandard air quality that exposed them to higher risks of asthma, respiratory ailments and cancers (Nuestras Raíces, 2009).

These results were used later to organize against the proposal for a solid waste transfer station into a neighbourhood already disproportionately impacted by poor air quality. Vital to organizing and educating the community against the solid waste transfer station was incorporating environmental concerns into their existing programmes, outreach and events. However, most crucial to the organization was the community gardens and urban farm which acted as sites for not only sharing information and concern among the community, but a space for organizing the community (Nuestras Raíces, 2009). Through these spaces, the organization, while unsuccessful in stopping the construction of the solid waste transfer station, was able to secure major commitments from the City Council to reduce the impact of the facility on the community (Ross, 2009).

The organizing around air-quality issues, especially the solid waste transfer station, highlights the tensions between co-evolving spaces. Local economic development officials conceptualized the solid waste transfer station as a major way to boost tax revenue and a job-creation mechanism (Ross, 2008b; Plaisance, 2009). In contrast, the residents of the community rejected the sacrifices to their well-being at the hands of the economic development opportunity that disproportionately placed them at risk. Moreover, these tensions reflect the broader struggles the community faces in its day-to-day dealings with the city. Many city officials, including the mayor, were unreceptive to the notion that environmental racism played any role in the health and economic disparities occurring in the city. Such hostilities date back to the founding of the first community garden site (Williams, 1992). Despite such hostilities, Nuestras Raíces was able to use its efforts to build alliances that ultimately led to the election of a new mayor who was more receptive to the community's concerns.

What becomes clear in the struggles of Nuestras Raíces is that alternatives to the dominant economic paradigm are not without conflict and setbacks. More importantly, it shows the ways in which community gardens are radical spaces that allow non-capitalist alternatives to flourish and serve as participatory venues for organizing for a just and sustainable community. Members involved in Nuestras Raíces have become empowered to shape the place around them and transform private spaces into public sites of resistance all through the simple act of growing food. Significantly, it allows people a way to change their local economy to serve human well-being rather than a profit margin. This reclaiming of the city and their metabolism allows residents of South Holyoke to create an alternative that seeks to place human well-being and a just and equitable food system and region ahead of economic development initiatives that would sacrifice residents' well-being and continue to perpetuate the nature–city divide in attempts to commodify nature and maintain exploitative socio-ecological relationships.

8.4.2 Gardening the Community

Gardening the Community (GTC) emerged in 2002 as a joint project of the Massachusetts Branch of the Northeast Organic Farmer's Association (NOFA) and grassroots community organizers in Springfield's Six Corners, a historically poor and African-American neighbourhood. Its modest beginnings were largely educational and involved the implementation of a small garden on an abandoned lot near one of the neighbourhood's elementary schools. Produce was sold at local markets and donated to local food pantries. Since that modest beginning, the organization has grown into a social and ecological force in the city, expanding to three garden sites (Fig. 8.1) and producing over 3300 lb of organically grown food for its farm stand, the neighbourhood farmers' market, and for delivery by bike to local residents (Gardening the Community, 2013). GTC now finds itself at the centre of a network of food activists, producers, researchers, volunteers and enthusiastic organic agriculture supporters, and is now poised

Fig. 8.1. Gardening the Community's Lebanon Street garden utilizes a rain garden and a diverse mixture of flora to control aspects of urban metabolism in a high-density mixed-use neighbourhood. This food-producing site also highlights the tenuous tenure rights as it is on loan from a local business. (Photograph by Brian Conz, August 2015.)

to expand production into a recently purchased quarter of hectare lot. It is a success story not without its ongoing challenges, but with a growing momentum and the support to go to scale and increase its impact in a city neighbourhood that has been desperate for positive change.

Six Corners is a high-poverty neighbourhood in several respects. The median household income is US$18,763. Almost 50% of the population lives below the poverty level and the correspondingly high unemployment rate was 18.5% for 2012. Roughly 12% of children born in the area in 2010 were low birth-weight babies (PVPC, 2014). Adding to an existing list of problems and challenges, the neighbourhood was devastated by a powerful tornado that ripped through it in June of 2011.

The larger setting in which Six Corners is situated, the city's Mason Square neighbourhood, is identified for its inadequate access to nutritious food by area food-justice activists and concerned community members. Indeed, the preponderance of fast food establishments and bodega-style corner stores, complemented by a single, small grocery, is suggestive of the familiar pattern of supermarket flight that has come to characterize inner city neighbourhoods, especially those inhabited by people of colour. Evidence of the links between this type of food environment and high rates of obesity and diabetes, and somewhat counterintuitively, food insecurity, abounds and the strategy that has emerged in Springfield, as elsewhere, involves a combination of urban policy interventions and educational initiatives to support food production (urban agriculture), greater access to fresh food (farmers' markets and full service supermarkets) and healthy eating (Raja *et al.*, 2008). GTC has been at the centre of these initiatives in one of Springfield's most marginalized neighbourhoods and is emerging as a galvanizing force in the greater Springfield area.

Alongside its efforts to promote food security in the community, GTC is an organization dedicated to youth development, sustainable community economies and social justice. In 2005, GTC's collaboration with a Springfield urban farm and homestead helped to boost the organization's sustainability profile and emphasis, by integrating ethics and principles of sustainability as part of GTC's mission. The organization implemented rainwater-harvesting techniques, emphasized human-powered transportation by delivering produce to markets and residents by bicycle, increasingly made use of salvaged materials for use in garden sites, and drew inspiration from the urban homesteaders' emphasis on barter and exchange. The technical urban farming and garden development skills deployed on the organization's sites have undergone continued refinement and deepening through regular collaboration with organizations such as NOFA, and with activists and practitioners from the Pioneer Valley's burgeoning local foods movement, contributing to the professional development of long-term project participants. Training is channelled into local people, with long-term investments in, and connections to, the community; and these community members continue to staff the projects, rather than outsiders with lots of farming expertise. This has been an important aspect of GTC's commitment to its home community (Belanger, 2014). Springfield youth have been some of the key beneficiaries of the educational component, with dozens coming through the programme and going on to college and beyond.

The network of organizations collaborating with GTC also has grown significantly and includes local universities where GTC draws upon resources in the form of volunteers and greenhouse space for vegetable starts. The organization's work spurred the formation of Springfield's Food Policy Council in 2010, which created an urban agriculture sub-committee and led to the updating, revision and adoption of the city's community gardening ordinance in 2012. In 2011, GTC began collaboration with Next Barn Over Farm and CSA, initiating their GTC Eats programme. Farms shares are delivered to a GTC garden in Springfield, where they are picked up or delivered by bicycle to participating members. Overall, GTC and its allies are asserting their metabolic right to the city. They utilize their community gardens as a catalyst to create a healthy and equitable community, which has created many challenges for the organization.

One of the common challenges faced by organizations like GTC is secure land tenure, a challenge GTC suffered in 2005 and again in 2007 when the city reclaimed lots that the organization had put into production. With a stroke of the pen, the Economic Development Department sacrificed tens of thousands of dollars in investment, food produced for residents and soup kitchens, and appropriated surplus from the volunteer labour that initially cleaned the abandoned run-down sites in order to put the land on the tax rolls (Paine, 2005). Since that period, the organization's three main lots have been stable, but on a year-to-year basis, provided by a local business (Fig. 8.1). In 2014, after a sustained fundraising campaign, negotiations with the city and a competitive bidding process, GTC acquired a quarter of a hectare of property in an ideal location within the Six Corners neighbourhood and close to its other gardens. Clearing for the site began in July and the challenging process of soil remediation was initiated.

The acquisition of this property represents an historic moment for the 13-year-old organization and a pivotal juncture for urban agriculture in Western Massachusetts' largest city, which has been plagued by capital flight and the associated poverty, high crime and urban decay typical of the postindustrial city at the turn of the 21st century. In one of the city's most underserved neighbourhoods, food justice has become a beacon of hope, galvanizing interest and action on social, economic and environmental issues. In a sense, the networks that helped produce this success – emerging from collaborations between concerned community members and organic farmers, urban homesteaders and social justice activists, youth leaders and college volunteers – can be seen as a product of a moral, alternative community economy that is giving birth to sets of relationships quite distinct from those of the conventional food system. To what scale, volume and effects on health this particular project may reach remains to be seen, but the consistent growth and enthusiasm this project has generated cannot be ignored. For the foreseeable future at least, GTC is likely to continue to make claims on urban space and metabolism, hoping to replicate, on some level, the scale of success of Nuestras Raíces and other urban agriculture projects. In doing so, they are growing a diverse community economy and demonstrating a real alternative to neoliberal economic models by leveraging human creativity, and social and ecological capital towards real community power and change. Addressing the issues of social injustice suffered by the black and Latino communities of Springfield through food systems work contributes in substantial ways to what Gibson-Graham (2008) call a 'politics of economic innovation', a place-based enactment of possibility that may be seen as 'lend[ing] credibility to the existence and continual emergence of "other economies" world-wide' (Gibson-Graham, 2008, p. 628).

8.5 More than Garden and Food

The community gardens and inner-city urban farms in Holyoke and Springfield not only play a vital role in providing affordable culturally appropriate nutritious food to neighbourhoods, but also serve as radical spaces to assert marginalized groups' right to the city and their metabolism. They are intergenerational spaces that have allowed residents to appropriate urban space in ways that challenge neoliberal local development initiatives in order to promote more equitable, just and sustainable socio-ecological relationships. At one level, these spaces reaffirm and celebrate the connection between community members, nature and food, while placing the community well-being over the accumulation of capital and development initiatives that seek to sacrifice such well-being. Consequently, Nuestras Raíces and Gardening the Community efforts do not end in the plots. Instead, the community gardens have become a vital starting point to tackle broader socio-economic injustices in the agri-food system and in their respective cities.

At a fundamental level, the community gardens are part of a larger autonomous foodscape that is open to the possibility of non-capitalist alternatives. They are relational spaces that leverage non-market mechanisms to create more just and equitable socio-ecological relationships to advance the well-being of the community. Whether it is Nuestras Raíces and GTC's attempts to tackle environmental racism, promote more sustainable food production, address affordable housing issues, secure land ownership, empower youth or uphold economic justice, these broader efforts seek to move beyond the provision of food. They dismantle the nature–city divide to reveal the exploitative socio-ecological relationships in an attempt to mobilize the community. While Nuestras Raíces and GTC have been successful in their efforts to reclaim their metabolic right to the city, their success has not been without tensions.

Both organizations struggle to balance their efforts to promote food sovereignty and equitable autonomous foodscapes with neoliberal development initiatives. GTC and Nuestras Raíces both implement youth empowerment and green job training and skills development programmes. Such efforts, along with the provision of food, can be viewed within a neoliberal framework as filling the gap in the social safety left by the erosion of the state and a reskilling of workers to fit an economic model that often has resulted in widespread disinvestment in their communities. At the same time, both organizations rely on private–public partnerships to establish funding underlying many of their programmatic areas. Yet as documented in this chapter, both organizations also are providing non-capitalist alternatives and resisting local economic development initiatives that do more harm than good to the communities' well-being.

8.6 Conclusion

Nuestras Raíces and Gardening the Community demonstrate how grassroots organizations and their allies are attempting to reassert their metabolic right to the city. This reassertion is no different from their counterparts around the globe. For it is the fundamental act of appropriating space to feed the heart and soul that makes it possible to create more just and equitable food systems, while placing human needs above economic development. By envisioning more just and equitable socio-ecological alternatives to the mainstream development practices, both organizations are creating radical spaces that are open to the possibilities of non-capitalist alternatives. These spaces are attempts to leverage local knowledge and heritage that has never known of any nature–city dichotomy, and in many ways share the rich cultural heritages surrounding food production and consumption found in the Global South. Moreover, through these protean spaces both organizations are able to leverage initiatives that attempt to place the well-being of the community at the centre of urban renewal in the Global North.

The quest to produce autonomous local spaces that lead to food sovereignty is a long road that is not without its complications. Both organizations walk a fine line in their attempts to balance their efforts to create a more sustainable and equitable food system with the neoliberal pressures embedded in the existing food system as well as within their cities. This is most evident in the partial successes and setbacks that both organizations have faced. Moreover, while some of the actions of Nuestras Raíces and Gardening the Community might underwrite the rolling back of the state and support mainstream economic development initiatives that place profit above people, it should not be forgotten that such attempts to create placed-based non-capitalist alternatives do not preclude a co-evolving relationship with current economic practices. By freeing ourselves from the all-encompassing logic of capitalism, it becomes possible to see how creative alternative local economies occurring around the globe are reshaping the city to reflect more just, equitable and sustainable socio-ecological relationships for future generations to come.

References

Agyeman, J. and McEntee, J. (2014) Moving the field of food justice forward through the lens of urban political ecology. *Geography Compass* 8(3), 211–220.

Alkon, A.H. (2013) The socio-nature of local organic food. *Antipode* 45(3), 663–680.

Alkon, A.H. and Agyeman, J. (2011) The food movement as polyculture. In: Alkon, A.H. and Agyeman, J. (eds) *Cultivating Food Justice: Race, Class and Sustainability*. MIT Press, Cambridge, Massachusetts, USA, pp. 1–20.

Allen, P. (2010) Realizing justice in local food systems. *Cambridge Journal of Regions, Economy and Society* 3, 295–308.

Allen, P. and Guthman, J. (2006) From 'old school' to 'farm-to-school': neoliberalization from the ground up. *Agriculture and Human Values* 23, 401–415.

Belanger, N. (2014) Alleviating hunger by increasing urban farm productivity. NOFA Newsletter, June 2014 (Northeast Organic Farming Association). Available at: http://www.nofamass.org/sites/default/files/newsletter-issue/2014_june_web.pdf (accessed 24 February 2015).

Berg, E., Elwood, A. and Macchiarolo, M. (2014) Food in the city, an old way in a new time: a process to assess land suitable for urban agriculture. Available at: https://issuu.com/conwaydesign/docs/springfield_final_report_issuu_vers/17?e=1127520/8234521 (accessed 25 February 2015).

Bluestone, B. and Harrison, B. (1982) *The Deindustrialization of America: Plant Closings, Community Abandonment, and the Dismantling of Basic Industry*. Basic Books, New York, USA.

Born, B. and Purcell, M. (2006) Avoiding the local trap: scale and food systems in planning research. *Journal of Planning Education and Research* 26, 195–207.

Burke, M. (2006) Group to help check toxic risks. *The Springfield Republican*, 5 December, B2.

Constantine, S.E. (1999) Holyoke youths aid in building new greenhouse. *The Springfield Republican*, 16 March, B2.

Domene, E. and Saurí, D. (2007) Urbanization and class-produced natures: vegetable gardens in the Barcelona Metropolitan region. *Geoforum* 38, 287–298.

Engel-Di Mauro, S. (2012) Urban farming: the right to what sort of city? *Capitalism Nature Socialism* 23(4), 1–9.

Federal Reserve Bank of Boston (2009) Towards a more prosperous Springfield, Massachusetts: project introduction and motivation. Community Affairs Discussion Paper, No. 2009-01.

Federal Reserve Bank of Boston (2010) Jobs in Springfield, Massachusetts: understanding and remedying the causes of low resident employment rates. Community Affairs Discussion Paper, No. 2009-05.

Federal Reserve Bank of Boston (2011) Housing policy and poverty in Springfield. Community Affairs Discussion Paper, No. 2011-01.

Forrant, R. (2003) The roots of Connecticut River valley deindustrialization: the Springfield American Bosch plant, 1940–1975. *Historical Journal of Massachusetts* 31(1), 90–106.

Forrant, R. (2005) Grinding decline in Springfield: is the finance control board the answer? *New England Journal of Public Policy* 20(2), 67–88.

Gardening the Community (2013) *2013 Annual Report*. Gardening the Community, Springfield, Massachusetts, USA.

Gibson-Graham, J.K. (1996) *The End of Capitalism (As We Knew It): A Feminist Critique of Political Economy*. Blackwell Publishing, Oxford, UK.

Gibson-Graham, J.K. (2006) *A Postcapitalist Politics*. University of Minnesota Press, Minneapolis, Minnesota, USA.

Gibson-Graham, J.K. (2008) Diverse economies: performative practices for 'other worlds'. *Progress in Human Geography* 32(5), 613–632.

Gottlieb, R. and Joshi, A. (2013) *Food Justice*. Massachusetts Institute of Technology, Cambridge, Massachusetts, USA.

Graham, J. (2001) Imaging and enacting noncapitalist futures: the community economies collective. *Socialist Review* 28(3), 93–135.

Graham, J. and Cornwell, J. (2009) Building community economics in Massachusetts: an emerging model of economic development? In: Amin, A. (ed.) *The Social Economy: International Perspectives on Economic Solidarity*. Zed Books, New York, USA, pp. 37–65.

Hartford, W. (1990) *Working People of Holyoke: Class and Ethnicity in a Massachusetts Mill Town, 1850–1960*. Rutgers University Press, New Brunswick, New Jersey, USA.

Harvey, D. (1996) *Justice, Nature and the Geography of Difference*. Blackwell Publishing, Malden, Massachusetts, USA.

Healy, S. and Graham, J. (2008) Building community economies: a postcapitalist project of sustainable development. In: Ruccio, D. (ed.) *Economic Representations: Academic and Everyday*. Routledge, New York, USA, pp. 291–314.

Heynen, N. (2006) Justice of eating in the city: the political ecology of urban hunger. In: Heynen, N., Kaika, M. and Swyngedouw, E. (eds) *In the Nature of Cities: Urban Political Ecology and the Politics of Urban Metabolism*. Routledge, New York, USA, pp. 129–142.

Heynen, N., Kaika, M. and Swyngedouw, E. (2006) Urban political ecology: politicizing the production of urban natures. In: Heynen, N., Kaika, M. and Swyngedouw, E. (eds) *In the Nature of Cities: Urban Political Ecology and the Politics of Urban Metabolism*. Routledge, New York, USA, pp. 1–20.

Jacobson-Hardy, M. and Weir, R.E. (1992) Faces, machines, and voices: the fading landscape of papermaking in Holyoke, Massachusetts. *The Massachusetts Review* 33(3), 361–384.

Kaika, M. and Swyngedouw, E. (2000) Fetishizing the modern city: the phantasmagoria of urban technological networks. *International Journal of Urban and Regional Research* 24(1), 120–138.

Keil, R. (2003) Urban political ecology. *Urban Geography* 24(8), 723–738.

Keil, R. (2005) Progress report – urban political ecology. *Urban Geography* 26(7), 640–651.

Klindienst, P. (2006) *The Earth Knows My Name: Food, Culture and Sustainability in the Gardens of Ethnic Americans*. Beacon Press, Boston, Massachusetts, USA.

Lauer, M.J. (1999) Carpenters donate labor to project. *The Springfield Republican*, 21 April, B3.

Marzan, G. (2009) Still looking for that else where: Puerto Rican poverty and migration in the Northeast. *Centro Journal* 21(1), 101–117.

Massey, D. (2005) *For Space*. Sage Publications, Thousands Oaks, California, USA.

McClintock, N. (2010) Why farm the city? Theorizing urban agriculture through a lens of metabolic rift. *Cambridge Journal of Regions, Economy and Society* 3, 191–207.

McClintock, N. (2014) Radical, reformist, and garden-variety neoliberal: coming to terms with urban agriculture's contradictions. *Local Environment* 19(2), 147–171.

Morales, J. (1986) *Puerto Rican Poverty and Migration: We Just Had to Try Elsewhere*. Praeger Publishers, New York, USA.

Nuestras Raíces (2009) Creating the community environmental health coalition. EPA CARE Level 1 Final Report.

Nuestras Raíces (2013a) *2013 Annual Report*. Nuestras Raíces, Holyoke, Massachusetts, USA.

Nuestras Raíces (2013b) *Planning for Our Future 2013–2016*. Nuestras Raíces, Holyoke, Massachusetts.

Oehler, K., Sheppard, S.C. and Benjamin, B. (2007) The economic impact of Nuestras Raíces on the city of Holyoke: current and future projections. Center for Creative Community Development, North Adams, Massachusetts, USA.

Paine, D. (2005) Youth protest sale of garden site. *The Springfield Republican*, 28 October, B2.

Plaisance, M. (2009) Trash facility to get hearing. *The Springfield Republican*, 9 August, C1.

Pothukuchi, K. and Kaufman, J.L. (2000) The food system: a stranger to the planning field. *Journal of the American Planning Association* 66(2), 113–124.

PVPC (2013) *The Pioneer Valley Regional Housing Plan*. Pioneer Valley Planning Commission, Springfield, Massachusetts, USA.

PVPC (2014) Data Atlas by Neighborhood, City of Springfield, MA. Pioneer Valley Planning Commission. Available at: http://www.pvpc.org/sites/default/files/Springfield%20Data%20Atlas%209-23-14-web-reduced.pdf (accessed 24 February 2015).

Raja, S., Born, B. and Russell, J.K. (2008) *A Planners Guide to Community and Regional Food Planning: Transforming Food Environments, Facilitating Healthy Eating*. American Planning Association, Chicago, Illinois, USA.

Ross, K. (2008a) Holyoke found prone to fumes. *The Springfield Republican*, 10 November, B2.

Ross, K. (2008b) Waste project spurs debate. *The Springfield Republican*, 21 December, C1.

Ross, K. (2009) Waste project clears hurdle. *The Springfield Republican*, 20 February, C1.

Sbicca, J. (2012) Growing food justice by planting an anti-oppression foundation: opportunities and obstacles for a budding social movement. *Agriculture and Human Values* 29, 455–466.

Sbicca, J. (2014) The need to feed: urban metabolic struggles of actually existing radical projects. *Critical Sociology* 40(6), 817–834.

Shillington, L.J. (2013) Right to food, right to the city: household urban agriculture, and socionatural metabolism in Managua, Nicaragua. *Geoforum* 44, 103–111.

Slocum, R. (2007) Whiteness, space and alternative food practice. *Geoforum* 38(3), 520–533.

Swyngedouw, E. (1996) The city as a hybrid – on nature, society and cyborg urbanization. *Capitalism Nature Socialism* 7(2), 65–80.

Swyngedouw, E. and Heynen, N.C. (2003) Urban political ecology, justice and the politics of scale. *Antipode* 35(5), 898–918.

Toensmeier, E. (2013) *Paradise Lot*. Chelsea Green Publishing, White River Junction, Vermont, USA.

US Census Bureau (2013) 2009-2013 American Community Survey 5-Year Estimates. Tables B03002, S1701 and S2201; generated by Timothy LeDoux using American FactFinder. Available at: https://factfinder.census.gov/bkmk/table/1.0/en/ACS/13_5YR/B03002/1600000US2530840|1600000US2567000 (accessed 24 February 2015).

Wachsmuth, D. (2012) Three ecologies: urban metabolism and the society-nature opposition. *The Sociological Quarterly* 53, 506–523.

Wilson, A.D. (2013) Beyond alternative: exploring the potential for autonomous food spaces. *Antipode* 45(3), 719–737.

Williams, S. (1992) *Seeds of Change: Community Gardening in South Holyoke, Massachusetts*. Division III examination thesis, School of Natural Science, Hampshire College, Amherst, Massachusetts, USA.

9 Sustainability's Incomplete Circles: Towards a Just Food Politics in Austin, Texas and Havana, Cuba

Jonathan T. Lowell[1]* and Sara Law[2]

[1]*University of Texas, Austin, Texas, USA;* [2]*Sustainable Food Center, Austin, Texas, USA*

9.1 Introduction

The purpose of this chapter is both to question the use of sustainability discourses in urban agriculture projects and to analyse how the discourse traverses over spaces and the bodies who occupy those spaces. As Alkon and Agyeman put it 'food is not only linked to ecological sustainability, community, and health but also to racial, economic, and environmental justice' (Alkon and Agyeman, 2011, p. 4). However, in theory and in praxis, 'sustainability' is largely talked about in terms of food production, to the detriment of food consumption (i.e. food security). More specifically, *how, where* and *by whom* the food is produced is privileged over *who* and to what extent food is accessible to *everybody*.

We turn to two places heralded for their cultures and systems of alternative agriculture: La Habana, Cuba and Austin, Texas. Faced with a food security crisis in the 1990s, Cuba turned to a diversified and mostly organic agricultural system in order to feed its habitants. Austin also has a well-established community garden programme and a lot of positive press for its sustainable agriculture programmes. The case studies we provide may seem different: the Austin case revolves around community resentment at the establishment and practices of urban farms, while the work in Cuba focuses on the inability of government-led sustainable production to meet the needs of many low-income communities. However, we have found some comparable dynamics. In both Austin and La Habana, the ways in which agricultural practices are deemed sustainable in the cultural and political discourses is predicated upon the ways in which food is produced, with less regard to continuing issues of food insecurity that urban agriculture has yet to redress. Still less attention is paid to the cultural politics of food: the ways in which race, class and gender can shape access to food choices, as well as have the capacity to shape political discourse around food. What we find in Austin and La Habana are similar sets of agricultural practices, as well as similar discourses towards their value for the community and the nation. We also find comparable outcomes in terms of the ways in which food can expose spatial and social divisions.

In this chapter, we interrogate the discourses and practices of sustainable agriculture in urban settings in a holistic way that brings histories of colonialism and segregation to the forefront. We point to deeper questions as to how urban agriculture impacts not only food provisioning but the social relations in which food production is embedded. What are the material and

---

*Corresponding author; e-mail: jonathan.t.lowell@utexas.edu

discursive roles of urban agriculture in terms of the socio-political agendas related to sustainability and mitigating food insecurity? How do factors such as race and spatial segregation contribute to the success and failure of urban agriculture projects?

Our methods included activist participant research, structured and unstructured interviews and discourse analysis. In Austin, the first author conducted a discourse analysis (Lees, 2004) on Austin newspapers such as the *Austin-American Statesman* and the *Austin Chronicle*, along with the meeting notes from the urban farm ordinances. The research in Cuba was conducted by the second author and included participant observation while walking the markets and gardens of Pogolotti and El Vedado and interviewing ten families that were involved with the Centro Memorial de Martin Luther King (CMMLK). Additionally, a discourse analysis was conducted on Cuban and US news sources, which included but were not limited to *Granma*, *Juventud Rebelde*, the *New York Times* and National Public Radio.

We root this paper in the premise of the 'sympathetic critique', which operates from an understanding that 'critique and critical inquiry that comes from a place of support and compassion for those doing the work of building subversive and interstitial food spaces' (Galt et al., 2014, pp. 138–139). This means acknowledging the various and serious constraints to creating alternatives and to also take a 'long-term, empathetic, structural view' (Galt et al., 2014, p. 139), that ultimately makes the work of all involved in remaking the food system better.

9.2 Tracing Sustainability in Discourse, Practice and Place

The standard starting point in tracing the diffusion of sustainability discourse and unpacking its various mobilizations is the Brundtland report, released in 1987, in which sustainability was defined as 'that which meets the needs of the present without compromising the ability of future generations to meet their own needs' (United Nations, 1987, p. 43). In discussing the ways in which sustainability has travelled and is mobilized, we wish in part to spend some time decentring what sustainability means from the perspective of the Global North. It was early in its development that Fidel Castro said at the Rio Earth Summit in 1992: 'Stop transferring to the Third World lifestyles and consumer habits that ruin the environment . . . Use science to achieve sustainable development without pollution. Pay the ecological debt. Eradicate hunger and not humanity' (Castro, 1992). Coming from the leader of one of the world's few non-capitalist nation-states, this statement reveals both the extent to which sustainability has travelled, while also revealing some of the tensions that the term has generated. In this context, Castro utilized sustainability as an anti-capitalist critique as well as a way to bolster his own regime in the wake of an economic depression (Gold, 2014). Others have also observed the latent radicalism of sustainability (Campbell, 1996; Davidson, 2009), while others claim that the underlying goal of sustainability is the sustaining of hegemonic capitalism (Swyngedouw, 2007; Tretter, 2013). The proliferation of sustainability projects and discourses led Campbell to claim that 'in the battle of big public ideas, sustainability has won: the task of the coming years is simply to work out the details, and to narrow the gap between theory and practice' (Campbell, 1996, p. 312). It has been nearly 20 years and the task has not proven simple.

The critiques of sustainability have come from two fronts, but both come down to this: while sustainability is often described as having three 'pillars' – environment, economy and society – the social is often neglected. Agyeman and Evans (2004) call for a 'just sustainability' that explicitly takes social justice into account. Others have found it problematic to consider the social a separate 'pillar' at all, as it renders the environment and the economy 'desocialized' (Psarikidou and Szerszynski, 2012). This points towards Swyngedouw's (2007) issues as well. For him, sustainability discourses are not just desocialized, but depoliticized as well. Everyone can be in favour of sustainability in principle without actually challenging any of the entrenched systems that have created our social and ecological problems. Marcuse makes a similar point when he says that the 'promotion of "sustainability" may simply encourage the sustaining of the unjust status quo' (Marcuse, 1998, p. 103) and that attempts to appeal to common interests can disguise conflicts of interest that have to be resolved through politics.

Like sustainability generally, urban agriculture is both radical and neoliberal (McClintock, 2014). On one hand, urban agricultural practice seeks to heal the 'metabolic rifts' in the food system by not depleting the nutrients of the social and de-commodifying land and labour (McClintock, 2010). While the benefits of local/sustainable agriculture are said to range from alleviating food insecurity in low-income neighbourhoods, providing economic development and green space, as well as a place for developing social capital and community (Pearson et al., 2010; McClintock, 2014), the heart of what makes them 'sustainable' lies in practices by which the farming is carried out. At least, this mostly holds true in agricultural projects in the Global North. With the prominent exception of Cuba, the literature on the Global South is much more focused on food insecurity and many urban producers in the Global South rely on synthetic fertilizers (Bryld, 2003). In many cities in the Global South, urban agriculture is widely a needs-based practice to supplement income, increase food security and even continue cultural traditions. With the rapid urbanization and influx of peoples from rural areas, urban agriculture tends to be for personal or family use. While class and gender have been the main source of analysis in these conversations, race seems to be widely excluded when discussing urban agriculture in the Global South (Hovorka, 2006). We intend to redress these issues by interweaving race in our case studies.

According to Slocum: 'we can better understand farming and provisioning, tasting and picking, eating and being eaten, going hungry and gardening by paying attention to race' (Slocum, 2011, p. 303). We bring the racialized class relations to the forefront in the discussion about sustainable agriculture. While there are several ways in which food and race intersect, we focus on the raw histories of spatial segregation that are embedded in both Austin and La Habana, as well as more broadly (Omi and Winant, 1994; Lipsitz, 2011). To a degree, this is a matter of using juxtaposition to expose paradox. Many of Austin's urban farms are located in a neighbourhood that has a deep history of marginalization and suffers from issues related to food access. Pogolotti, the first workers' neighbourhood of La Habana, is also embedded in a history of government failure and ambivalence that has relegated them socially and spatially. By asking who and what sustainable agricultural practices are for, we are able to interrogate the diverse array of responses and attitudes to food.

Finally, we want to speak to the non-monolithic nature of sustainability as a set of discourses and practices. While it may be ubiquitous, it is certainly not homogeneous. Rather, it refracts in ways similar to overall geographic unevenness reflective in other social situations. A similar set of discourses can serve different functions at different scales and locations. Fidel Castro can speak at the Rio Summit and deliver a Brundtland report-inspired message about the need to rein in the exploitation of the earth's resources, but tailor it to his Cuban context to bring out sustainability's latent critique of capitalism, while simultaneously folding it into Cuba's nationalist project (Gold, 2014). Similarly, agricultural practices deemed sustainable serve different, yet overlapping, functions. In Austin, they are folded into development, gentrification, weirdness, progressivism. In La Habana they were born out of an immediate necessity, a crisis that was the result of the shifting geopolitics.

9.3 Space and Food Politics in Austin, Texas

We begin our discussion of Austin by talking about a compost pile. Inspired by Michael Pollan's critique of industrial agriculture (Vickery, 2014), the farm's owner set up shop in the early 2000s in the Eastside of Austin and installed a black soldier fly compost system as part of her chicken-raising operation. After slaughtering a chicken, the scraps of unused flesh and feathers would be placed in the pile where soldier flies would eat them and the little fly grubs that emerged from the pile would then be eaten by the next batch of chickens. It was a lovely and innovative closed-loop cycle that kept all the material on-site, in accordance with the principles of sustainable agriculture. The soil enriched by the compost pile produced fresh organic vegetables that were sold on-site or to local restaurants. The farm's owner felt the farm to be part of a moral obligation to act against global warming

(Gándara, 2012). However, the compost pile and the farm do not exist in a vacuum, but within a variety of socio-spatial contexts operating at different scales. The farm's neighbours, for example, began to complain about the smell emanating from the compost and eventually called the city authorities to complain (Gándara, 2012). Some were also upset by the way in which Haus-Bar and other urban farms had become event spaces that attracted traffic and noise to their neighbourhood. Govalle neighbourhood exists in a rapidly gentrifying district in which high-end restaurants were quickly replacing the long-time cantinas, yet one in which many low-income residents lacked access to fresh produce either from local sources or from chain grocery stores. The city in which the farm is situated, Austin, Texas, prides itself on its progressive image and policies, making 'sustainability' the guiding frame of reference for its most recent master plan (City of Austin, 2012). Yet, Austin also is a city deeply segregated racially and economically, a historical by-product of a city policy that only provided social services to people of colour if they lived in certain sections of town (Busch, 2011). The farm is also within a country in which commercial small-scale agriculture has been eroded for decades, yet has started to see a groundswell of interest in remaking the food system, that is, how we grow, distribute and consume food, and how we dispose of its waste (Allen, 2004).

It is in these contexts that Austin's urban farms, and local food system as a whole, are embedded. In fact, many of the same general dynamics play out in the food system (Guthman, 2008; Alkon, 2012). Austin has a celebrated food scene, with several farmers' markets dotting the city, and 4000 restaurants (Marty, 2015), dozens of which are part of a burgeoning scene marked by fusion and creativity displayed by well-known chefs such as Paul Qui, and commitment to buy locally sourced ingredients. This scene is regularly celebrated in publications such as the *Austin Chronicle* and *Edible Austin*, yet at the same time it forms part of the gentrifying landscape of consumption of Austin's changing eastside, replacing the cantinas that served the Latino working-class communities for decades previously (Ward, 2009). This displacement process is compounded by the fact that many of these communities struggle with food insecurity, lacking both financial and geographic access to either full-service grocery stores or local farmer's markets (Sustainable Food Center, 1995; Banks, 2011). Despite its reputation as a locavore haven, the total amount of direct farmer to consumer sales in Austin's MSA is about a quarter of the national average (TXP, Inc., 2013) and current production of vegetables in Austin's municipality does not even register in the agricultural economy data (Banks, 2011, p. 10). This is in part due to the loss of regional farmland as the city continues to grow, as well as lack of access and participation among the city's low-income and people of colour communities (Sustainable Food Center, 1995; Banks, 2011; Vickery, 2014).

9.3.1 Public arguments over sustainability

In the months following the neighbours' initial complaints about HausBar farm, the local press followed the story closely and various actors made their voices known through social media and op-eds. Both sides of the debate felt attacked and vilified. A collective of urban farms in the Govalle neighbourhood created a website (http://www.austinurbanfarms.org), to make the case that their farms deserved public support. That case centred on using only organic methods and being good 'stewards of the land' by collecting rainwater and reusing materials. At the heart of this appeal is the production methods used as what is of value to the community. Sixty local chefs penned a letter to the city council in support of urban farms. One was quoted in the weekly alternative paper: 'I definitely support urban farms 100% percent [sic]. The produce and product they provide the city of Austin and from the education aspect, as well. If people approach it with an open mind, they will support urban farms' (Toon, 2013a). Neighbourhood advocates countered that 'The objection is not to urban farming but to commercialization of the single-family zoning in general and to the slaughtering of animals in particular, which presents health hazards to the neighbourhood and is inappropriate to the area' (Austin American-Statesman, 2013).

The public debate led to the Sustainable Food Policy Board (SFPB) being appointed to draft a more clear and elaborate urban farm ordinance (Vickery, 2014). SFPB hosted four public engagement meetings on various topics surrounding the urban farm ordinance, with sessions dedicated to issues of animal-raising, site requirements, and issues of sustainable waste management. It culminated in a town hall meeting, in which the recommendations were presented and open for public comment. Dozens of people showed up to these meetings, a mixture of direct stakeholders, city staffers and those who were interested in local food issues. Conversations mostly revolved around what the best practices should be for an urban farm, such as water and soil management, as well as the policies that would best enable urban farms to be economically viable. For example, the working group defined sustainable waste management as 'a close circle requiring as little waste disposal away from the site as possible and the reuse of as much material as possible already on the site' (Sustainable Food Policy Board, 2013). While members of the SFPB proclaimed the process to be one of robust public engagement, others disagreed. One community organizer was quoted as saying:

> The fact that persons of color proclaimed that they felt uncomfortable or unsafe speaking about race, privilege, or gentrification in a white-dominated space, the UFO [urban farm ordinance] public meetings, should serve as a loud horn with glaring and flashing red lights that something was wrong. (Toon, 2013b)

Neighbourhood activists complained that none of their recommendations to the board, including outlawing slaughtering animals for commercial use and banning commercial agriculture on single family zoning, made it into the document that was to be presented to the city council (Gándara, 2013).

The public process reached a head when the issue made it to the city council in November of 2014. Neighbourhood advocates reiterated that they were not opposed to urban farms in theory but were concerned with having commercial and undesirable land uses imposed in their neighbourhoods.

They criticized the process of pursuing the ordinance by having the SFPB stacked with farming advocates, for not providing an inclusive space to talk about issues of gentrification, food justice and white privilege. Some expressed outrage that none of their recommendations made it into the document. The farms were taken to task for not accepting food benefits for low-income consumers. Several speakers brought up the history of East Austin's segregation. Urban farm advocates made an appeal to the idea of 'community' by providing locally grown food and green space to their surrounding neighbours. Urban farmers, they claimed, are facing an uphill battle against an industrial food system, and that what they were doing was a labour of love; no one was getting rich from farming. After showing a video of some local residents stating in Spanish that they did not have any problems with farms, and after two African-American children, with adopted white parents, testified to loving HausBar farms in their neighbourhood, some people in the gallery got upset. 'The tokenization here is just really outrageous', a woman called out from the gallery, eventually being removed from chambers.

The final ordinance that the city council passed maintained that urban farms could be established across all zoning types, but prohibited slaughtering animals for commercial use. Urban farm advocates decried the decision as a move 'away from sustainability' (Toon, 2013b). Barger was quoted in the local paper, stating: 'I truly hate to see Austin moving backwards when it claims to want to set the bar for sustainability for the whole country' (Toon, 2013b) and felt it was unfair to frame the practice as one that subjected poor residents of East Austin to industrial/commercial uses. Following the council's decision, there was cautious optimism on both sides. One community organizer who spoke against the ordinance said: 'with both sides claiming victory in the decisions voted on by Council, we can pivot back to creating dialogue about race, gentrification, climate change, and food justice, and identify other points of unity' (Toon, 2013b), while one of the prominent members of the SFPB expressed hope that the community could come together.

9.3.2 Sustainability and its limits

This case is a good example of what Marcuse means by limiting the scope of how sustainability is used (Marcuse, 1998). Instead of referring to long-term environmental planning, its gradual assimilation into all matters of public policy has rendered it unopposable. As such, those that are able to wear sustainability as a cloak are protected and those that raise concerns about a set of practices deemed sustainable are dismissed. It would be simplistic to position the two opposing camps as being pro- and anti-urban farm (as neighbourhood advocates were always quick to point out); rather, it is a debate about how to integrate them into the urban fabric. However, the two sides used very different language and appeals in talking about the issue and they had a difficult time communicating. Both pointed to this debate being in a context of a larger struggle: the broken food system on one side and the gentrification and displacement of low-income people of colour on the other.

Without wanting to denigrate the urban farms of Austin too much – we do feel that they have mostly good intentions – it is important to attend to the fissions that were laid bare throughout the controversy. For if, as McClintock (2010) suggests, urban agriculture is an imperfect attempt at healing 'metabolic rifts', we must examine where and why it plays a role in division. The racialized social division in Austin and the rampant gentrification taking place in many of its neighbourhoods are much larger issues than urban agriculture alone can either be responsible for or address positively. That being said, the debates surrounding urban farming in Austin reveal the limits of sustainability as a set of discourses and practices. When the social relations in which the practices are embedded are fraught with unequal power dynamics and painful histories, more work needs to be done to go beyond the notion that 'fresh, organic, and local' is the solution.

9.4 Space and Food Politics in La Habana

In La Habana, the story begins with a plot of land roughly 6 miles outside the city centre and a state-funded project championed by President José Miguel Goméz (Scarpaci et al., 2002). In the early 1900s, the administration attempted to build workers' housing for the urban poor but, due to bad planning and 'flimsy materials', only 950 units were completed out of the 2000 promised by 1913. This early housing plan quickly deteriorated and Pogolotti soon became labelled as a slum, 'notorious for outbreaks of infectious diseases' (Scarpaci et al., 2002, p. 58). While Pogolotti was not the only neighbourhood in La Habana that experienced poor planning and government neglect, it serves as a crucial case study to analyse how cultural marginalization and infrastructural decay surpasses temporal scales and bleeds into present-day relationships with food and food practices. Forty years after the creation of Pogolotti, political and economic pressures in Cuba initiated a series of events that led to the Cuban Revolution in 1959. The Revolutionaries' attempt to overthrow the Batista regime was marked by near-suicidal attacks, rebel soldiers and guerilla fighting, but their real success was the quick formation of the new State (Pérez, 2010).

Part of this success came from Fidel Castro's ability to generate an inspiring and nation-building discourse for Cuban citizens. Throughout the early years of the Revolution, and even during the economic collapse in 1991, the regime's capacity to maintain its nationalist message has been key to its ideological longevity. Similar to how environmentalists invoke figures like John Muir and Aldo Leopold as forebears to their movement, Cuban leaders look to José Martí, the father of the Cuban revolution, as the inspiration of sustainability. While his writings explicitly reference environmental themes in connection to political struggle, Fidel Castro reinterprets Martí's writings to create a platform for sustainable development as an ideological battlefront against imperialism (Gold, 2014). More recently, Raul Castro has modified the Cuban discourse of sustainability by speaking more in terms of efficiency and austerity (Gold, 2014). Gold argues that this tactical move not only reinforced the practical and progressive move to sustainable agriculture but also served as an 'official revolutionary discourse (that) has channeled popular global notions of sustainability in order to capture international funds, realign itself with regional partners, and redefine the terms of the essential revolutionary battles' (Gold, 2014, p. 406). The discourses of sustainability are important to how the Cuban state projects itself both to the outside world and to its own citizens. As a consequence, issues like food access to certain members of Cuban society and the politicization of food across lines of race, class and gender are often erased or ignored. In what follows, we lay out the ways in which sustainable producers and production are privileged and how the attendant cultural politics around consumption are left out or swept aside.

9.4.1 The increase of state-led production

With the collapse of the Eastern Bloc and Soviet Union in 1989 and 1991, Cuba was suddenly without its principal trading partners. Unable to export its sugar cane or import its food as before, meeting its food demands became Cuba's top priority (Torres et al., 2010). A new model was created for a more sustainable agricultural system, setting goals like diversifying crops, only utilizing organic fertilizers and creating food self-sufficiency (Chan and Freyre, 2012). Before 1989, the government limited urban food production to backyards. However, as the crisis set in, they relaxed their ordinances. Urban gardens began to emerge throughout La Habana, through the use of local resources and relatively low-input tools for production to alleviate food shortages. As the movement grew and diversified, the government provided state support in the form of technological and information services and established state-run gardens (Altieri et al., 1999).

The GNAU (Grupo Nacional de Agricultura Urbana) represents the state-led formalization of the once grassroots movement and provided the language that would define Urban Agriculture in Cuba. The GNAU's definition of Urban Agriculture includes 'all agricultural lands within certain distances of cities and towns having populations in excess of 1,000 persons' (Koont, 2011). Consequently, any form of agriculture in the La Habana provinces would be registered as urban agriculture by the Cuban government, which may skew our perceptions of how ubiquitous the movement actually is in La Habana. Additionally, the Urban Agriculture Department under the Minister of Agriculture sought to secure land rights, granting all citizens usufruct rights to unused land (Altieri et al., 1999; Torres et al., 2010). However, the urban agricultural movement is more than just an attempt to resolve food insecurity in and around cities. Fuster Chepe (2006), director of ACTAF (Asociación Cubana de Técnicos Agrícolas y Forestales), attempted to organize the movement in urban and peri-urban areas, provide support and training for neighbourhood production and make a clearer link from the fields to the home (Koont, 2011).

To a certain extent these objectives worked. Altieri et al. (1999) reported that in 1996 the urban farms of La Habana provided the city's urban population with 8500 tons of agricultural produce, 4 million dozens of flowers, 7.5 million eggs, and 3650 tons of meat. There was a high concentration of production from some of Cuba's most important agricultural areas like La Habana, Matanzas, Camaguey and Pinar del Rio. A survey carried out by Torres et al. (2010) revealed that market vendors were frequently purchasing produce from growers in their own districts, probably indicating a blossoming urban agricultural movement in the early years of the Special Period and a need to stay 'local' due to the high price of oil, as well as crumbling infrastructure across the nation.

Raul Castro has many times taken the party line that privileges production over consumption (Gold, 2014). This captures the problematic sentiment that is shared not only in cities like La Habana but also in the Global North. By only analysing sustainability from the standpoint of increased production, we ignore and marginalize the racialized, gendered and classed politics behind food consumption (Guthman, 2008; Harper, 2011; Slocum, 2011). While the stateled emphasis on production ameliorated the situation for thousands of Cubans, it was not an equally distributed remedy. Those who lived further away from the city centre, such as the residents of Pogolotti, were already experiencing 80 years of infrastructural deterioration, and food access became increasingly challenging as the Special Period heightened.

9.4.2 Consumption politics and continuing hauntings

Although things have changed somewhat over the last few decades, both materially and culturally, Pogolotti remains spatially and socially marginalized. Located in the district of Marianao, Pogolotti is seen as a 'low culture' neighbourhood by many Cuban elites. They are far removed from the tourist district in La Habana Vieja (lit. 'Old Havana'), making it difficult to work in the most lucrative industry in Cuba. The majority of Pogolotti residents are Afro-Cuban, who are generally excluded from participating in the tourist economy. As a result of this physical distance and racial marginalization, many residents are unable to supplement their state wage and find that food can often be an expensive expenditure.

In Pogolotti, it was often repeated that, 'things are getting better, but prices are too high'. Of the lauded *organopónicos* in La Habana, only one is located in the vicinity of Pogolotti. The rest of the produce was brought in from urban farms and distributed to small vendors, which increased the price due to transportation costs. One white male community member said he would spend up to 3 hours grocery shopping each morning to find the quality and, more importantly, quantity of produce necessary to sustain his family of four. In these interviews with residents of Pogolotti, conversations would inevitably circle back to life during the Special Period, revealing the haunting memory of a time not so far gone. The legacies of Cuba's past in relationship to food politics are not limited to the Special Period (Gordon, 2008). It is a compounding of the violence of colonialism and imperialism, the socio-political agendas of government leaders, the subtle and unspoken racism, that together create an environment where the practice of who consumes what is highly politicized. This is not just restricted to the conversation of food, though it clearly affects it.

However, there are various ways in which residents of Pogolotti exercise agency in making food consumption choices. Some would spend hours in the morning to avoid the heat, jumping from vendor to vendor to find the right amount and highest quality of produce. Often, single-parent households or families with more means preferred the *carretilleros*, or produce carts, that would be trolleyed down the neighbourhood. The consensus was that these *carretilleros* might be pricier, but the quality of vegetables was higher, their arrangement was visually appealing and, perhaps most significantly, they were convenient.

Additionally, there were specific vendors or markets where the residents would not shop. One worker of the Centro Memorial de Martin Luther King said, 'I like the meat on 51st Street because it is fresh and they don't use chemicals. I like to be environmentally friendly.' A professor from the Centro said that when she and her mother go grocery shopping they look for the *platanito de verdad*, or a real banana. She explained that some fruits like mango, avocado and papaya were injected with an artificial preservative called *flor de ines*. Many residents of Pogolotti had a social and environmental conscience when buying groceries and would often refuse to consume what they perceived to be non-sustainable produce. These consumers had specific practices to how they approached grocery shopping, whether it was based on price, convenience or quality. By highlighting their conscious consumption practices, we hope to problematize the notion that only the elites 'vote with their fork'. Additionally, it is important to include the residents of Pogolotti as conscious consumers, in order to undermine the notion that Afro-Cubano communities are only producers but never consumers of the product (Pieterse, 1995). While the residents of Pogolotti may not have as much buying power, their shopping habits reflect a larger national discourse about sustainability and sustainable agricultural systems.

During the height of the Special Period, as a way to mitigate the economic crisis, the state opened the country to tourist ventures. Tourism can be interpreted as an act of consumption in the way that it consumes spaces, peoples and culture. In Cuba, this type of consumption has manifested as a tension between what spaces Cubans are allowed to enter, foods they are allowed access to, and in what capacity they are permitted to interact with foreigners. Examining the food consumption practices in tourist enclaves reveals a 'two-tiered food system' (Allen, 2004, p. 109), in which those who are able to afford it most benefit from reforms made in the food system. For example, a traditional Cuban dish *Ropa Vieja* (lit. 'old clothes'), is now only found in *paladares* (private Cuban restaurants) instead of being served in the household. Beef, a chief ingredient, has been rare in Cuba for decades now, and what was once a typical household dish is now a delicacy available only to elites and foreigners. This analysis is not to privilege the act of consumption or the consumer as an 'agent of choice', but rather as a lens to interrogate the racial, gendered and classed politics behind food access and distribution in marginalized communities.

9.5 Conclusion

We would like to conclude by doing a SWOT analysis of urban agriculture in La Habana and Austin. SWOT (Strengths, Weaknesses, Opportunities and Threats) is an exercise common in activist/community organizing circles as well as other organizational settings. It is meant as an assessment and analysis of the various components of a system or organization, thereby identifying strategies for how best to move forward. This is supposed to be done in a collaborative group setting, with the various actors involved contributing. So, this must be read as a partial – meaning both incomplete and biased – attempt at developing a conversation.

The *strengths* of urban agriculture in both La Habana and Austin include relatively secure land tenure and discerning customers. In Austin, most of the urban farmers owned their own land, a rarity in UA. The Cuban government also worked to provide access to all unused land for agricultural purposes. As was highlighted above, Cuba has a rich food tradition and, despite the constraints upon residents of Pogolotti, many are quite discerning and exercise agency in making their food choices. What this points to is perhaps a broad interest in the food system as a whole; caring about where food comes from.

However, this leads to considering the *weaknesses* we have encountered. For one, food prices are too high in relation to income levels. As Torres et al. (2010) reported, Cubans on average spend close to one-third of their income on food. In Pogolotti, it may be even higher (Law, 2015). The second weakness, and perhaps the more pernicious one, is the lack of engagement with the racialized cultural politics of food politics. In Austin, these issues were vocalized by long-time residents and other actors, while in La Habana, they remain muted and hidden under the surface. However, in both cases, if the issues of race, marginalization and segregation continue, without policy changes to improve people's material conditions, or without public and discursive acknowledgement of past and continuing injustices, efforts to reform the food system will fall short.

Moving forward, there are several possible *opportunities*. In Austin, one of the criticisms levelled at the urban farms was their inability to accept food benefits, which, with the partnerships between the city and the Sustainable Food Center, they have begun to do. As such, the farms can potentially better serve the community. The city also has hired a full-time food policy manager, who has been working very hard to link up the different segments of the food system to keep the volunteer-run Sustainable Food Policy Board going. Moreover, a shift in the city's electoral politics from an at-large system, in which nearly all elected officials came from a few powerful neighbourhoods, to one of broad geographic representation, creates more opportunities for a politically engaged city. While the divisions over urban farms are real and indicative of deeper social divisions across the city, it at least points to a politically engaged populous. Since President Barack Obama announced the new chapter of relations in Cuban–US diplomacy, there has been endless speculation on what this could mean for the island. While there are some concerning implications for opening of trade relations, the articles portray a refreshing reality of Cuban–US relations. John Kavulich, president of the US–Cuban Trade and Economic Council, cautions: 'What people tend to forget is it's not what the US wants to do to for Cuba. It's about what Cuba feels is in its interest' (Barclay, 2014). This is a clear divergence of discourse from even 20 years ago, in which Cuba was portrayed as without agency.

Many *threats* and challenges lie ahead though for urban agriculture. The most obvious is perhaps that the industrial food system is still well entrenched across the globe. More specifically to our case studies, the continuing gentrification of Austin's eastside neighbourhoods and the intensifying crisis around housing affordability across the city create an added layer of complexity. Not only are black and brown communities being pushed out of their neighbourhoods, but many farmers are potentially losing their land to out-of-state developers. Recently, Austin's metropolitan area was designated the most economically segregated in the country (Florida and Mellander, 2015). Also, among the ten fastest-growing cities in the US, it is the only city to see a net loss in the African-American population from 2000 to 2010 (Tang and Ren, 2014). In Cuba, it remains an open question what the impact of the 'thawing' of relations will have on the Cuban population. In March 2015, the Associated Press published a hopeful piece on the promise of increasing agricultural trade relations, after a bipartisan group of senators introduced a bill to end the embargo, with clear farm and business backing, and to explore the 'untapped market' (Associated Press, 2015). The potential re-introduction of US large-scale agriculture could be a threat to the current Urban Agricultural system; however, it is still unclear.

Through these two cases, we have demonstrated the problematic dynamics that arise when sustainability seems to frame the beginning and end of conversations. However, we also feel that the concept of sustainability should be extended by thinking through the broad circularity that sustainability discourses supply. It is important not just to think about the ground from which food is produced, but also to trace the often marginalized and segregated spaces through which food travels and the implications it leaves behind.

References

Agyeman, J. and Evans, B. (2004) 'Just sustainability': the emerging discourse of environmental justice in Britain? *Geographical Journal* 170(2), 155–164. DOI: 10.1111/j.0016-7398.2004.00117.

Alkon, A.H. (2012) *Black, White, and Green: Farmers Markets, Race, and the Green Economy*. University of Georgia Press, Athens, Georgia, USA.

Alkon, A.H. and Agyeman, J. (2011) Introduction. In: Alkon, A.H. and Agyeman, J. (eds) *Cultivating Food Justice: Race, Class, and Sustainability*. MIT Press, Cambridge, Massachusetts, USA.

Allen, P. (2004) *Together at the Table: Sustainability and Sustenance in the American Agrifood System*. Published in cooperation with the Rural Sociological Society. Pennsylvania State University Press, University Park, Pennsylvania, USA.

Altieri, M.A., Companioni, N., Cañizares, K., Murphy, C., Rosset, P., Bourque, M. and Nicholls, C.I. (1999) The greening of the 'barrios': urban agriculture for food security in Cuba. *Agriculture and Human Values* 16(2), 131–140. DOI: 10.1023/A:1007545304561.

Associated Press (2015) Cuba Looks North to U.S. Farmers For Help With Food Crisis. *AGWEB*. Available at: https://www.agweb.com/article/cuba-looks-north-to-us-farmers-for-help-with-food-crisis-naa-associated-press/ (accessed 21 April 2017).

Austin American-Statesman Editorial Board (2013) Protect integrity of neighborhoods. *Austin American-Statesman*, 16 October.

Banks, K. (2011) *Central Texas Foodshed Assessment*. Sustainable Food Center, Austin, Texas, USA.

Barclay, E. (2014) What the change in U.S.–Cuba relations might mean for food. Available at: http://www.npr.org/sections/thesalt/2014/12/18/371478629/what-the-change-in-u-s-cuba-relations-might-mean-for-food (accessed 18 November 2016).

Bryld, E. (2003) Potentials, problems, and policy implications for urban agriculture in developing countries. *Agriculture and Human Values* 20(1), 79–86. DOI: 10.1023/A:1022464607153.

Busch, A. (2011) *Entrepreneurial City: Race, the Environment, and Growth in Austin, 1945–2011*. PhD Dissertation, University of Texas at Austin, Austin, Texas, USA.

Campbell, S. (1996) Green cities, growing cities, just cities? *Journal of the American Planning Association* 62(3), 296.

Castro, F. (1992) *Discurso pronunciado en Río de Janeiro por el Comandante en Jefe en la conferencia de Naciones Unidas sobre Medio ambiente y desarrollo* [in Spanish]. United Nations Conference on Environment and Development. Available at: http://www.cuba.cu/gobierno/discursos/1992/esp/f120692e.html (accessed 18 November 2016)

Chan, M. and Freyre, E.F. (2012) *Unfinished Puzzle: Cuban Agriculture: The Challenges, Lessons and Opportunities*. Food First Books, Oakland, California, USA.

City of Austin (2012) Imagine Austin Comprehensive Plan. City of Austin. Available at: http://www.austintexas.gov/imagineaustin (accessed 18 November 2016).

Davidson, M. (2009) Social sustainability: a potential for politics? *Local Environment* 14(7), 607–619. DOI: 10.1080/13549830903089291.

Florida, R. and Mellander, C. (2015) *Segregated City: The Geography of Economic Segregation in America's Metros*. Martin Prosperity Institute, Toronto, Canada.

Fuster Chepe, E. (2006) Diseño de La Agricultura Urbana Cubana [in Spanish]. *Agricultura Orgánica* 12(2), 6.

Galt, R.E., Gray, L.C. and Hurley, P. (2014) Subversive and interstitial food spaces: transforming selves, societies, and society–environment relations through urban agriculture and foraging. *Local Environment* 19(2), 133–146. DOI: 10.1080/13549839.2013.832554.

Gándara, R. (2012) Urban farm, neighbor collide in East Austin. *Austin American-Statesman*, 17 December, p. A01. Austin, Texas, USA.

Gándara, R. (2013) City Council to consider urban farm ordinance heads to City Council. *Austin American-Statesman*, 17 October, p. B2. Austin, Texas, USA.

Gold, M. (2014) Peasant, patriot, environmentalist: sustainable development discourse in Havana. *Bulletin of Latin American Research* 33(4), 405–418. DOI: 10.1111/blar.12175.

Gordon, A. (2008) *Ghostly Matters: Haunting and the Sociological Imagination*, 2nd edn. University of Minnesota Press, Minneapolis, Minnesota, USA.

Guthman, J. (2008) 'If they only knew': color blindness and universalism in California alternative food institutions. *The Professional Geographer* 60(3), 387–397. DOI: 10.1080/00330120802013679.

Harper, A.B. (2011) Vegans of color, racialized embodiment, and problematics of the 'exotic'. In: Alkon, A.H. and Agyeman, J. (eds) *Cultivating Food Justice: Race, Class, and Sustainability*. MIT Press, Cambridge, Massachusetts, USA.

Hovorka, A.J. (2006) The no. 1 ladies' poultry farm: a feminist political ecology of urban agriculture in Botswana. *Gender, Place and Culture* 13(3), 207–225. DOI: 10.1080/09663690600700956.

Koont, S. (2011) *Sustainable Urban Agriculture in Cuba*. University Press of Florida, Gainesville, Florida, USA.

Law, S.E. (2015) *Retroactively Rewriting the Revolution: The Discursive Mobilization of Sustainability in La Habana, Cuba*. University of Texas Press, Austin, Texas, USA.

Lees, L. (2004) Urban geography: discourse analysis and urban research. *Progress in Human Geography* 28(1), 101–107.

Lipsitz, G. (2011) *How Racism Takes Place*. Temple University Press, Philadelphia, Pennsylvania, USA.

Marcuse, P. (1998) Sustainability is not enough. *Environment and Urbanization* 10(2), 103–112. DOI: 10.1177/095624789801000201.

Marty, E. (2015, January) *Sustainable Food System Update*. Presented at the Sustainable Food Policy Board monthly meeting, Austin, Texas, USA.

McClintock, N. (2010) Why farm the city? Theorizing urban agriculture through a lens of metabolic rift. *Cambridge Journal of Regions, Economy and Society*, rsq005. DOI: 10.1093/cjres/rsq005.

McClintock, N. (2014) Radical, reformist, and garden-variety neoliberal: coming to terms with urban agriculture's contradictions. *Local Environment* 19(2), 147–171. DOI: 10.1080/13549839.2012.752797.

Omi, M. and Winant, H. (1994) *Racial Formation in the United States: From the 1960s to the 1990s*, 2nd edn. Routledge, New York, USA.

Pearson, L.J., Pearson, L. and Pearson, C.J. (2010) Sustainable urban agriculture: stocktake and opportunities. *International Journal of Agricultural Sustainability* 8(1–2), 7–19. DOI: 10.3763/ijas.2009.0468.

Pérez, L.A. (2010) *Cuba: Between Reform and Revolution*, 4th edn. Oxford University Press, New York, USA.

Pieterse, J.N. (1995) *White on Black: Images of Africa and Blacks in Western Popular Culture*. Yale University Press, New Haven, Connecticut, USA.

Psarikidou, K. and Szerszynski, B. (2012) Growing the social: alternative agrofood networks and social sustainability in the urban ethical foodscape. *Sustainability: Science, Practice, and Policy* 8(1), 30–39.

Scarpaci, J.L., Segre, R. and Coyula, M. (2002) *Havana: Two Faces of the Antillean Metropolis* (rev. edn). University of North Carolina Press, Chapel Hill, North Carolina, USA.

Slocum, R. (2011) Race in the study of food. *Progress in Human Geography* 35(3), 303–327.

Sustainable Food Center (1995) *Access Denied: An Analysis of Problems Facing East Austin Residents in Their Attempts to Obtain Affordable, Nutritious Food*. Sustainable Food Center, Austin, Texas, USA.

Sustainable Food Policy Board (2013) Backup materials and findings. City of Austin, 24 June. Available at: http://www.austintexas.gov/edims/document.cfm?id=191680 (accessed 18 November 2016).

Swyngedouw, E. (2007) Impossible 'Sustainability' and the postpolitical condition. In: Krueger, R. and Gibbs, D. (eds) *The Sustainable Development Paradox: Urban Political Economy in the United States and Europe*. Guilford Press, New York, USA.

Tang, E. and Ren, C. (2014) *Outlier: The Case of Austin's Declining African-American Population*. Institute for Urban Policy Research & Analysis, Austin, Texas, USA.

Toon, A. (2013a) Urban farm debate intensifies. *Austin Chronicle*, 17 October. Available at: http://www.austinchronicle.com/daily/food/2013-10-17/urban-farm-debate-intensifies (accessed 18 November 2016).

Toon, A. (2013b) The farm report: both sides win, lose. *Austin Chronicle*, 6 December. Available at: http://www.austinchronicle.com/news/2013-12-06/the-farm-report-both-sides-win-lose (accessed 18 November 2016).

Torres, R.M., Nelson, V., Momsen, J.H. and Niemeier, D.A. (2010) Experiment or transition? Revisiting food distribution in Cuban agromercados from the 'Special Period.' *Journal of Latin American Geography* 9(1), 67–87.

Tretter, E. (2013) Contesting sustainability: 'SMART Growth' and the redevelopment of Austin's Eastside. *International Journal of Urban and Regional Research* 37(1), 297–310. DOI: 10.1111/j.1468-2427.2012.01166.x.

TXP, Inc. (2013) *The Economic Impact of Austin's Food Sector*. City of Austin, Austin, Texas, USA.

United Nations (1987) *Our Common Future*. Oxford University Press, New York, USA.

Vickery, K. (2014) Barriers to and opportunities for commercial urban farming: case studies from Austin, Texas and New Orleans, Louisiana. Master's Thesis, University of Texas at Austin, Austin, Texas, USA.

Ward, T. (2009) East Austin Mosaic: The Shape and Color of Gentrification. Unpublished manuscript. Available at: https://utexas-ir.tdl.org/handle/2152/6096 (accessed 18 November 2016).

10 A Political Ecology of Community Gardens in Australia: From Local Issues to Global Lessons

Jason A. Byrne,[1]* Catherine M. Pickering,[1] Daniela A. Guitart[2] and Rebecca Sims-Castley[3]

[1]*Environmental Futures Research Institute, Gold Coast, Queensland, Australia;* [2]*Griffith School of Environment, Gold Coast, Queensland, Australia;* [3]*Independent scholar*

10.1 Introduction

The local impacts of global urbanization (e.g. dwindling green spaces, food insecurity, land shortages, loss of biodiversity) have triggered resurgent interest in various forms of urban agriculture (Godfray *et al.*, 2010; Evers and Hodgson, 2011). In many rapidly growing cities across the Global North (GN) and Global South (GS), residents are clamouring for better access to places to grow safe and healthy food, for spaces that foster social inclusion, and improved environmental quality (Guitart *et al.*, 2015). Urban cultivation initiatives are often framed around the social benefits of local food growing and typically seek to be 'sustainable' (Chapters 8 and 9, this volume). These twin goals have important implications for land-use planning and policy, implications that we address in this chapter.

The scholarly literature suggests that urban agriculture initiatives may be driven by six threats to food security:

1. rapid urban population growth (Barthel *et al.*, 2015);
2. climate change impacts (e.g. floods, drought);
3. energy vulnerability (e.g. peak oil);
4. phosphorous shortages (Godfray *et al.*, 2010);
5. the rising dominance of multi-national food corporations (Shiva, 2008; Cordell *et al.*, 2009; Sage, 2012);
6. increasing use of genetically modified food (Monteiro and Cannon, 2012).

These threats have refocused the attention of scholars and practitioners alike upon the vulnerability of urban food systems (Allen, 2010) and the need for food justice (Agyeman and McEntee, 2014).

Community gardens are a form of urban agriculture, often reported to have food, social and environmental benefits. Such gardens are found in cities in both the GN and GS, and their popularity is burgeoning (Corrigan, 2011; Turner *et al.*, 2011). Community gardens are said to contribute to both social inclusion (Glover *et al.*, 2005a, b; Guitart *et al.*, 2012) and environmental quality (Ferris *et al.*, 2001). Yet many of the ostensible benefits of community gardens remain untested (Guitart *et al.*, 2014, 2015). Moreover, the scholarly literature has reported conflicts between land-use planners, decision makers, community gardeners, and non-profit organizations in the development, management

---

*Corresponding author; e-mail: jason.byrne@griffith.edu.au

and administration of community gardens (Staeheli *et al.*, 2002). Key problems include insecure tenure, public liability, social exclusion and insufficient public participation in decision making (Guitart *et al.*, 2015; also Chapter 5, this volume). Such problems can harm garden viability and entrench environmental injustice (Chapter 8, this volume).

There are many examples (e.g. from New York and Los Angeles) where scholars have reported that community gardens have been shut down, and gardeners forcibly evicted, because municipal governments prioritized land and property development over food production and community well-being (Staeheli *et al.*, 2002; Irazabal and Punja, 2009). There are also examples from cities such as Oakland, California, Berlin, Germany, and Nairobi, Kenya, where concerns have been raised about the suitability of urban land for food cultivation, due to potential soil contamination (McClintock, 2012; Säumel *et al.*, 2012; Gallaher *et al.*, 2013). These are important matters for land-use planners and policy makers. As Allen (2010, p. 297) reminds us: 'rather than assuming that local food systems are necessarily socially just' and sustainable, planners need to better understand the 'economic, political, [environmental] and cultural forces' that shape urban cultivation. In short, there is currently a failure to incorporate the findings from much community garden research into land-use planning policies, processes, and decision making. Conversely, many scholars have also failed to consider the importance of governance and institutional processes (such as planning) for community garden research.

An urban political ecology approach has the potential to redress some of these problems, by illuminating when, 'where and what kinds of gardens have taken root and for whom', and what benefits they provide (McClintock, 2014, p. 163). Better understanding of the interactions between historical, social, economic, political and environmental factors that potentially drive community garden development is the first step towards improving land-use policy and decision making (L'Annunziata, 2010; Galt, 2013). In this chapter, we report the findings of a study of community gardens in Brisbane and Gold Coast cities, Australia, using an urban political ecology approach. In the research we discuss here, we have sought to answer five questions:

1. are there different types of community gardens in the study area;
2. what motivations have underpinned garden development;
3. which communities do the gardens serve;
4. do the gardens provide social and/or ecological benefits; and
5. if so, for whom?

The chapter begins by concisely assessing the utility of political ecology for examining community garden development specifically (and urban cultivation generally), before presenting findings from our empirical research, and discussing their implications for cities in the GN and GS.

10.2 The (Urban) Political Ecology of Urban Community Gardens

The term political ecology refers to a theoretical frame that has been used by geographers, anthropologists and sociologists to investigate nature–society interrelationships. Political ecology originated in the work of Blaikie and Brookfield (1987), whose pioneering research sought to better understand the interconnected causes of poverty and land degradation (Chapter 1, this volume). Looking beyond biophysical explanations, they recognized the importance of the interplay of historical, socio-economic, political and ecological factors in configuring environmental outcomes (Agyeman and McEntee, 2014). Specifically, Blaikie and Brookfield showed how agricultural practices such as the management of soil, water and livestock, together with labour practices, are nested within a broader political-economy of power relations, including institutions such as land-use planning. This perspective has informed subsequent political ecology research.

There is a substantial body of research about the political ecology of various types of rural agriculture, but comparatively less has been written about the political ecology of urban agriculture (hereafter urban cultivation), and urban community gardens specifically (L'Annunziata, 2010 is a notable exception). The focus of much of the urban political ecology research has been on networks of knowledge generation, technologies of food production and distribution, food consumption practices, and the regulation of 'agricultural activities' (Galt, 2013). Although particular attention has been directed at the role of structures, actors, social networks and institutions in determining social and environmental outcomes in food production, urban political ecology research on agricultural systems in the so-called first world, especially 'alternative food networks' such as community gardens, is still inchoate (Galt, 2013). However, this situation is rapidly changing, as urban political ecologists take up the challenge of researching and theorizing urban cultivation, including urban food systems, community supported agriculture, alternative food networks, and food security (Agyeman and McEntee, 2014; McClintock *et al.*, 2016).

Capitalist imperatives for profit maximization and/or land enclosure can result in forms of land tenure and management, and social relations that exploit both people and ecosystems (Galt, 2013). The extent to which the state intervenes in local labour and property markets (e.g. via land-use policy and planning) can produce divergent environmental outcomes (Galt, 2013; McClintock, 2014). Shifts in local economic conditions due to rapid urbanization, market deregulation, and/or changes to land-use policy can either foster or undermine local food production (McClintock, 2014). Domene and Saurí (2007) for example, have observed that land-use planning policies in the densely populated city of Barcelona, Spain, intended to address declining environmental aesthetics accompanying rapid urbanization, led to the destruction of informal communal gardens developed by marginalized and impoverished residents. Recent Australian studies suggest that compact and denser forms of urban development, intended to accommodate rapid population growth, can hinder urban cultivation (Evers and Hodgson, 2011). Importantly, Larder *et al.* (2014) have shown that the viability of urban cultivation hinges upon gardeners' value and belief systems; supportive state and local government policies; and sympathetic planning regulations and local government strategies.

Urban political ecology analyses are often attentive to multiple dimensions of environmental change, including:

1. metabolic shifts accompanying urbanization;
2. the uneven distribution of environmental benefits and harms;
3. how social and natural capital intersect to (re)configure environmental outcomes (Chapter 14, this volume).

Countervailing factors can improve environmental outcomes arising from different types of urban agricultural practices. On the one hand, urban political ecologists have found that strong social movements, alternative economies, grassroots resistance, participatory governance, and the valorization of local environmental knowledge and practices can foster food cultivation (Agyeman and McEntee, 2014; Heynen, 2014), with implications for food security and social equity. On the other hand, market-driven planning policies, government subsidies for agri-business, economic restructuring driven by neoliberal ideology, the corporatization of local government, rapid population growth, and the deregulation of local property markets can hinder local food production (McClintock, 2014). Many of these socio-ecological dynamics are observable in the development and maintenance of community gardens.

Three broad themes can be discerned from the community garden literature, which are relevant to a political ecology of urban cultivation:

1. community gardens may contribute to a broadly defined 'sustainability' agenda, for example, by reducing 'food miles' or composting putrescible waste (Galt, 2013; Heynen, 2014; also Chapter 9, this volume);
2. the 'environmental justice' functions of community gardens are important for the well-being of marginalized and vulnerable residents, for instance, by building self-reliance and strengthening social networks (Agyeman and McEntee, 2014);
3. community gardens reflect institutional dynamics, because they are highly responsive to different types of governance (e.g. neoliberalism) (Moore, 2006; McClintock, 2014).

Local policies and practices thus have the capacity to either invigorate or extirpate community gardening. In reality, these three themes are interrelated; community gardens are complex socio-ecological entities (Shillington, 2013; Chapter 19, this volume).

Despite the broad consensus that community gardens are beneficial urban spaces, planning for community gardens in countries in the GN such as the UK, US and Australia has been haphazard, especially in areas undergoing rapid urbanization. While local authorities may be quick to promote these gardens, seldom do land-use policy and planning instruments provide specific guidance about land area requirements, site coverage, optimal plot numbers, key facilities, or requirements for solar access (DeKay, 1997). Indeed, Lawson (2004) has argued that the relationship between land-use planning and community gardens has historically been deeply ambivalent – ranging from neglect, to encouragement, to tight regulation, to hostility. For community gardens to succeed in the long term, we need a better understanding of the role of social, ecological, political and institutional factors shaping their development and management (McClintock et al., 2016; Passidomo, 2016; Spilková and Vágner, 2016). This chapter takes up that task.

10.3 Methods

Our research has sought to understand whether community gardens in Brisbane City and Gold Coast City, Australia, provide social and/or sustainability benefits, and if so, for whom (as discussed in Section 10.2). Specifically, we sought to systematically assess:

1. the factors driving the development of community gardens in the study area (e.g. motivations for garden development);
2. the characteristics of communities served by the gardens;
3. the social and environmental functions provided by the gardens;
4. the role of land-use policy in fostering or curtailing garden development.

Data were collected about:

1. socio-demographic and psychometric characteristics (e.g. age, sex, organization type, gardeners' motivations, etc.);
2. gardening practices used and the philosophies informing those practices;
3. the plant types grown in the gardens;
4. surrounding property values.

Two methods were used to collect the data: a desk-top survey of community gardens followed by standardized-questions, in-person interviews with garden managers; field assessments of garden plots and plants. Statistical tests were performed to test for relationships between the variables and spatial analyses were undertaken to assess socio-demographics, property values and population densities surrounding the community gardens.[1] The research was limited to food-producing community gardens rather than ornamental or native-plant gardens. Data collection was also limited to gardens located within the Brisbane City and Gold Coast City municipal boundaries.

10.3.1 Study area

South East Queensland (SEQ) is one of Australia's fastest-growing metropolitan regions (Fig. 10.1). Rapid growth in the region has reduced natural areas and fostered unhealthier sedentary lifestyles, imposing significant social, economic and environmental pressures on the populace (Guitart et al., 2015). Brisbane (2011 pop. 1.04 million) and Gold Coast (2011 pop. 494,501) cities are respectively the first- and second-largest municipalities in Australia (by population) and contain most of the SEQ region's population. The two cities have responded to growth pressures through land-use policies that promote urban consolidation, i.e. encouraging smaller lot sizes, increasing residential density, and promoting urban infill (Guitart et al., 2014). The result has been reduced backyard garden size and an increase in apartment living. The ability of many residents to experience nature and grow their own food has declined.

Partly in response to these growth pressures, Brisbane and Gold Coast cities (like others in Australia) have fostered community garden development through supportive local policies. Gold Coast City Council has a guideline called a 'start-up kit' that is linked to Council's Corporate Plan (2009–2014), and which articulates objectives for sustainability, social inclusion, and physical and mental health. The document, which is administered by the city parks department, sets out a process

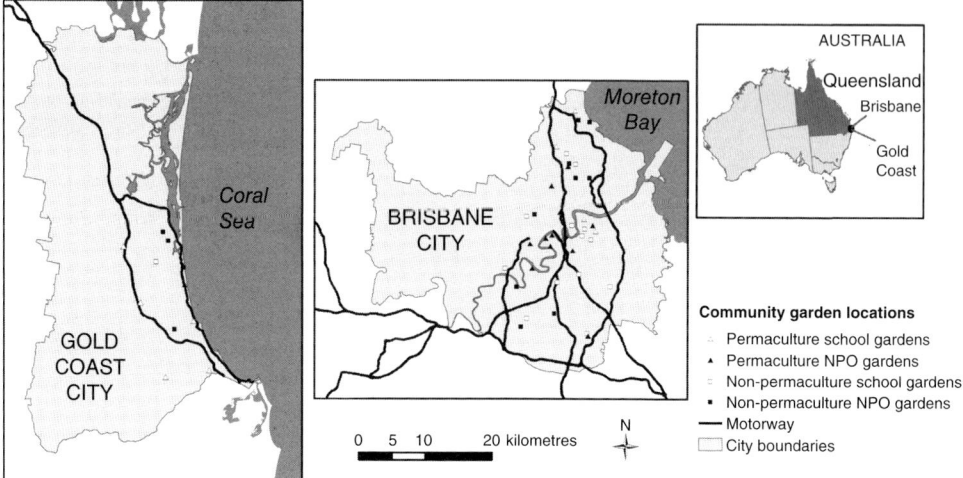

Fig. 10.1. Community Gardens in South East Queensland, Australia.

of applying for approval, garden design, meeting procedures, performance benchmarking, compliance procedures and reporting (Gold Coast City Council, n.d.). Brisbane City Council has similar but more extensive guidelines, administered by the community services department, which also include guidance for site assessment, occupational health and safety, and financial management (Brisbane City Council, 2013). Brisbane City Council also provides up to AU$2500 grants for garden start-up or for upgrading facilities.

But little is known about the characteristics of community gardens in these municipalities, including the type of organizations managing the gardens, garden ages and sizes, motivations underpinning garden development, security of land tenure, existence of membership programmes and associated costs, and the factors that configure community garden location and longevity. In other words, local governments have been encouraging community gardens, at least in principle, but we do not know if their policies are working as intended. Even less is known about gardening practices and agro-biodiversity within and between gardens, whether gardens foster social inclusion and address environmental justice issues, and whether gardening practices impact local ecosystems and agro-biodiversity. It is difficult to sustain land-use policies promoting community gardens in the absence of such information.

10.3.2 Desk-top study

To identify food-producing community gardens in the study area, Brisbane and Gold Coast City Council websites were systematically searched. A list of gardens was compiled, and a web search was undertaken to determine whether identified gardens had their own websites, and the type of information available on those websites. The search terms were 'community', 'garden', 'Brisbane', and 'Gold Coast'. The first 3–5 Google result pages were examined, and all relevant websites were visited. An initial list of 40 community gardens was generated. A 'snowball' sampling technique was then used to expand the list, as the web searches could underestimate the existence of gardens where organizations did not have web presence. Key informants including the community garden officers from Brisbane and Gold Coast City Councils were contacted, and community garden general meetings were attended to identify additional potential informants. A map was generated using Google Earth to show informants the location of all known community gardens and to cross-validate existing gardens. Informants could see known gardens and identify missing ones. This map was later shown to garden managers for the same purpose, and was later formalized via a Geographic Information System (GIS) for spatial analysis. We recognize that gardens without a web presence or that were unknown to informants would not have been picked up by this approach.

A database of all known community gardens within the study area was created; it included the garden name, address (physical and postal), and garden managers' names and contact details (phone number, email). A total of 56 community gardens were located across Gold Coast and Brisbane, although three gardens were no longer operating (having closed after the catastrophic Brisbane floods in early 2010).

10.3.3 Garden manager interviews

Several studies have used surveys to obtain sociodemographic information about community gardens (Guitart et al., 2012), but comparatively fewer have addressed the interconnections between social/environmental justice and ecological sustainability. We developed a new survey instrument for this purpose, which was approved by the home institution's human ethics committee (ENV/12/11/HREC), was pilot-tested, and was then modified based on responses. The instrument was administered via semi-structured, in-person interviews.

The survey instrument consisted of three sections. General data about the community gardens were obtained through questions in part one, assessing garden age, management (organization type, individual or communal system, land ownership and motivation), membership system (number of members, cost of membership, waiting lists, cultural backgrounds) and facilities present in the gardens (e.g. water and electricity sourcing, worm farm, compost, solar panels, nursery, toilets). Part two examined gardening practices using a composite scale derived from a review of the literature (e.g. chemical use in gardening philosophy, soil nutrient improvement and pest management) (Guitart et al., 2015). Gardening philosophy (e.g. organic vs conventional) was cross-validated with actual practices using a 'permacultureness index', assessed via site visits. The final section of the instrument collected quantitative data about the diversity and composition of food plants grown in the gardens.

Data was collected about types of plants grown rather than the species grown, because one species can produce a range of different types of foods and gardeners are typically more familiar with common names than their scientific counterparts. Similarly, data was not collected at the level of plant varieties because gardeners would likely be unfamiliar with the correct variety name for the plants they grew, and might use a single name for several different varieties. We devised a list of 347 possible plant types that could be grown in community gardens in the study area, categorized by their most common use (e.g. fruit, vegetable, herbs, spices, nuts, grains, legumes, natives and flowers). Garden managers were asked to mark each type of plant currently in the garden and/or grown in the last 12 months on the list provided. Plants missing from the list but found to be growing in the gardens were added when observed.

Garden managers were initially contacted via telephone or email to verify garden status (i.e. presence of plants and ongoing activity) and confirm participation. Potential participants were given an information sheet and asked to sign an informed consent. Garden managers could answer the survey with another garden member, to bolster information accuracy. The questions were administered in person. Interviews lasted approximately 30–45 minutes. In some schools, teachers or principals lacked sufficient time to complete all the questions during the site visit, so a pre-paid envelope was provided to enable them to return their responses by mail. Three school community gardens did not participate. A total of 50 community gardens were surveyed; each garden was visited to record its exact location and area and to identify plants grown in the garden. The perimeter of each garden was walked, and a Global Positioning System (GPS) was used to calculate garden area. Photographs were taken and the presence of facilities and gardening practices were cross-validated.

10.4 Data Analysis

All information gathered from the interviews was entered into two Microsoft Excel datasets: one about gardening characteristics and practices, the other about plant types grown. The plant database was expanded to include additional information about each type of plant grown, including the species, family, primary use, life form and its growth form. This information was sourced from books used to compile plant lists and from two taxonomic websites (Guitart *et al.*, 2015). The total number of plant types (plant richness) was calculated for each garden. All data was transferred to SPSS (version 22) statistical package for further analysis.

To determine whether the size of gardens varied with when they were established, linear regressions were performed in SPSS. Linear regressions were also used to test the effect of garden area and age on plant-type richness. The first garden established in 1974 in Brisbane, by far the largest at 19,200 m^2, was excluded from regression analyses because it was an outlier. A series of One-way Analysis of Variance (ANOVA) were performed to compare school and non-profit (NPO) gardens, and separately, between permaculture and non-permaculture gardens. Plant-type richness, life form and growth form were assessed. Plant densities were calculated (number of plant types divided by garden area), and were compared between school and NPO gardens, and permaculture and non-permaculture gardens using ANOVA.

Garden addresses were geocoded and a map was generated by creating a point for each garden polygon (derived from the GPS) and their centroids. In cases where a garden had more than one polygon allocated to it (e.g. the gardens were distributed throughout different spaces in some schools), a single point was allocated between polygons. Spatial analysis was undertaken using ArcGIS 10.1 to ascertain relationships between socio-demographic characteristics of garden catchments and garden characteristics. A garden catchment was defined by maximum walking distance to the garden (Gorham *et al.*, 2009) representing an 800-metre (½-mile) buffer. All remaining residential areas falling outside the garden neighbourhoods are hereafter referred to as 'non-garden' areas.

We then assessed property values, population density and relative disadvantage (measured using the Australian Bureau of Statistics Socio-Economic Index for Areas or SEIFA index) for areas within and outside garden catchments. Property value analysis employed the latest Queensland Valuation and Sales data (Queensland Government, 2014b) and property parcel layer from the Queensland Digital Cadastral Database (Queensland Government, 2014a). Unimproved land valuation figures for freehold properties were used. Large variation in property sizes precluded normalization of values. Mean property value per square metre (AU\$/m^2) for garden vs non-garden areas within Brisbane and the Gold Coast precincts were then compared.

To assess comparative disadvantage, the most recent population figures and SEIFA index were obtained at the Statistical Area level 1 (SA1) scale from the 2011 Census database (Commonwealth of Australia, 2013b). Population density was calculated by dividing the total population by the area of the SA1s inside and outside the garden neighbourhoods. Socio-economic disadvantage was represented by scores from the Index of Relative Socio-economic Disadvantage (IRSD), one of the indexes from SEIFA (Commonwealth of Australia, 2013a). The score of the index is derived for every SA1, using variables such as low income, low educational attainment, unemployment, and car ownership. In this case, the decile score was used, which divides the distribution of SEIFA scores into ten equal groups. Decile 1 contains the bottom 10% of SA1s, Decile 2 contains the next 10% of SA1s, and so on.

Exploratory data analysis was completed for all described parameters using descriptive statistics (measures of kurtosis and skewness) and Q–Q plots in SPSS. Where assumptions of normality were not met, a natural log transformation (ln[x+1]) was applied to the data and a standard Student's *t*-test carried out to compare the means between the two independent samples. In cases where transformation of the data did not achieve a normal distribution (i.e. the property value dataset), a non-parametric Mann–Whitney test was undertaken. Analyses were completed in SPSS and significance was set at the 0.05 level.

10.5 Results

Results are structured according to garden characteristics (types), motivations underpinning garden development, communities served by the gardens, and environmental functions and benefits of the gardens (including food provision).

10.5.1 Garden characteristics

Most community gardens in the Brisbane and Gold Coast cities are relatively new; 30 of the 50 gardens were established post-2007 (Fig. 10.2). The oldest and largest Brisbane garden was established in 1974 (19,200 m²) while the most recent was established in 2011 (340 m²). The oldest Gold Coast garden (1225 m²) was established in 1998, while the most recent was established in 2011 (250 m²). Collectively the 50 gardens occupied 57,000 m² of land, ranging in size from 6 m² to 3080 m²; the average size was 1139 m² ± 384 m². There were no significant differences in the size of the gardens depending on age, even after excluding the very large Northey Street City Farm from analysis.

Two types of organizations were found to operate the gardens: non-profit organizations (NPO) and schools (46%) (Table 10.1). The NPOs (similar to non-government organizations in North America) included 14 volunteer groups, six government-funded community support groups, two churches, two learning disability centres, one early parenting centre, one senior centre and one police-citizens youth club. Of these gardens, 20 were located on public land, whereas seven were located on private land. For the school gardens, 22 were in government-funded public schools, while only one was in a private school. Most of the land on which gardens were located (84%) was publicly owned by state (e.g. Department of Education) or local government, suggesting a high level of government support for community gardens.

Membership systems differed between the gardens. Non-profit gardens (NPO) were based on voluntary (casual and organized) and paid memberships. Teachers and students ran school gardens, however, some also involved parent committees. All of the school gardens, and most of the NPO gardens (60% of non-profits) did not have private plots allocated to individuals; instead they shared the work, and allocated produce from the whole garden among members. For the school gardens, produce was used in teaching (82%), sold in tuck shops (school canteens) (48%), sold to staff and students (22%), or was taken home by staff and/or students (56%) (produce was used in multiple ways so percentages do not add up to 100%).

10.5.2 Motivations underpinning garden development

The most common motivations for developing the community gardens were education (50%), community building (50%), and environmental sustainability (34%). Garden managers also mentioned increased health (18%), living skills (10%) and contact with nature (6%). Other motivations included local food production, urban revitalization, subsistence, leisure, exercise and

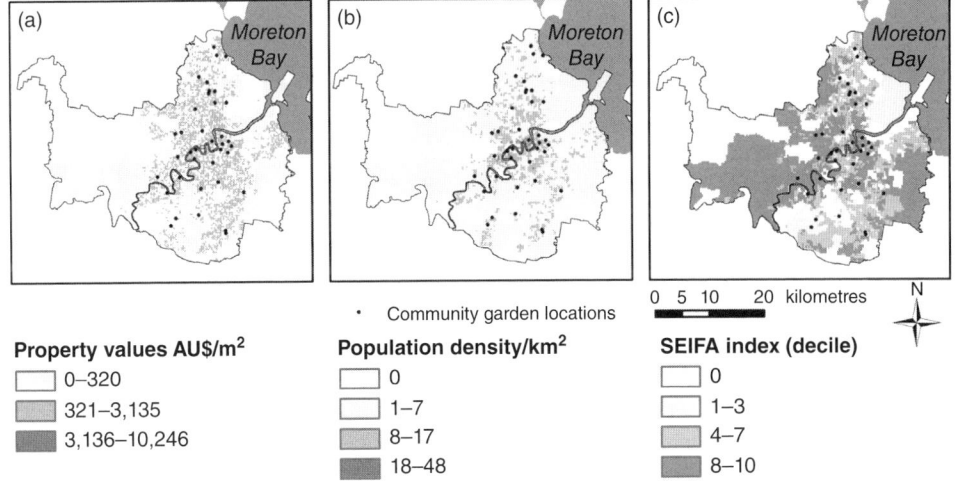

Fig. 10.2. (a) Property values, (b) population density and (c) SEIFA index for neighbourhoods surrounding community gardens in Brisbane City.

Table 10.1. Characteristics of 50 community gardens surveyed in Brisbane and Gold Coast cities.

Variables	Total		School		NPO		Permaculture		Non-permaculture	
Garden location	n	%	n	%	n	%	n	%	n	%
Brisbane City	38	76	17	74	21	77	15	72	23	79
Gold Coast City	12	24	6	26	6	33	6	28	6	21
Total	50	100	23	100	27	100	21	100	29	100
Land ownership										
Public	42	84	22	95	20	74	17	80	25	86
Private	8	12	1	5	7	26	4	20	4	14
Individual plots										
No	39	78	23	100	16	60	18	86	21	73
Yes	11	22			11	40	3	14	8	27
Motivation(s)[1]										
Education/research	25	50	20	40	5	10	9	18	6	12
Community building	25	50	5	10	20	40	13	26	12	24
Environmental sustainability	17	34	8	16	9	18	8	16	9	18
Health	9	18	7	14	2	4	4	8	5	10
Living skills	5	10	2	4	3	6	0	0	5	10
Contact with nature	3	6	2	4	1	2	3	6	0	0
Other	38	76	18	36	20	40	20	40	18	36
Cultural backgrounds										
Australian	76	29	44	30	32	28	32	34	44	26
Aboriginal	17	7	14	10	3	3	7	8	10	6
Torres Strait Islander	9	3	7	5	2	2	4	4	5	3
Other Australians	50	19	23	16	27	23	21	23	29	17
Asian	71	27	36	25	35	30	22	24	49	29
Maori	20	8	14	10	6	5	7	8	13	8
Pacific Islander	18	7	11	8	7	6	5	5	13	8
African	17	7	10	7	7	6	5	5	12	7
European	31	12	15	10	16	14	13	14	18	11
Middle Eastern	15	6	9	6	6	5	3	3	12	7
Latin American	9	3	5	3	4	3	5	5	4	2
North American	4	2	2	1	2	2	1	1	3	2

[1]Note: Percentages add up to more than 100 because multiple selections were possible.

spirituality. Motivations differed between school and non-profit gardens; schools emphasized education (40%), whereas non-profit gardens emphasized community building (40%). Permaculture gardens were primarily motivated by community building (26%), followed by education (18%) and environmental sustainability 16%), whereas non-permaculture gardens were primarily motivated by community building (24%) and sustainability (18%) (Table 10.1).

10.5.3 Communities served by the gardens

Garden membership was culturally diverse. The highest proportions of gardeners were Asians (27%), followed by non-indigenous Australians (19%), Europeans (12%) and Aboriginal Australians (7%). School gardens were significantly more culturally diverse than non-profit gardens (One-Way ANOVA, $F=5.369$ $P=0.025$). School gardens reported significantly higher percentages of Australian Aboriginals (10%), and Pacific Islanders (8%) (chi-square test, $P=0.007$). Non-profit gardens reported higher percentages of Asians (30%), Europeans (14%) and non-indigenous Australians (23%). Some differences are also observable between permaculture and non-permaculture gardens, with higher percentages of Asians (29%) in non-permaculture gardens and higher percentages of non-indigenous Australians in permaculture gardens.

Spatial analysis revealed uneven distribution of the gardens and hence garden benefits.

Gardens tend to be concentrated in inner Brisbane close to the Brisbane River and nearer the coastline on the Gold Coast (Fig. 1). Gardens were also located in areas with higher land values (Figs 10.2a and 10.3a; Table 10.2). The mean land value for properties proximate gardens in Brisbane was AU\$725 per m^2, which was on average AU\$153 more than the land value of properties in the rest of the city ($U=-96.012$, $P < 0.001$). For the Gold Coast there was an even greater difference, with the value of properties proximate gardens averaging AU\$860 per m^2, AU\$300 higher than for the rest of the city ($U=-70.223$, $P < 0.001$) (Table 10.2).

The gardens also tended to be located in higher-density areas. The Brisbane average was 8.3 people/km^2 close to the gardens, vs 7.4 people/km^2 for the rest of the city ($t=-4.454$, $P < 0.001$) (Figs 10.2b, 10.3b, Table 10.2). Similarly for the Gold Coast, the average was 2.7 people/km^2 close to the gardens, vs 2.4 people/km^2 for the rest of the city (but this was not significant, $t=-1.933$, $P=0.053$). However, local communities within garden catchments were comparatively more disadvantaged in both Brisbane and Gold Coast closer to the gardens (Figs 10.3c and 10.4c; Table 10.2). Interaction effects may explain this finding.

10.5.4 Environmental functions

The gardening philosophy, facilities and practices of the community gardens appear to be predominantly sustainable (i.e. ecologically viable over the long term). For example, almost half of the gardens (42%) reported that permaculture was their principal gardening philosophy, with 12 (57%) non-profit gardens and nine (43%) school gardens practising permaculture. Permaculture gardens used environmentally beneficial practices, such as homemade fertilizers, companion planting, crop rotation and planting to attract beneficial insects (Guitart et al., 2015). Many garden facilities supported a sustainable approach to gardening, with 92% of gardens possessing compost bin(s), 74% with worm farm(s) and 52% having a nursery or propagation area on site. Nearly all of the gardens had facilities for building social capital, including shared tool sheds (90%), kitchens (58%), and meeting areas (56%) (Table 10.3). Such philosophies and facilities are similar in the Global South. Very few gardens used commercial synthetic pesticides, fungicides and herbicides (10%).

The gardens were highly botanically diverse, similar to findings from research on home gardens (Galluzzi et al., 2010; WinklerPrins and de Souza, 2010). A total of 317 different types

Table 10.2. Facilities present in the 50 community gardens surveyed in Brisbane and Gold Coast.

Facilities	Total n	Total %	School n	School %	NPO n	NPO %	Permaculture n	Permaculture %	Non-permaculture n	Non-permaculture %
Compost	46	92	21	42	25	50	19	38	27	54
Communal tool shed	45	90	20	40	25	50	20	40	25	50
Rainwater tanks	44	88	20	40	24	48	18	36	26	52
Toilets	40	80	19	38	21	42	16	32	24	48
Worm farm	37	74	20	40	17	34	17	34	20	40
Wheelchair access	33	66	12	24	21	42	14	28	19	38
Kitchen	29	58	15	30	14	28	14	28	15	30
Meeting room/shed	28	56	16	32	12	24	14	28	14	28
Nursery/propagation area	26	52	13	26	13	26	18	36	8	16
Library with gardening books	25	50	16	32	9	18	9	18	16	32
Barbeque	25	50	10	20	15	30	8	16	17	34
Play area for children	25	50	17	34	8	16	9	18	16	32
Fence	20	40	11	22	9	18	7	14	13	26
Public art	20	40	8	16	12	24	8	16	12	24
Educational signs	18	36	7	14	11	22	10	20	8	16
Solar panels	13	26	9	18	4	8	5	10	8	16
Garden materials storage shed	9	18	2	4	7	14	4	8	5	10
Cob oven	6	12	2	4	4	8	5	10	1	2

128 J.A. Byrne et al.

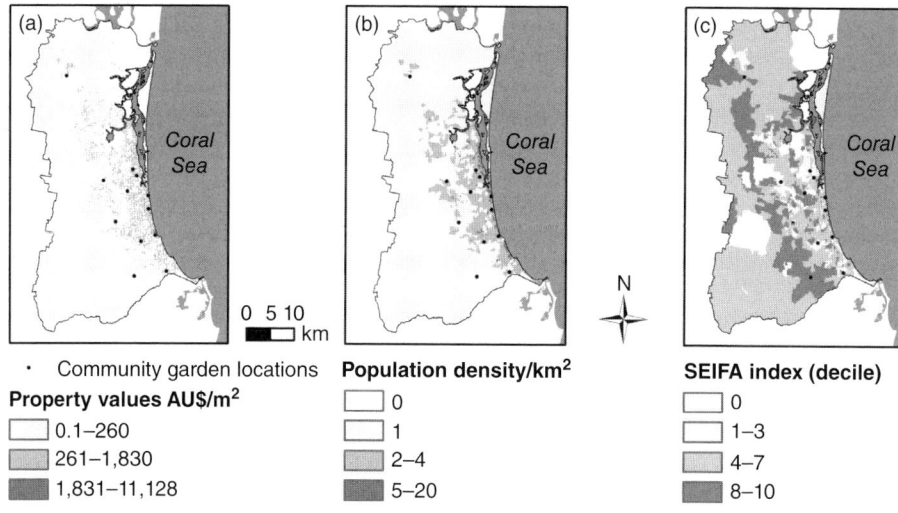

Fig. 10.3. (a) Property values, (b) population density and (c) SEIFA index for neighbourhoods surrounding community gardens in Gold Coast City.

Table 10.3. Land values, population density and disadvantage for Brisbane and Gold Coast cities.

Location	Parameter	Statistic	Garden	Non-garden
Brisbane	Property value (AU$/m^2)	N	63,297	360,922
		Mean	724.806	573.174
		SD	582.001	533.092
		Range (min.)	0.221	0.002
		Range (max.)	62,589.7	27,190
Brisbane	Population density (No./km^2)	N	568	2290
		Mean	8.312 (2.085)	7.400 (1.971)
		SD	5.670 (0.539)	6.014 (0.572)
		Range (min.)	0.019	0.005
		Range (max.)	48.306	154.44
Brisbane	SEIFA index (IRSD decile)	N	568	2290
		Mean	6.77 (1.958)	7.09 (2.025)
		SD	2.741 (0.479)	2.481 (0.400)
		Range (min.)	1 (0.693)	1 (0.693)
		Range (max.)	10 (2.397)	10 (2.397)
Gold Coast	Property value (AU$/m^2)	N	16,528	103,757
		Mean	859.816	559.541
		SD	1079.091	1392.275
		Range (min.)	0.53	0.099
		Range (max.)	27,484.373	221,352.553
Gold Coast	Population density (No./km^2)	N	165	1 106
		Mean	2.569 (1.135)	2.288 (1.050)
		SD	1.915 (0.535)	1.919 (0.526)
		Range (min.)	0.014 (0.014)	0.003 (0.003)
		Range (max.)	10.799 (2.468)	19.682 (3.029)
Gold Coast	SEIFA index (IRSD decile)	N	165	1106
		Mean	4.78 (1.666)	5.67 (1.825)
		SD	2.291 (0.439)	2.298 (0.408)
		Range (min.)	1 (0.693)	1 (0.693)
		Range (max.)	10 (2.397)	10 (2.397)

of plants were grown across the 50 different gardens, including 294 different types of food plants, representing at least 230 species. On average, there were 72 ± 6 different types of plants grown per garden; the least diverse had seven plant types, whereas the most diverse had 224 different plant types. The most common plants were a mixture of food, herb and companion plants; tomato, lettuce, parsley, basil, pumpkin, chives, marigolds, lemon grass, cucumber, passion fruit, rocket, rosemary, sweet potato, broccoli, mint, papaya, sunflower, capsicum (bell pepper) and nasturtium were reported in more than 70% of the gardens. There were differences among gardens, with 137 types of plants occurring in five or fewer gardens, and 52 plants unique to a single garden.

10.6 Discussion

Comparatively less attention has been given to urban political ecologies of cultivation in the Global North. We still lack the kind of robust empirical data that can inform land-use policy to guide various forms of urban cultivation, including community gardens. Our study of community gardens in Brisbane and Gold Coast cities, Australia, begins to redress this lacuna. We have found that community gardening is growing in these cities; 30 gardens (60%) have been established since 2007, reflecting national and international trends (Commonwealth of Australia, 2013a). Community gardens in Brisbane and Gold Coast cities appear to be well supported by local government via land provision, in contrast to many cities in the Global North (Chapters 3 and 4, this volume). It is common for Australian local governments to allow community gardens within parkland, although it is contested, as some argue that this practice represents de facto enclosure of public land (Guitart et al., 2015).

The two local authorities we examined (Brisbane and Gold Coast) provide direct support in the form of training and assistance, such as start-up grants, support for grant-writing, and expert input into garden design and maintenance. The Queensland State Government also supports community garden initiatives via the Department of Education, which encourages the development of school-based community gardens. And the state government fosters community garden development more broadly, through a regional statutory land-use planning instrument, which promotes garden development as a component of community resilience (Guitart et al., 2015). Recently though, the two local governments have stepped away from targets for new garden development (which were once as ambitious as 1000 new gardens) and have curtailed funding. Initially in response first to fiscal austerity that emerged following the Global Financial Crisis (e.g. budget deficits and rising infrastructure costs); funding cuts also reflect a growing perception that community gardens may burden municipal coffers.

Many community gardens elsewhere in the world have attracted some government support (Voicu and Been, 2008) including cities in the USA, such as New York (Staeheli et al., 2002), where 96% of community gardens are located on publicly owned land. Community gardens are also located on publicly owned land in the United Kingdom, Sweden, South Africa and Canada, and similar support exists within Australia. Government support is thus an important factor underpinning the long-term viability of community garden development, and has become an important consideration in urban political ecology analyses (Chapter 19, this volume).

Organizations managing gardens also appear to play a strong role in both the long-term success of community gardens and the benefits they provide. Education, community cohesion and environmental sustainability, for example, appear to play a strong role in motivating garden development. These were the primary motivations driving garden development in our study. Community gardens in schools in Brisbane and the Gold Coast were used for teaching science, environmental studies and nutrition and, in many instances, were part of the school curriculum, corroborating previous research (Somerset et al., 2005). Community building, on the other hand, was the primary motivation for non-profit organizations in our study area, which is also the most commonly cited motivation in the literature (besides fresh food consumption) (Glover et al., 2005a, b). Corroborating the literature, other common motivations we found included environmental sustainability, improving health, and enhancing social inclusion and preserving cultural practices (Chapter 5, this volume).

Unlike our study, global research has found that poverty alleviation and reducing crime are also strong motivations for garden development (Guitart et al., 2012). International studies, particularly those in the USA, have found strong associations between the location of community gardens and spatial and social disadvantage (Chapters 5, 6 and 8, this volume). Unlike those studies, we found no direct connection between garden location and concentrated poverty. Most community gardens in Brisbane and Gold Coast cities were not in suburbs with higher levels of social disadvantage, suggesting that in South East Queensland, community gardens do not appear to provide environmental equity benefits. Just two garden managers listed 'supplementing incomes' as a motivation underpinning garden development. Nonetheless, we found that the areas within garden catchments appeared to have higher levels of relative disadvantage than outside the catchments, an apparent paradox that requires further investigation.

International research has found that some community gardens in the Global South also fulfil both subsistence and livelihood functions, enabling gardeners to sell their produce for profit (Wade, 1987). This was not a common practice for Brisbane or Gold Coast community gardens. One explanation for this difference is that socio-economic disadvantage in Australia tends to be concentrated in the outer-ring suburbs. Community gardens in Brisbane and Gold Coast cities were disproportionately located in inner-ring suburbs, areas dominated by comparatively wealthier people. Gardens thus appear to serve social, recreational and educational functions, as well as providing opportunities to encounter nature, rather than subsistence functions, which are commonly provided by gardens located in areas of concentrated poverty in both the Global North and the Global South (Chapters 8 and 14, this volume). In other words, these gardens may act as a form of middle-class welfare (not unlike the findings in Chapter 4, this volume).

Our spatial analysis has also revealed that community gardens in the two cities tended to be located in areas with higher land values. This finding corroborates research by Voicu and Been (2008), who found that community gardens can increase surrounding property values. But the spatial effect between property values and community gardens in our study area is ambiguous. The higher property values around the gardens in our study could be partially attributable to the fact that the gardens are located on school and parklands. Some research in Australia and internationally has shown that property values tend to be higher around schools and parks (Crompton, 2005), but park quality must also be considered.

Cultural diversity appears to be another important land-use planning consideration, when assessing community garden benefits. In contrast to the literature, we found higher cultural diversity in school gardens compared to non-profit gardens. This may reflect the nature of government-funded, but free to the public, schools, which are ethno-racially diverse (Mansouri and Jenkins, 2010), or it could be an artefact of the greater knowledge of school garden managers about the ethno-racial composition of child gardeners. We could not assess the socio-demographic characteristics of individual gardeners to cross-check the managers' data, due to limited time and funding.

Our findings about cultural diversity contrast with global studies, which have found peoples' cultural values can explain some of the differences in agricultural biodiversity among gardens (Galluzzi et al., 2010). For example, when differences were examined in what was grown among three gardens in Toronto, Canada, with culturally homogeneous members (e.g. Sri Lankan, African and Chinese), plants grown were related to gardeners' country of origin (Baker, 2004; Chapter 19, this volume). We found that Brisbane and the Gold Coast garden memberships tended to be multicultural, with lower levels of segregation – and hence no effect between garden socio-demographic characteristics and plants grown. This finding suggests that the gardens we studied may be fulfilling an important function in promoting social inclusion.

10.7 Conclusion

Urban political ecology can provide a better understanding of the intertwined factors that lead to socio-economic marginalization and environmental degradation. Community gardens offer fertile terrain for urban political ecology research. The growing importance of community gardens globally suggests that they may play important roles in mitigating some impacts of environmental change. But their benefits may disproportionately accrue to some populations and not others – a potential environmental inequality. Surprisingly little research has assessed whether community gardens are meeting the expectations enshrined in land-use policy. In this chapter, we have argued that planners need grounded research to inform land-use policy making, and to align appropriate land use and environmental planning objectives with community needs.

International studies have found that some community gardens have been developed to combat entrenched poverty and social disadvantage – and to address concomitant health impacts. We found no evidence for this in our study. We did, however, find that community gardens in the study area may be providing other important benefits, including improving nutrition, conserving agricultural biodiversity, and fostering environmental education. These gardens may also perform 'compensation functions' due to the loss of residential backyards accompanying rapid population growth in the study area. We found statistically significant associations between the location of the community gardens and residential densities in Brisbane City.

Residents of higher-density suburbs in most Australian cities tend to be comparatively wealthier than their outer suburban counterparts. Our results suggest that property values appear to be higher in close proximity to community gardens. But this observation must be tempered by our finding that in the study area relative disadvantage is also somewhat higher closer to the gardens. This paradoxical finding suggests that some people might be drawn to the gardens for their subsistence benefits – though garden managers did not confirm this supposition. Future political ecology research should better assess the impact of community gardens on the value of adjacent properties, quantifying the magnitude of effect of the gardens, independent from the effect of proximity to parkland or schools. This would provide a clearer picture for planners and policy makers about the sociospatial dynamics of community gardens. Rising land values (as a form of eco-gentrification) could potentially threaten the long-term viability of the gardens, as has been demonstrated by international studies (Staeheli et al., 2002). Land-use policy will thus need to manage the potentially regressive impacts of community gardens on property values. Ignoring these effects could undermine the food security benefits of community gardens, and impede the success of urban cultivation, an unpalatable scenario.

Acknowledgements

Special thanks to Dr Christina Qi Li for her contribution in developing and testing the spatial analysis methods, and to Dr Guy Castley for his statistical analysis advice and mapping assistance. The authors also wish to thank community garden managers for their participation.

Note

[1] A range of statistical tests were undertaken using SPSS (version 22), including linear regressions, analysis of variance (ANOVA), non-parametric Mann–Whitney tests, chi-square tests, t-tests and Q–Q plots.

References

Agyeman, J. and McEntee, J. (2014) Moving the field of food justice forward through the lens of urban political ecology. *Geography Compass* 8, 211–220.

Allen, P. (2010) Realizing justice in local food systems. *Cambridge Journal of Regions, Economy and Society* 3, 295–308.

Baker, L. (2004) Tending cultural landscapes and food citizenship in Toronto's community gardens. *Geographical Review* 94, 305–325.

Barthel, S., Parker, J. and Ernstson, H. (2015) Food and green space in cities: a resilience lens on gardens and urban environmental movements. *Urban Studies* 52, 1321–1338.

Blaikie, P.M. and Brookfield, H.C. (1987) *Land Degradation and Society*. Methuen, London, UK.

Brisbane City Council (2013) *Brisbane Community Garden Guide*. Department of Community Services, Brisbane City Council, Brisbane, Australia. Available at: https://www.brisbane.qld.gov.au/sites/default/files/community_garden_guide_aug_2013.pdf (accessed 14 November 2016).

Commonwealth of Australia (2013a) SEIFA 2011. Available at: http://www.abs.gov.au/websitedbs/censushome.nsf/home/seifa2011?opendocument&navpos=260 (accessed 10 March 2014).

Commonwealth of Australia (2013b) Socio-economic indexes for areas. Available at: http://www.abs.gov.au/websitedbs/censushome.nsf/home/seifa?opendocument&navpos=260 (accessed 10 March 2014).

Cordell, D., Drangert, J.-O. and White, S. (2009) The story of phosphorus: global food security and food for thought. *Global Environmental Change* 19, 292–305.

Corrigan, M.P. (2011) Growing what you eat: developing community gardens in Baltimore, Maryland. *Applied Geography* 31, 1232–1241.

Crompton, J.L. (2005) The impact of parks on property values: empirical evidence from the past two decades in the United States. *Managing Leisure* 10, 203–218.

DeKay, M. (1997) The implications of community gardening for land use and density. *Journal of Architectural and Planning Research* 14, 126–149.

Domene, E. and Saurí, D. (2007) Urbanization and class-produced natures: vegetable gardens in the Barcelona Metropolitan region. *Geoforum* 38, 287–298.

Evers, A. and Hodgson, N.L. (2011) Food choices and local food access among Perth's community gardeners. *Local Environment* 16, 585–602.

Ferris, J., Norman, C. and Sempik, J. (2001) People, land and sustainability: community gardens and the social dimension of sustainable development. *Social Policy and Administration* 35, 559–568.

Gallaher, C.M., Kerr, J.M., Njenga, M., Karanja, N.K. and WinklerPrins, A.M.G.A. (2013) Urban agriculture, social capital, and food security in the Kibera slums of Nairobi, Kenya. *Agriculture and Human Values* 30, 389–404.

Galluzzi, G., Eyzaguirre, P. and Negri, V. (2010) Home gardens: neglected hotspots of agro-biodiversity and cultural diversity. *Biodiversity and Conservation* 19, 3635–3654.

Galt, R.E. (2013) Placing food systems in first world political ecology: a review and research agenda. *Geography Compass* 7, 637–658.

Glover, T., Parry, D. and Shinew, K. (2005a) Building relationships, accessing resources: mobilizing social capital in community garden contexts. *Journal of Leisure Research* 37, 450.

Glover, T., Shinew, K. and Parry, D. (2005b) Association, sociability, and civic culture: the democratic effect of community gardening. *Leisure Sciences* 27, 75–92.

Godfray, H.C.J., Beddington, J.R., Crute, I.R., Haddad, L., Lawrence, D., Muir, J.F., Pretty, J., Robinson, S., Thomas, S.M. and Toulmin, C. (2010) Food security: the challenge of feeding 9 billion people. *Science* 327, 812–818.

Gold Coast City Council (n.d.) *Community Gardens Start Up Kit*. Gold Coast Parks, Gold Coast City Council, Gold Coast, Queensland, Australia. Available at: http://www.goldcoast.qld.gov.au/documents/bf/community-gardens-kit.pdf (accessed 14 November 2016).

Gorham, M., Waliczek, T., Snelgrove, A. and Zajicek, J.M. (2009) The impact of community gardens on numbers of property crimes in urban Houston. *HortTechnology* 19, 291–296.

Guitart, D., Pickering, C. and Byrne, J. (2012) Past results and future directions in urban community gardens research. *Urban Forestry and Urban Greening* 11, 364–373.

Guitart, D.A., Pickering, C.M. and Byrne, J.A. (2014) Color me healthy: food diversity in school community gardens in two rapidly urbanising Australian cities. *Health and Place* 26, 110–117.

Guitart, D.A., Byrne, J.A. and Pickering, C.M. (2015) Greener growing: assessing the influence of gardening practices on the ecological viability of community gardens in South East Queensland, Australia. *Journal of Environmental Planning and Management* 58, 189–212.

Heynen, N. (2014) Urban political ecology I: the urban century. *Progress in Human Geography* 38, 598–604.

Irazabal, C. and Punja, A. (2009) Cultivating just planning and legal institutions: a critical assessment of the South Central Farm struggle in Los Angeles. *Journal of Urban Affairs* 31, 1–23.

L'Annunziata, E. (2010) Following the plant: the political ecology of a Hmong community garden. *Humboldt Journal of Social Relations* 33, 97–134.

Larder, N., Lyons, K. and Woolcock, G. (2014) Enacting food sovereignty: values and meanings in the act of domestic food production in urban Australia. *Local Environment* 19, 56–76.

Lawson, L. (2004) The planner in the garden: a historical view into the relationship between planning and community gardens. *Journal of Planning History* 3, 151–176.

Mansouri, F. and Jenkins, L. (2010) Schools as sites of race relations and intercultural tension. *Australian Journal of Teacher Education* 35, 8.

McClintock, N. (2012) Assessing soil lead contamination at multiple scales in Oakland, California: implications for urban agriculture and environmental justice. *Applied Geography* 35, 460–473.

McClintock, N. (2014) Radical, reformist, and garden-variety neoliberal: coming to terms with urban agriculture's contradictions. *Local Environment* 19, 147–171.

McClintock, N., Mahmoudi, D., Simpson, M. and Santos, J.P. (2016) Socio-spatial differentiation in the sustainable city: a mixed-methods assessment of residential gardens in metropolitan Portland, Oregon, USA. *Landscape and Urban Planning* 148, 1–16.

Monteiro, C.A. and Cannon, G. (2012) The impact of transnational 'Big Food' companies on the South: a view from Brazil. *PLoS Medicine* 9, e1001252.

Moore, S. (2006) Forgotten roots of the green city: subsistence gardening in Columbus, Ohio, 1900–1940. *Urban Geography* 27, 174–192.

Passidomo, C. (2016) Community gardening and governance over urban nature in New Orleans's Lower Ninth Ward. *Urban Forestry and Urban Greening*. DOI: 10.1016/j.ufug.2016.01.001.

Queensland Government (2014a) Property boundaries Queensland. Available at: https://data.qld.gov.au/dataset/property-boundaries-queensland (accessed 14 November 2016).

Queensland Government (2014b) Property sales and valuation products and services. Available at: https://www.qld.gov.au/environment/land/title/valuation/property-sales (accessed 14 November 2016).

Sage, C. (2012) The interconnected challenges for food security from a food regimes perspective: energy, climate and malconsumption. *Journal of Rural Studies* 29, 71–80.

Säumel, I., Kotsyuk, I., Hölscher, M., Lenkereit, C., Weber, F. and Kowarik, I. (2012) How healthy is urban horticulture in high traffic areas? Trace metal concentrations in vegetable crops from plantings within inner city neighbourhoods in Berlin, Germany. *Environmental Pollution* 165, 124–132.

Shillington, L.J. (2013) Right to food, right to the city: household urban agriculture, and socionatural metabolism in Managua, Nicaragua. *Geoforum* 44, 103–111.

Shiva, V. (2008) *Soil, Not Oil: Climate Change, Peak Oil and Food Insecurity*. Zed Books, London, UK.

Somerset, S., Ball, R., Flett, M. and Geissman, R. (2005) School-based community gardens: re-establishing healthy relationships with food. *Journal of the Home Economics Institute of Australia* 12, 25–33.

Spilková, J. and Vágner, J. (2016) The loss of land devoted to allotment gardening: the context of the contrasting pressures of urban planning, public and private interests in Prague, Czechia. *Land Use Policy* 52, 232–239.

Staeheli, L., Mitchell, D. and Gibson, K. (2002) Conflicting rights to the city in New York's community gardens. *GeoJournal* 58, 197–205.

Turner, B., Henryks, J. and Pearson, D. (2011) Community gardens: sustainability, health and inclusion in the city. *Local Environment* 16, 489–492.

Voicu, I. and Been, V. (2008) The effect of community gardens on neighboring property values. *Real Estate Economics* 36, 241–283.

Wade, I. (1987) Community food production in cities of the developing nations. *Food and Nutrition Bulletin* 9, 29–36.

WinklerPrins, A.M.G.A. and de Souza, P.S. (2010) The diversity and circulation of plants in urban homegardens, Santarém, Pará, Brazil. *Boletim do Museu Paraense Emílio Goeldi – Ciências Humanas* 5(3), 493–507.

11 Urban Agriculture as Adaptive Capacity: An Example from Senegal

Stephanie A. White*
Michigan State University, East Lansing, Michigan, USA

11.1 Introduction

This chapter discusses how resilience theory can help to better qualify and situate urban agriculture (UA) in relation to city food systems and urban food security. Specifically, it demonstrates how UA can be regarded as a food practice that builds urban adaptive capacity in a number of ways and across a range of scales. It then uses these conceptual frames to examine urban agriculture in M'Bour, Senegal, drawing attention to:

1. how city processes and space, or urban assemblages, are implicated in producing various food vulnerabilities and resiliencies;
2. how food practices are leveraged to survive and thrive in dynamic and variable urban environments.

Although the research specifically addresses the food environment in M'Bour, Senegal, the conceptual framing is generalizable, and is intended to reveal the diversity of food environments and the contingent ways people experience them.

Urban agriculture is a common component of urban food systems in the GS (and increasingly in the GN) and is practised in various ways at various scales depending on the goals, opportunities and constraints of urban cultivators (Chapter 1, this volume). UA is also commonly embraced by development organizations as a measure to improve food security in urban households. Recent research by the African Food Security Network (AFSUN) (e.g. Crush *et al.*, 2010; Frayne *et al.*, 2010) and Webb (1998, 2011) convincingly casts doubt on the facile association made between urban agriculture and urban food security, while Hovorka (2006) observes that simplistic prescriptions for urban agriculture risk reproducing social and economic marginalization, conditions that cause people to be vulnerable to food insecurity. Yet there is also recent evidence to support the view that urban agriculture can enhance household food and nutrition security, and is a critically important source of both material and non-material benefits for some low-income city residents (Battersby-Lennard and Haysom, 2012; Gallaher *et al.*, 2013; Riley and Legwegoh, 2014). Clearly, the link between urban agriculture and improved food security is not straightforward. Improved methods for planning how it might be better used to enhance food security and for assessing its potential in context and for particular populations are needed.

A shortcoming of much UA research is the tendency to treat urban farms and gardens in isolation, decontextualized from the workings of

---
*E-mail: whites@msu.edu

the city (i.e. urban assemblages) and the various ways people experience urban life, including the ways they engage with food environments. Consequently, the resulting literature tends to take an advocative position on the value of UA (Mougeot, 2006; Veenhuizen, 2006; Karanja and Njenga, 2011), or an adversarial one (Webb, 1998, 2000, 2011; Crush and Frayne, 2011), based primarily on UA's material contributions to households, i.e. its contributions to households are significant versus its contributions are negligible. A bias towards focusing on the productive dimensions of UA misses the social and place-specific features of the practice, and thus remains ignorant of its causative factors, which may be 'multiple, diverse, and dispersed' (Jasanoff *et al.*, 1997, p. 2066). Indeed, 'urban assemblage' theories remind us that the search for *meaning* must not only consider the practice or outcome of the practice (the thing), but also the values and relationship that generate the practice or thing (Jacobs, 2012). Without an examination of its connections to 'the assemblages of urban life', the similarities that transcend urban agriculture may obscure the differences that can provide insight about why it is being done in particular places and by particular people, as well as what the effects are on bodies, cities and urban food security.

In Chapter 2 of this volume, White and Hamm put forth the notion that locating and exploring UA as an *urbanistic* practice can improve the nuance with which it is understood. The premise of this chapter, which is in keeping with the overall orientation of this edited volume, is that theoretical rigour is enhanced by framing UA *in relation to* the local food provisioning system. In particular, this chapter asserts that UA is better understood as an *adaptive practice* that persists because it enhances food system resilience by helping people to 'absorb shocks and perturbations and adapt to change' (Berkes *et al.*, 2003, p. 14). Using the concepts of resilience, vulnerability and adaptive capacity is useful because they draw attention to the diverse and contingent socio-ecological interactions that comprise urban food systems, and thus enable analyses that move past the simplistic equation that producing food=greater food security. In the context of emergent and highly decentralized food systems, like those typically found in the GS, the concepts are also helpful in identifying and valorizing the many ways in which people adapt to and manage vulnerabilities, variability and disruption in food environments. This is important because all too often, the highly individualized and diversified food activities that comprise food systems in, for example, Africa, are uniformly referred to as 'informal', which hides critical dimensions of self-organization and adaptive management that could otherwise help to improve food security in both the GS and GN.

11.2 Conceptual Framework – Vulnerability, Adaptive Capacity and Resilience in Emergent Urban Food Systems

Each UA site is at once a contained system in which people interact with the environment to produce food, and a system that is connected to the larger urban food system. Situating UA within urban food systems has its conceptual roots in Berkes *et al.* (2003), which frames socio-ecological systems as embedded processes within progressively larger socio-ecological systems, where 'phenomena at each level of the scale tend to have their own emergent properties, and different levels may be coupled through feedback relationships' (Berkes *et al.*, 2003, p. 6).

In the literature on socio-ecological systems, resilience, vulnerability and adaptive capacity are system properties. Adger (2000) characterizes the conceptual relationship between vulnerability and resiliency as loosely antonymic. Vulnerability refers to the degree to which a system or individual is susceptible to harm or unable to cope with change or perturbations in the socio-ecological system, while resilience captures the ability of people and socio-ecological systems to tolerate those perturbations and to adapt to change without losing fundamental functions of the system (Adger and Brown, 2009; Preston and Stafford-Smith, 2009; Barthel and Isendahl, 2013). The capacity to incorporate change to maintain function of the system, i.e. to build resilience, is referred to as 'adaptive capacity'. Adaptive capacity is scale-dependent, contingent, and reliant on access to material and social resources. At a societal level, adaptive capacity is dependent on collective ability, action and consensus (Brooks and Adger, 2005).

Understood with the concepts of vulnerability, adaptive capacity and resilience, city food provisioning systems exhibit numerous properties of a resilient socio-ecological system that *works to provide an outcome of food security*. This last point is important, because as Levin et al. (1998) note, 'resilience makes no distinctions, preserving ecologically or socially undesirable situations as well as desirable ones' (p. 225). In other words, despite how 'resilience' is used in contemporary discourse to generally indicate something good, resilience theory is agnostic about the goodness or badness of a system. As it applies to food systems, however, the normative outcome of 'food security' is a necessary criteria because, as Hodbod and Eakin (2015) observe, 'at a global scale and the indivisible scale of the individual human, there must be adequate production and distribution of food to maintain all human life; no regime shift that compromises that core function can be morally permitted' (p. 477).

Urban food systems in the GS have typically emerged in relation to highly variable and chronically uncertain food environments, with very minimal centralized municipal management or support. In such conditions, it is unsurprising that people would evolve multiple pathways to ensure food access and would maintain a diversity of food practices from which to draw, even if those food practices produce little in the way of the overall food share (Lourenço-Lindell, 1995). Food is available from a mix of diverse sources including markets, roadside stands, home gardens and livestock production, as well as through informal cultural practices. Movement and distribution of food is widely decentralized, socially self-organized, emergent, locally interdependent, and serves as a source of income for large numbers of people. Such qualities understood through the conceptual frameworks of resiliency and vulnerability suggest that they are highly adaptive and flexible systems, with multiple 'moving parts' that perform redundant functions and thus are capable of dealing with surprises and perturbations.

11.2.1 Vulnerability

While urban food provisioning systems are resilient in a number of ways, there is a high degree of vulnerability to food insecurity in urban environments. In contrast to 'poverty', which blurs the distinctions of *how* poverty occurs, the concept of vulnerability compels attention to the contingent and located ways in which risk is differentially experienced (Turner et al., 2003). The concept of vulnerability helps to disaggregate these processes, and thus compels more specific and grounded explanations, by recognizing that people experience urban social and spatial processes differently, which, in turn, affects their ability to cope, i.e. their adaptive capacity. In cities, in addition to income and the fragility of livelihood, spatial location, gender, crime, length of time in city, and social networks have important effects on the ways in which one is able to provision the household (Battersby, 2012).

11.2.2 Adaptive capacity

Perceived through the lens of resilience, UA can be understood as one adaptive management practice that deals with uncertainty by diversifying food sources and increasing the numbers of food access points around the city (Lourenço-Lindell, 1995). As an adaptive practice that intersects with different kinds and levels of vulnerability, people derive different benefits from doing it. That is, the extent to which, and how, UA provides resilience varies with the ways in which people experience food vulnerabilities. This is because people's adaptive capacity, as an effect of their vulnerability, varies according to their level of exposure and sensitivity to the factors that cause vulnerability (Turner et al., 2003; Adger, 2006; Preston and Stafford-Smith, 2009). For example, people may experience similar levels of poverty, but some may be more sensitive to food vulnerabilities because they have recently immigrated to a city and lack social networks. While they may have access to a plot of land to farm, a lack of social networks limits their ability to draw on material resources, as well as the socio-ecological memory, or 'framework of accumulated experience', associated with urban farming in that particular place (Folke et al., 2003). In turn, the inability to leverage those social and material resources limits the adaptive capacity of UA. Likewise, limited adaptive capacity minimizes the extent to which those new immigrants can improve resilience of the household food provisioning system, and thus has a negligible, or no, effect on household food resiliency. Such an understanding that the value of UA can be *leveraged* through the factors associated with adaptive capacity requires that research into urban farming considers not only the material benefits of urban farming, but also the social and ecological contexts in which it is practised.

11.2.3 Resiliency

'Food resiliency' refers to the ways that food provisioning systems can cope with perturbations and continue to operate without losing fundamental functions. Increased attention to household food provisioning strategies can provide lessons for improving food resiliencies at a wider scale. This means that it is necessary to draw attention to the underlying principles of resilience behind seemingly chaotic or inconsequential food provisioning strategies.

11.3 Characterization of Study Area Food System and Methods

M'Bour, Senegal, is a rapidly growing, ethnically diverse, increasingly sprawling coastal town with a population of around 200,000 located about 80 km south of the capital city, Dakar. Many of M'Bour's citizens make their lives, either partially or wholly, with unregulated, or 'informal', economic activities, many of them food-related. Much of the housing development in M'Bour is also informal.

Life in M'Bour is characterized by chronic uncertainty, marked by both small and large perturbations. Daily power and water outages, petrol and cooking gas shortages, severely depleted fish stocks, and increases in the price of flour and other basic necessities are common occurrences, with causative factors at multiple scales. Food, however, is readily, albeit differentially, available. In contrast to the capital-intensive and centralized food distribution and exchange systems in the GN, food provisioning in M'Bour is highly decentralized, with large portions of the population engaged in the food trade and/or food production. Most households do not have refrigerators, nor large amounts of cash on hand, so that small and frequent purchases of ingredients are the norm. Although the market is the primary source of food, food is also available throughout the city from small to medium-sized markets or kiosks, roadside stands, home gardens and livestock production. Women, especially, are highly involved with decentralizing the food system, and often make the trip to the central market, buy a number of food ingredients that are in high demand, and bring them back to their neighbourhoods to sell at increased prices. The neighbours who purchase these ingredients do not have to pay to travel to and from the market, so this arrangement benefits both parties. In addition, neighbourhood kiosks and individuals often extend credit, which allows people to cope with uncertain and uneven income streams. Other women sell fruits, vegetables and milk through door-to-door sales, while men were observed selling eggs, fish and goat's cheese. Urban livestock production is more common than gardening or farming, and observed animals included chickens, rabbits, turkeys, goats, pigs, sheep and cattle. Many households maintain fruit trees, such as papaya or mango, and people protect and wild-harvest from a variety of trees, both in and around the household, for their useful products, such as Moringa (drumstick tree; *Moringa oleifera*).

The following empirical findings are drawn from a larger qualitative study on small-scale dry season urban cultivation (White, 2014). Methods included interviews, participant observation, photographs of research sites, and document analysis. The forms of cultivation include micro-gardening (simple form of hydroponics), ornamental plant production, fruit tree production, and vegetable production. Interviews were conducted with eight men and ten women at 14 different cultivation sites, and eight officials (seven men, one woman) representing five government bureaus.[1] Two semi-structured interviews with each cultivator, recorded several months apart, focused on livelihood strategies and practices, outputs of gardens, individual life histories, economic challenges, urban governance, and hopes for the future (Seidman, 2012).

11.4 Findings: Adaptive Capacity M'Bour's Food Provisioning Environment

UA is one adaptive practice in complex food networks and offers an entry point for discerning food provisioning strategies and informal institutions that produce food resiliencies. Findings are organized into four sections that explore how the practices associated with UA build adaptive capacity to deal with change and uncertainty. The first three sections address how resilience is built through socio-economic and ecological diversity, nutrient recycling and social institutions. The final section discusses the contingent and particular ways that urban cultivators experience vulnerability in their efforts to food provision their households.

11.4.1 Practices of ecological and socioeconomic diversity

Diversity helps to buffer the system by spreading risk and addressing critical points of instability. Two ways that UA diversifies the system are discussed: (1) income and agroecosystem; (2) food access.

As farmers diversified their income sources, they also diversified the urban agroecosystem. None of the primary farmers in the study relied wholly on UA for their livelihood, and UA did not often rank as the most important source of household income. Although UA has been associated with desperation, performed by only poor urban residents, this research found that UA is viewed as a pragmatic secondary, or even tertiary, way to earn an income. Lamine,[2] who worked intermittently as a brick builder, sums up the general attitude of many farmers:

> You won't leave your [informal day work] because that's the work that is more important to you. And then you think, 'Ah, I'm not working today . . . let me just go ahead and plant a garden. If I am working, I go to work. If I don't have work, I'll just do some farming here.' That's what's most common here.

Moustapha noted that, 'Farming doesn't interfere with my other work; you can do everything you need to do between 4 and 6 pm.' As a guard, Samba works 12 hours a day, and makes less than 50,000CFA (~US$100) per month, which is not enough to support his family. Farming supplements his income significantly, and he was pondering leaving his formal employment to devote more time to it, saying he could earn at least twice what he was earning as a guard. Because it is self-directed and relies on minimal or often readily available inputs, farming 'fits' with the circumstances of people who need to earn an extra income to make ends meet, or who would like to improve their household resilience through the accumulation of capital (Hansen and Vaa, 2004; Owusu, 2007).

None of the farmers in the study grew just one crop, and several of them diversified by planting both annuals and perennials. Growing both annuals and perennials provided more opportunities to earn an income. Farmers who grew mint, a perennial crop, were able to harvest from these plots several times a week and earn at least several hundred CFA (~US$0.5) each time. Annual crops were harvested less often, but brought in larger amounts of cash. All farmers who were farming on vacant lots ($n=7$), except for one, were cultivating mango, papaya or banana trees, despite not having secure or long-term access to that land. All farmers at the other cultivation sites ($n=7$) were also growing these and other fruit trees, including grapefruit and lime. Polycultures are generally recognized to improve resilience, not only for the diverse products/sources of income they provide to farmers, but because diversity works as a buffer against perturbations that might affect some elements of the system and because they take advantage of diverse ecosystem niches, thus efficiently using space (Altieri, 1989; Colding et al., 2003; Gliessman, 2007). The diversity of the urban agroecosystem was served both by a 'socio-ecological memory' and by innovations from outside the system. A number of farmers sourced favoured seed varieties from their home villages, which are widely dispersed around Senegal, while others bought imported European varieties from local seed shops. Thus, the agroecosystem, via the practices and preferences of farmers, was constantly undergoing small adjustments to deal with change.

A number of farmers remarked on the emotional and sensory effect of gardens. Aba, for example, noted that when she walks around a garden she feels calm, and that she enjoys being in the garden after it has been watered because it is cooler than the ambient temperature. Mohammad said that looking at plants 'clears your eyes up', while Andre said that when he plants something and sees it sprout, it pleases him. Andre also observed, 'If you go somewhere without plants, you can't relax. If you go somewhere with plants, you can relax and your heart becomes cool.' These responses suggest that gardens provide people with a respite from the heat and bustle of the city, which are under-recognized necessities within an economic development agenda that prioritizes economic growth (see also Chapter 12, this volume).

Urban agriculture diversifies and decentralizes food access. As cities grow, the food system must adjust. As new people arrive in M'Bour, even temporarily, 'points of instability' in the food provisioning systems are created, i.e. gaps in servicing food needs. One of the advantages of a self-organized and emergent food system, carried out by large segments of the urban population, is that it can move nimbly and flexibly to fill these critical points of instability (Berkes et al., 2003). In this process, the food provisioning system is constantly renewed, or replenished, with new ideas, new relationships, new networks of exchange and increased food exchange points. Urban farmers play a role in this process because their production activities generally occur in neighbourhoods where increased food access points are required.

In M'Bour there is one central market that services the entire town, which is spread out over about 22 km². Because most people must make small and frequent purchases, due to both lack of access to refrigeration and generally small incomes, daily trips to the market are the norm. Every morning, residents from all over M'Bour make their way to the market. Many walk or take horse-drawn carts, or some other form of public transport. Many people note the expense of travelling to the market each day, and for those living on the outskirts of town, it represents a significant time expenditure as well. Many people, especially women, said that the existence of just one market was unsatisfactory, inconvenient and onerous. In the rainy season, it can even become dangerous, as carts become stuck in the mud, and people have to disembark so that horses can pull the carts free.

Productive gardens in the study were generally located several kilometres away from the market and were valued by neighbours as being a convenient alternative. On several occasions, as interviews were being conducted with farmers, neighbours arrived to purchase fresh produce. During an interview with Mohammad, a woman restaurateur stopped by to buy bissap (*Hibiscus sabdariffa*) and okra, which, as Mohammad explained, she did quite often. On another occasion, two neighbourhood women arrived to buy lettuce. When asked why she buys it from Mohammad, one woman explained that it is fresh and of good quality, and that she knows what she is getting, unlike in the market. Mohammad explained that this happens often, and that people commonly buy tomatoes and peppers at the same time. Samba noted how it benefits his neighbours when he sells food to them 'because if I sell a head of lettuce for 100CFA (~US$0.2), at the market it costs 150'. Lamine described how his wife could cut mint and sell it around the neighbourhood and make as much as 3000CFA (~US$6) in a short amount of time.

Urban farmers develop multiple relationships in order to distribute food (see also Chapter 7, this volume). Such practices enable flexibility on the part of both producer and agent of distribution. In addition to the individual sales noted above, farmers develop working relationships with *bana bana*, i.e. the market women who walk around neighbourhoods and buy produce in bulk to re-sell in the market or elsewhere in the city. Of the seven farmers cultivating on vacant lots, six of them sold to *bana-bana*, while the seventh sold what he produced in a small shop in close proximity to his field. In addition, three of the farmers lived in an area of M'Bour that is at the crossroads to Saly Portudal, a thriving tourist town that has a large expatriate population. In those cases, farmers had the advantage of being able to sell some of their produce at even more outlets *and* at an increased price. Thus, in this case, because of the spatial location, the flexibility of the informal and emergent food provisioning system enabled farmers to gain a bit of extra income from those who could afford it. Such a system, in which there are many connections among a high number of individuals, and in which people are free to make connections with as many people as they can, reflects the principle of 'appropriate connectedness', which, in practice, means 'collaborating with multiple suppliers and multiple outlets, including consumers, rather than just one' (Cabell and Oelofse, 2012, p. 6). Such a high degree of connectedness enables flexibility and responsiveness to change, which helps to ensure that the food systems can shift to fill in 'critical points of instability' that come with urban growth and change (Gunderson and Holling, 2002; Berkes *et al.*, 2003, p. 6).

11.4.2 Nutrient and resource (re)cycling in the city agroecosystem

In many African cities, urban primary production is common, so much so that thinking of the city as an integrated farm, with multiple farmers, helps to understand how productive resources are cycled, and how people draw on these resources in the production and exchange of food. A farm is integrated when the outputs of one productive activity are used as the inputs for another, and when crop and livestock production is carried out via complementary practices (Agbonlahor *et al.*, 2003). A major quality of resilient systems is that the 'system functions as much as possible within the means of the bioregionally available natural resource base and ecosystem services' (Cabell and Oelofse, 2012, p. 4). Although the concept and practice of integration usually refers to the farm level, applying it to a landscape level helps to see the ways in which production practices across the city are integrated and resilient. In addition, better understanding how productive resources are recycled demonstrates how urban farming is an *urban* practice conditioned by, and intersecting with, urban processes.

In M'Bour, farmers depend heavily on the natural resources of a city agroecosystem. Soils in M'Bour are mainly composed of sand and, thus, have little nutrient or water-holding capacity. Classified as Arenosols on the World Soils Map, these soils have greater than 70% sand and less than 15% clay. To make them viable as a growing medium, cultivators must amend them with considerable amounts of organic material, which serves as a nutrient source and helps to mitigate water loss. Warm ambient temperatures and daily watering that encourage rapid decomposition of organic material requires farmers to amend soil regularly.

Manure is readily available and is the most common soil amendment. In some cases, it is necessary to pay for the manure, and in other cases, it is possible to get it for free. How someone accesses manure, and the kinds and quality of manure available to them, depends on where they live and whom they know. Lamine, for example, never paid for horse manure, which he said was because he lived in an area of the city where it was not in high demand. For those farmers closer to the hotels, where manure was used to fertilize lawns, it was necessary to pay around 2000CFA (~US$4) for a load of manure. Often, chicken manure mixed with wood shavings, which are used as bedding in urban chicken production, is available. These shavings are acquired by farmers from city furniture makers or wood workers. Chicken manure is widely recognized to provide a quick boost of nitrogen, but all farmers noted that it should be used with caution since it can 'burn' the plants. Samba noted that the best fertilizer is a mix of wood-shavings and chicken manure that comes from coops that had not been cleaned in a long time. The several dumps around the city were also mined for compost by some farmers, though this practice was used more often by hotels. Two farmers in the study also used fish waste to enrich the soil, which is a plentiful but under-used resource in M'Bour.

11.4.3 Food as a means of building adaptive capacity in social systems

Food practices are embedded within social institutions, broadly conceived here to include 'habitualized behavior and rules and norms that govern society' (Adger, 2000, p. 348). In M'Bour, there is a significant social component to food practices, which not only works to mitigate food vulnerabilities, but also serves to build social relationships, and thus strengthens the collective capacity of people to cope with stress and change (Adger, 2000; Lourenço-Lindell, 2001). Through their cultivation practices, farmers were improving their relationships with others in the community by giving away a portion of their production to others in the community, which served not only to improve their ability to make claims on other households in times of need, but also improved overall community cohesiveness. This quality of communities is recognized as an important prerequisite and determinant of social development and resilient societies (Buchmann, 2009; Cabell and Oelofse, 2012).

In general, respondents agreed that major determinants of well-being, referred to as *nattangué*, are interpersonal relationships, both within and between households (Galaskiewicz, 1979; Lourenço-Lindell, 2001; Swift, 2006). Lamine explained it this way:

> Relationships between neighbors are more important than relationships with your family. I'm here today. My family isn't here . . . they are out in the village. If something were to befall us here at the house, like my father died, before anyone back in the village knew, everyone here on this corner would know . . . everything that should be done will be done before my family gets here.

Moustapha, who said, 'There is no one on this corner who can say I haven't given them something', gives away a portion of his harvest before he sells anything. When asked whether the garden was necessary for his livelihood, Mohammad said that he could survive without it, but that 'there are always people depending on you. There are always people who have needs. If it is just you, you can get what you want very quickly . . . but you have to be able to give.' In rural areas, these bonds result in large part from kinship ties and longstanding inter-family relationships. In urban areas, especially among recent migrants, these ties must be created anew and the outputs of gardens help to create those ties (Gallaher et al., 2013).

Giving away a portion of the harvest fulfils a religious mandate, called *asaka* by Muslims, and practised by Catholics as well. For farmers who did not have a large income, the ability to give away a portion of harvests enabled them to fulfill this important cultural duty and create bonds with neighbours. In addition to this socio-spiritual aspect of UA, there was also an eco-spiritual dimension for some farmers, which linked the cultivation of social relationships with the cultivation and care of the soil. Mohammad explains:

> If someone comes to visit me, I like to give them mint or salad. This is 'doing good' . . . I worked this soil until I was tired, put my money into and then I gave it to you . . . God will pay me, because I am giving life to things of the world. I am making these things live. It is good. You are rehabilitating the soil. And, you know, when you see this, you are happy.

11.4.4 Vulnerability and the limits to urban agriculture's adaptive capacity

City processes influence individual or household vulnerability. Understanding these processes can provide insight about how to address the differential aspects of adaptive capacity.

In general, farmers had limited access to material and knowledge resources, which was due to multiple reasons, including (1) the inability to connect with other farmers because of their distance from each other, and (2) the lack of extension services. Farmers in the study were spread out over an area of roughly 22 km^2. Although urban farming is relatively common, farmers are not necessarily in close proximity to each other. As a result, the pool of expertise they can tap is limited. On a number of occasions, farmers encountered problems, but lacked a means of researching the problem to find a solution. Moustapha, for example, did not know the cause of the yellowing and stunting of his bitter tomato, and thought it might be due to overwatering. In Babacar's field, a large plot of bissap was afflicted with yellow spots, which rendered the leaves unsaleable. He thought it might be due to an insect. When a leaf sample was later shown to a horticulturalist, it was found to be a pathogenic fungus that is prevented by avoiding wetting the leaves when watering. Souleyman, the poorest and most vulnerable farmer in the research group, had the space to farm, but was new to farming and lacked basic farming skills. Due to his level of poverty, he had very little room for error and would have benefited immensely from the expertise of other farmers or trained professionals.

Farmers clearly had shared management challenges that could have been resolved by a more robust exchange across the city. For example, termites and whitefly were a problem mentioned by almost every farmer. Moustapha observed that applying compost to his field (a mix of manure and fish waste) instead of raw manure reduced termite damage. He also observed that watering at night reduced whitefly damage. Beyond the scientific expertise that it offers, extension services might help to contribute to resilience by serving as a conduit for new and diverse innovation, an important dimension of resilience (Folke et al., 2003) and as a sort of hub for farmers to exchange information and material resources, thus improving human capital (Buchmann, 2009; Cabell and Oelofse, 2012).

A number of structural issues also affected the ways in which farmers were able to profit from farming. Fatou explained that her ability to 'survive the city' was hindered, in part, by regulations issued by the mayor's office. Fatou lives on the outskirts of the city, where vacant land is still relatively plentiful. When she first arrived in M'Bour in 1998, she was able to grow millet, but in 2005 or 2006, municipal officials told her to stop:

> I used to farm a lot of millet out here, but they've prohibited it because they say that it increases how hot it is and increases the number of mosquitoes ... When we first came, there was no one here and I could farm all of this. They say you can only farm corn, beans or bissap. Not millet. They say the millet is just for people living in the rural areas.

Mosquito prevention is a major public health concern, which is reflected in M'Bour's Règles d'Hygiène Publique (Public Health Rules). That document states that the cultivation of plants that harbour mosquitoes is prohibited. Putting aside the distinction made by municipal officials between millet and other rainfed field crops, there is evidence that UA increases the incidence of malaria, but primarily in relation to irrigation and resistance to insecticides, neither of which is a factor in low-input, rainfed agriculture, millet or otherwise (Afrane et al., 2004; Dongus et al., 2009; Antonio-Nkondjio et al., 2011). A more tolerant approach would consider the risk associated with difficult socio-economic conditions in relation to the risk of malaria, as well as work with farmers to minimize mosquito breeding habitats (De Silva and Marshall, 2012).

Thus far my findings have focused largely on inter-city dynamics, but food environments in any particular place have causative factors at multiple scales. In M'Bour, farmers who chose to devote a large portion of their small spaces to onions were particularly disadvantaged during the year that the research was conducted, due to onion import/exchange practices. In recent years, farmers have been encouraged to grow onions by the Senegalese government. In an effort to ensure that local producers receive a good price, import restrictions are typically placed on Dutch onions from April to August. In 2011, however, the foreign onions kept arriving well into April, which meant that Senegalese farmers could not get a good price for their onions. Early in the season, Lamine had hoped to be able to stockpile his onions and sell them during the rainy season, from July to September, when the price can climb as high as 15,000CFA/sack (~US$30). The previous year, he harvested 28 sacks and sold them for 10,000CFA (~US$20) each, which allowed him to buy 30 chicks that he raised and then sold to earn an even greater profit. In 2011, however, the foreign onions kept arriving, which led some people to say that the Senegalese government never imposed the restriction, but which may have been caused by onion importers stockpiling onions, and then releasing them onto the market during the closed period (Agritrade, 2013). Because there was a lack of day work, Lamine had to sell three bags of onions in late March to buy rice and oil, and was able to get 9000CFA/bag (~US$18). On 13 April, he sold four more bags, and was only able to get 7000CFA/bag (~US$14). By 20 April, the price was reportedly down to 3000CFA/bag (~US$6), and there were so many onions on the market that they were rotting in their sacks. The online site Agritrade suggested that this was due to local traders stockpiling the imported onions and then releasing them onto the market during the import ban (Agritrade, 2013).

Space was also a factor in creating vulnerabilities. The majority of economic activity associated with food is centrally located, but many of the farmers live in the less densely settled areas of the city. Farmers who lived in areas that permitted easier access to tourists and expatriates were advantaged in at least two ways. First, there were more bana-bana with whom they could choose to do business, which means farmers had more negotiating power. Second, they had more access to tourists and expatriates, who were able to pay more for produce.

Thus, vulnerability is associated with living within particular socio-ecological contexts, and is dependent on the capacity of a group or individual 'to respond to external stresses that may come from environmental variability or from change imposed by economic or social forces outside of the local domain' (Adger and Brown, 2009, p. 109). Vulnerability is situated and contingent, and the factors can vary considerably from household to household. Understanding how farmers are vulnerable, however, gives a starting point for thinking about how to improve the adaptive capacity of urban farming.

11.5 Conclusion: Applying the 'Adaptive Capacity' Lens to Planning Future Food Systems

> When analyzing resilience . . . it makes sense to address the local level and build linkages to other scales . . . Resilience thinking helps the researcher to look beyond the static analysis of social systems and ecological systems, and to ask instead questions regarding the adaptive capacity of societies and their institutions.
> (Berkes *et al.*, 2003, p. 115)

Food is a primary resource need for people. Without a constant, well-distributed supply, society suffers. In much of the GS, the responsibility to produce and distribute that resource lies with a decentralized network of people, held together by formal and informal relationships. This emergent, self-organized system has as its primary concern to get food to people on a consistent basis, and urban residents have developed adaptive practices that can flexibly deal with critical points of instability that would otherwise create more vulnerability. In urban environments, those reorganizations are shaped by specifically urban processes with social, spatial, political and economic dimensions. In addition to the challenges posed by rapid urbanization, urban food provisioning faces climate change, increasing energy prices, and the volatility of global food markets, as evidenced by the 2008 food crisis. These complexities forewarn of even more challenging food provisioning environments in urban places in the coming years and call for new ways of thinking that grapple with such unprecedented conditions. An understanding of food provisioning and exchange practices as comprising the food system's adaptive capacity can help to inform approaches that are responsive to locally articulated conditions and consider both social and ecological uncertainty.

Notes

[1] Bureaus represented included the mayor's office, the prefect's office, the urban planning office, the office of decentralization and local development, and the rural development office, which manages the micro-gardening programme.
[2] All names are pseudonyms to protect identity.

References

Adger, W.N. (2000) Social and ecological resilience: are they related? *Progress in Human Geography* 24, 347–364.
Adger, W.N. (2006) Vulnerability. *Global Environmental Change* 11, 268–281.
Adger, W.N. and Brown, K. (2009) Vulnerability and resilience to environmental change: ecological and social perspectives. In: Castree, N., Demeritt, D., Liverman, D. and Rhoads, B. (eds) *A Companion to Environmental Geography*. Blackwell Publishing Ltd, Oxford, UK, pp. 109–122.
Afrane, Y.A., Klinkenberg, E., Drechsel, P., Owusu-Daaku, K., Garms, R. and Kruppa, T. (2004) Does irrigated urban agriculture influence the transmission of malaria in the city of Kumasi, Ghana? *Acta Tropica* 89, 125–134.
Agbonlahor, M., Aromolaran, A. and Aiboni, V. (2003) Sustainable soil management practices in small farms of southern Nigeria: a poultry-food crop integrated farming approach. *Journal of Sustainable Agriculture* 22, 51–62.
Agritrade (2013) 'Oversupply' on Senegalese onion markets sees prices plummet. Technical Centre for Agriculture and Rural Cooperation. Available at: http://agritrade.cta.int/Agriculture/Commodities/Horticulture/Oversupply-on-Senegalese-onion-markets-sees-prices-plummet (accessed 21 October 2013).

Altieri, M. (1989) *Agroecology: The Science of Sustainable Agriculture (Part 1)*. Westview Press, Boulder, Colorado, USA.

Antonio-Nkondjio, C., Fossog, B.T., Ndo, C., Djantio, B.M., Togouet, S.Z., Awono-Ambene, P., Costantini, C., Wondji, C.S. and Ranson, H. (2011) *Anopheles gambiae* distribution and insecticide resistance in the cities of Douala and Yaounde (Cameroon): influence of urban agriculture and pollution. *Malaria Journal* 10, 154–154.

Barthel, S. and Isendahl, C. (2013) Urban gardens, agriculture, and water management: sources of resilience for long-term food security in cities. *Ecological Economics* 86, 224–234.

Battersby, J. (2012) Beyond the food desert: finding ways to speak about urban food security in South Africa. *Geografiska Annaler: Series B, Human Geography* 94, 141–159.

Battersby-Lennard, J. and Haysom, G. (2012) *Philippi Horticultural Area*. AFSUN and Rooftops Canada Abri International, Cape Town, South Africa.

Berkes, F., Colding, J. and Folke, C. (2003) Introduction. In: Berkes, F., Colding, J. and Folke, C. (eds) *Navigating Social-Ecological Systems: Building Resilience for Complexity and Change*. Cambridge University Press, West Nyack, New York, USA, pp. 1–28.

Brooks, N. and Adger, W.N. (2005) *Assessing and Enhancing Adaptive Capacity*. Cambridge University Press, Cambridge, Massachusetts, USA.

Buchmann, C. (2009) Cuban home gardens and their role in social-ecological resilience. *Human Ecology* 37, 705–721.

Cabell, J.F. and Oelofse, M. (2012) An indicator framework for assessing agroecosystem resilience. *Ecology and Society* 17(1), article 18. Available at: http://www.ecologyandsociety.org/vol17/iss1/art18 (accessed 23 November 2016).

Colding, J., Elmqvist, T. and Olsson, P. (2003) Living with disturbance: building resilience in social-ecological systems. In: Folke, C., Berkes, F. and Colding, J. (eds) *Navigating Social-Ecological Systems: Building Resilience for Complexity and Change*. Cambridge University Press, West Nyack, New York, USA, pp. 113–185.

Crush, J.S. and Frayne, G.B. (2011) Urban food insecurity and the new international food security agenda. *Development Southern Africa* 28, 527–544.

Crush, J.S., Hovorka, A.J. and Tevera, D. (2010) *Urban Food Production and Household Food Security in Southern African Cities*. African Food Security Urban Network (AFSUN), Cape Town, South Africa.

De Silva, P.M. and Marshall, J.M. (2012) Factors contributing to urban malaria transmission in sub-Saharan Africa: a systematic review. *Journal of Tropical Medicine* 2012, 10.

Dongus, S., Nyika, D., Kannady, K., Mtasiwa, D., Mshinda, H., Gosoniu, L., Drescher, A.W., Fillinger, U., Tanner, M., Killeen, G.F. *et al*. (2009) Urban agriculture and *Anopheles* habitats in Dar es Salaam, Tanzania. *Geospatial Health* 3, 189–210.

Folke, C., Colding, J. and Berkes, F. (2003) Synthesis: building resilience and adaptive capacity in social-ecological systems. In: Berkes, F., Colding, J. and Folke, C. (eds) *Navigating Social-Ecological Systems: Building Resilience for Complexity and Change*. Cambridge University Press, West Nyack, New York, USA.

Frayne, B., Pendleton, W., Crush, J., Acquah, B., Battersby-Lennard, J., Bras, E., Chiweza, A., Dlamini, T., Fincham, R. and Kroll, F. (2010) The state of urban food insecurity in southern Africa. *Urban Food Security Series No. 2*. Queen's University and AFSUN Cape Town, Kingston and Cape Town South Africa.

Galaskiewicz, J. (1979) *Exchange Networks and Community Politics*. Sage Publications, Beverly Hills, California, USA.

Gallaher, C.M., Kerr, J.M., Njenga, M., Karanja, N.K. and WinklerPrins, A.M. (2013) Urban agriculture, social capital, and food security in the Kibera slums of Nairobi, Kenya. *Agriculture and Human Values* 30(3), 389–404.

Gliessman, S.R. (2007) *Agroecology: The Ecology of Sustainable Food Systems*. CRC Press, Boca Raton, Florida, USA.

Gunderson, L.H. and Holling, C.S. (2002) *Panarchy: Understanding Transformations in Systems of Humans and Nature*. Island, Washington, DC, USA.

Hansen, K.T. and Vaa, M. (eds) (2004) *Reconsidering Informality: Perspectives from Urban Africa*. Nordiska Afrikainstitutet, Uppsala, Sweden.

Hodbod, J and Eakin, H. (2015) Adapting a social-ecological resilience framework for food systems. *Journal of Environmental Studies and Sciences* 5(3), 474–484.

Hovorka, A.J. (2006) Urban agriculture: addressing practical and strategic gender needs. *Development in Practice* 11, 51–61.

Jacobs, J. (2012) Urban geographies I: still thinking cities relationally. *Progress in Human Geography* 36(3), 412–422.

Jasanoff, S., Colwell, R., Dresselhaus, M.S., Goldman, R.D., Greenwood, M.R.C., Huang, A.S., Lester, W., Simon, A.L., Linn, M.C., Lubchenco, J. *et al*. (1997) Conversations with the community: AAAS at the millennium. *Science* 278, 2066–2067.

Karanja, N. and Njenga, M. (2011) Chapter 10: feeding the cities. In Starke, L. (ed.) *State of the World: Innovations that Nourish the Planet*. Worldwatch Institute, New York, USA.

Levin, S.A., Barrett, S., Aniyar, S., Baumol, W., Bliss, C., Bolin, B. and Sheshinski, E. (1998) Resilience in natural and socioeconomic systems. *Environment and Development Economics* 3(2), 221–262.

Lourenço-Lindell, I. (1995) The informal food economy in a peripheral urban district: the case of Bandim District, Bissau. *Habitat International* 19, 195–208.

Lourenço-Lindell, I. (2001) Social networks and urban vulnerability to hunger. In: Tostensen, A., Tvedten, I. and Vaa, M. (eds) *Associational Life in African Cities: Popular Responses to the Urban Crisis*, Nordiska Afrikainstitutet, Stockholm, Sweden, pp. 30–45.

Mougeot, L.J.A. (2006) *Growing Better Cities*. International Development Research Centre, Ottawa, Canada.

Owusu, F. (2007) Conceptualizing livelihood strategies in African cities: planning and development implications of multiple livelihood strategies. *Journal of Planning Education and Research* 26, 450–465.

Preston, B.L. and Stafford-Smith, M. (2009) *Framing Vulnerability and Adaptive Capacity Assessment*. Discussion paper, CSIRO, Australia.

Riley, L. and Legwegoh, A. (2014) Comparative urban food geographies in Blantyre and Gaborone. *African Geographical Review* 33(1), 52–66.

Seidman, I. (2012) *Interviewing as Qualitative Research: A Guide for Researchers in Education and the Social Sciences*. Teachers College Press, Columbia University, New York, USA.

Swift, J. (2006) Why are rural people vulnerable to famine? *IDS Bulletin* 37, 41–49.

Turner, B.L., Kasperson, R.E., Matson, P.A., McCarthy, J.J., Corell, R.W., Christensen, L., Eckley, N., Kasperson, J.X., Luers, A. and Martello, M.L. (2003) A framework for vulnerability analysis in sustainability science. *Proceedings of the National Academy of Sciences USA* 100, 8074–8079.

Veenhuizen, R.V. (ed.) (2006) *Cities Farming for the Future: Urban Agriculture for Green and Productive Cities*. RUAF Foundation, IDRC and IIR, Manila, Philippines.

Webb, N.L. (1998) Urban agriculture. *Urban Forum* 9, 95–107.

Webb, N.L. (2000) Food-gardens and nutrition: three Southern African case studies. *Journal of Family Ecology and Consumer Sciences/Tydskrif vir Gesinsekologie en Verbruikerswetenskappe* 28, 62–67.

Webb, N.L. (2011) When is enough, enough? Advocacy, evidence and criticism in the field of urban agriculture in South Africa. *Development Southern Africa* 28, 195–208.

White, S.A. (2014) Cultivating the city: exploring production of place and people through urban agriculture. Three studies from M'Bour, Senegal. PhD dissertation, Michigan State University, East Lansing, Michigan, USA.

12 Intersection and Material Flow in Open-space Urban Farms in Tanzania

Leslie McLees*
University of Oregon, Eugene, Oregon, USA

12.1 Introduction

Researchers estimate that 40% of the urban population in sub-Saharan Africa relies on food cultivation in cities (Foeken, 2006). Translating that figure to Dar es Salaam, a city of an estimated population of 4 million (UN-Habitat, 2010), approximately 1.6 million people are involved in some manner of urban food production. Understanding how these spaces exist and persist, however, requires understanding them more than just as sites of food production. People, materials, money and even the sites of the farms themselves flow and intersect through time and space to create specific places and processes. This chapter examines how utilizing movement, flow and intersection as heuristic devices (Simone, 2005, p. 13) on the open-space farms of Dar es Salaam bridges the analytical gap between urban agriculture in the Global South and Global North. This analysis highlights the possibilities and creativities within urban space by understanding cities beyond a developmentalist approach that frames movement and creativity (or informality) as proof of failed cities (Robinson, 2002; Myers and Murray, 2006). Instead, I focus on how people and materials flow through the city in ways that disrupt the narratives of marginalization and powerlessness that frame cities in the Global South.

I start by providing a brief orientation to this approach, followed by background on open-space farming, and the methodology I utilized in this study. The empirics of this chapter will focus on flows of individuals and social relations to demonstrate the utility of tracing how all of these intersect and are manifest on and beyond open-space farms. First, I present the movements of farmers, exploring how they started farming, where they go, and how farms operate in their own life trajectory. I then examine farms as nodes in daily social life for people both on and off the farm, focusing on the seemingly simple interactions that provide information, companionship, pride, and knowledge transfer. I conclude with how this perspective can facilitate an understanding of urban agriculture in sub-Saharan Africa beyond the focus on poverty and marginalization. Through this, it is possible to develop a global approach to urban agriculture, appreciating the similarities and distinctions between how the practice is manifest in different places.

This approach to the city situates the urban as a process, emphasizing the roles of intersections, movements and flows in the continual processes of remaking that defines, and is distinct in, different cities. It is understanding the urban as a site of conjunction for seemingly endless possibilities (Simone, 2005, p. 9). Instead of focusing on informality as failure, I seek to examine the city

---

*E-mail: lmclees@uoregon.edu

through the eyes of people who consciously navigate places with intentions of creating connections or manipulating spaces to their benefit. I mobilize this approach through the interrogation of a specific space, not only to demonstrate the fluidity of cities, but to emphasize the integration of these spaces within city-making and re-making, and to demonstrate the very real impact these processes have on people's lives. This work contributes towards methods of urban analysis that can focus on the urban as a process, rather than the outcome of determined events, and as such, help move the analysis of urban agriculture beyond the South/North analytical divide that often frames informality, and urban agriculture in the Global South, as failed urban development.

12.2 The City as Flow and Intersection

I use the terms flow, movement and intersection not as categories, but as heuristic starting points, following Simone (2005, p. 13). These processes describe the varying capacities of people and spaces to operate together to help make sense of what otherwise appear as disparate and irrational dimensions of urban life. The motive is to understand how cities 'work' and how people articulate and disarticulate from each other, from places, and from institutions, to create their own urban futures (Simone, 2004, 2010). Simone attempts to capture the nuanced and ephemeral components of cityness, how those are made real, and how people use spaces and movement to make claims and connections that will help secure their livelihoods and their identity. Simone rarely focuses on specific spaces *per se*, but rather the ways people move through and use cities. I modify this to focus on a particular type of space in order to specifically grasp the roles that farms, and the people who move through them, play in the larger urban system.

Simone's work complements other frameworks emerging in critical urban studies, emphasizing the role of assemblage in urban life. Urban actors draw upon innumerable networks and objects in their daily lives to secure their livelihoods and their urban existence. One cannot understand a process (or space) without recognizing the connections, however fleeting, that actors create, maintain and leave (McFarlane, 2011a). Assemblage also highlights how people draw upon diverse practices and resources to constitute their whole experience. Analysing just one component of daily life might not provide a complete picture of the importance of that resource or practice.

Scholars have deliberated over the usefulness of assemblage in urban studies, arguing that it diverts focus from how cities are shaped through capitalism (Brenner *et al.*, 2011). Yet it is precisely this critique that postcolonial urban scholars seek to up-end. The political economic focus predestines cities such as Dar es Salaam for poverty and crisis because of the implicit comparison to more developed cities (Robinson, 2002, 2006; Roy, 2011). The focus on power (imposed by the state, by capitalism, etc.) that emerged in Western urban theory does not fully explain the subtleties, variations and expressions in power in different situations and spaces. Assemblage incorporates more marginalized knowledge of urban function (McFarlane, 2011b), making it appealing to critical urban theorists and activists in regional contexts throughout the world.

Simone draws upon ideas of assemblage in much of his work, focusing specifically on how marginalized actors desire to belong and define the terms of their own belonging, often in makeshift ways (Simone, 2004, p. 9). People embody the need of being what they need to be in any given moment or place – a reaction to the perpetual emergency and emergence that defines daily life for many urban residents in Africa (Simone, 2004, pp. 4–5). Simone uses the idea of emergency to convey the circumstances in which many people function in their daily lives: employment is insecure or impossible, food prices rapidly fluctuate, and other expenses (bribes, school fees, water) can wipe out a family's income very quickly. People must be ready to make connections and utilize them in ways that secure their livelihoods.

In this chapter, I employ Simone's work on movement, flow and intersection to understand how a place, an open-space farm in the city, reflects spatial and temporal relations between people and urban spaces. I highlight fluidity as a way of making spaces in the city. In Simone's words:

Intersection is about people and ways of doing things coming down to a crossroads, not knowing what else is going to be there, and no one being able to completely dominate what takes place there... Whatever happens, people coming to the crossroads are changed... The key is how spaces get turned into crossroads-points and experiences of intersection. (Simone, 2010, p. 191)

[F]or those who aspire to be something more than they are in the present, the objective is not to tie themselves down to prevailing notions about what can be taken into account, what makes sense, or what is logically possible... In the very lack of things seeming settled, people keep open the possibility that something more palatable to their sense of themselves might actually be possible. (Simone, 2010, pp. 260–261)

Emphasis on intersection, movement and flow is an effort to valorize the agency and constructive powers of urban subjects (Simone, 2005, 2008). These processes define cityness: 'a thing in the making ... the capacity of its different people, spaces, activities, and things to interact in ways that exceed any attempt to regulate [or analyse] them' (Simone 2010, p. 3). Rather than taking these processes for granted, or ignoring them because they seem too intangible, they should be privileged for their ability to disrupt urbanism that often fails the very people it purports to help. And it should also be recognized that these processes exist in cities around the world, not just in the seemingly chaotic cities of the Global South.

12.3 Urban Open-space Farming in Dar es Salaam

Foeken estimates that 90% of the leafy green vegetables consumed in Dar es Salaam are grown in the city (2006). While many people farm on home gardens or on peri-urban areas, open-space farms exist in the urban built environment near roadsides, river valleys, or other interstitial spaces where people can access land through takeover or negotiation (Dongus, 2001; McLees, 2011). Farmers represent a range of socio-economic classes, from the poorest surviving day to day, to those who can send their children to college in South Africa. However, open-space farming remains legally ambiguous. While the government recognizes urban agriculture as important for urban economies and food security (URT, 1997, 2000), no space has been zoned for this purpose. Farmers operate within the informal economy, selling leafy greens to individuals who purchase their daily needs, to vendors in the markets across the city, to restaurants, and to hotels in the city centre. Open-space farming remains visible in the urban landscape and represents an important component of urban nutrition.

Open-space cultivation has been controversial, especially after several sensationalist news articles in the late 2000s claimed that farmers were profiting while using polluted water and causing cancer in consumers (Andrew, 2008; Kato, 2010). In Kariakoo, the main market in the city, no vendors would admit to selling greens from Dar es Salaam, though farmers claimed they sold many of their vegetables to those vendors. While pollution goes unregulated, most farmers are aware of the issue and look for alternatives. However, many farms are located in floodplains or close to the water table, and floods can cause contamination.

Urban agriculture has been present throughout sub-Saharan Africa since the colonial era (Leslie, 1963; Armstrong, 1986; Page, 2002). However, most research in the region has focused on urban agriculture as a response to neoliberal adjustments in the 1980s and 1990s (Slater, 2001). This research highlighted the challenges of farming in cities and the ways that the activity subsidized dramatic reductions in government support and employment (Briggs and Mwamfupe, 2000; Egziabher et al., 1994), and it served as a platform for advocacy on behalf of farmers. More recent work has focused on urban agriculture as a part of networks of social capital, urban environments, or belonging (Freeman, 1991; Slater, 2000, 2001; Rakodi, 2002; Gallaher et al., 2013; Chapter 14, this volume). Some of this work has inspired my own approach, but I want to focus on the urban spaces as a whole and how they intersect with individual and social lives, rather than focusing on the farms themselves. My approach strives to situate farms as an integral part of the city's trajectory, rather than an exception to urban development.

12.4 Methods

Following calls by Crang (2003) and Latham (2003) to diversify qualitative approaches to focus on materiality and practice, I utilized several methods to capture the ephemerality, nuance and reality of how urban farms operate in the city. These methods included participant observation with farmers, agricultural extension workers, and an NGO. I also conducted interviews with farmers and government officials and developed photo-voice (see below) and mental-mapping projects with farmers. This array reflects my own desire to examine how methods can answer these questions through postcolonial inquiry that shifts the research away from the framework of the researcher and towards allowing people to tell their own stories about urban processes (McLees, 2012).

The initial round of data collection consisted of semi-structured interviews with 82 farmers from nine open-space farms. These focused on farmers' personal histories, why they moved to the city, how they began farming, and the challenges they face. I have moderate fluency in Kiswahili, but interviews were conducted with a translator, and her voice is reflected in the quotes below to recognize her role in interpreting farmers' stories (Temple and Young, 2004; Larkin et al., 2007). My team spent significant amount of time on farms, getting to know farmers, and earning their trust. As a result, interviews were relaxed, explored tangents, and were wide-ranging.

I incorporated photo-voice and mental maps to allow farmers to illustrate their daily activities and spaces, because interviews can be less effective at incorporating motivations, purpose, and unconscious meaning of people's actions (Dirksmeier and Helbrecht, 2008). Photo-voice allows participants to generate different ideas, by relying on the most powerful sense for most people: sense of sight (Darbyshire et al., 2005). When people see their pictures, they may become excited, proud, angry or tell a story. They make new associations, initiate dialogues between individuals and social worlds, and take the researcher to places where they otherwise could not go (Young and Barrett, 2001; McIntyre, 2003; Nowell et al., 2008; Packard, 2008). Twenty farmers from four farms participated in this project. They were given a camera for a week and asked to take pictures of their daily lives. They were then asked to write captions and participate in focus groups. The photos and descriptions, and responses to them, make this a useful method to approach the intangible, yet real, experiences in the farmers' daily lives.

Mental mapping allows participants to map their own spaces, privileging spatial practices that are often hidden by land-use research and top-down planning policies (Elwood, 2006; Brennan-Horley and Gibson, 2009; Brennan-Horley, 2010). Mental maps add to the richness of interview data by allowing respondents to express spaces in a creative way, and participants can illustrate new ways of describing the places they occupy (Darbyshire et al., 2005). Maps reveal how people relate to spaces within cities and demonstrate less-measurable forms of experience (Green et al., 2005; Pavlovskaya, 2009). In this project, 15 farmers were asked to draw maps, which we then discussed with them to understand how the social and personal landscapes are integrated with the physical (Brennan-Horley et al., 2010), and to reveal the relationships on and with these spaces.

12.5 Intersections and Flows Through Urban Farms

In the remainder of this chapter, I examine the flows and movements of people and social interactions to understand how farms function in the city. This process will demonstrate how spaces are produced at the junction of specific actors, ecologies, urban development processes and economies. Tracing the flows and intersections that create urban open-space farms highlights structural forces people navigate, but also the nuances, creativities and meanings embedded in these processes and to which people react. All of the farmers I introduce below have had their names changed to protect their identity.

12.5.1 Three life stories

Each farmer has a story about how they began working on the particular farm where I met them. I present three of these to illustrate the pathways and forces that people navigate, how their lives intersect with open-space farms, and how these experiences have in turn influenced their urban lives. These stories also demonstrate the ways that farms are integrated into people's broader urban experiences.

Mary comes from Lindi Region and went to school until class seven, the end of primary school and free education in Tanzania (at 12 years old). She had a cousin who lived in Masaki (a city ward), who wanted to pay for Mary to go to typing school. When Mary arrived in the city he tried to seduce her, and he stopped paying for her schooling when she refused him. She went to live with an uncle who had just moved to Msasani, but the uncle's wife did not like her. She found a job as a house girl in Oyster Bay, but the woman of the house paid her only 7500 Tsh (about US$1) per month, saying that Mary was getting free board and food. Mary had a child by this time and she needed more income to support herself and her daughter. She was eventually able to find a job with a family adjacent to Drive-In Farm. The woman of this house encouraged her to grow plants in an open area next to the house to earn extra money. The area was a rubbish dump, and Mary cleared some of it to grow *mchicha* (a local spinach), which she quickly sold, and she was able to send her child to school. When her mother died, Mary had to visit her village for 2 months. When she returned to the city, her former employer had moved away. Her plots had remained, and she decided to expand and live by growing and selling *mchicha*.

Mary is now the leader of the Drive-In farming group. She has also used an area of the farm to cook and sell lunches. This has been of limited success, however, and Mary remains relatively poor. She hopes to make more money by growing more profitable vegetables (peppers and tomatoes), but she doesn't know what else she could do. She believes she will depend on her children, who she hopes have better lives. Yet she enjoys farming because of the independence she has, and she does not have to work for people who treat her poorly. She is incredibly proud of her work and often (though not always) enjoys the challenges of farming. She has a reputation of being sensitive, fair and accommodating, which is why the farmers elected her to group leader. That, too, is challenging for her, but she enjoys the leadership position and knows it gives her good experience in case another opportunity comes her way.

David comes from the Kilimanjaro region, going to school until class seven, when his family ran out of money. He found work as a tree planter in Tanga, a town north of Dar es Salaam. He did not earn much money, and decided to attend trade school learning how to make hats, shoes and belts. He moved to Dar es Salaam to earn more money, and at first he made a good income working at an Indian company making special-order leather goods. However, the emerging second-hand clothes market resulted in fewer orders, as people bought the cheaper used products. About this time, he and his wife moved to the Msimbazi River Valley in the middle of the city and built a house in the informal settlements near the valley floor. Msimbazi lies in the middle of the densely built-up city, but remains undeveloped due to regular flooding along the valley floor. David and his wife kept chickens and sold eggs for a while, but eventually became friends with people farming in the valley, and when plots became available, they were able to secure several for planting vegetables.

David has been successful and is very proud of what he has accomplished. He bought a gas-powered water pump, which he rents out to other farmers. He showed farmers how to dig canals between wells, meaning they did not have to walk so far to carry water. He also showed them how to mix powdered pesticides with water to spray on the crops, rather than throwing the powder directly on them. His innovations led the 50 farmers that formed a group to elect him as leader. He believes that in the future he will be a 'big farmer', a phrase used by people who want to buy some land in the peri-urban areas where they can grow higher-value crops, such as fruit.

Cindy lives and works on the Oyster Bay Farm, located on the Oyster Bay Police Station land. She was born in Ruvuma in southwestern Tanzania. She went to primary and 3 years of secondary school, when one of her parents died and the family could no longer pay her school fees. She met a policeman who lives in Dar es Salaam, and eventually moved to be with him. She cooked chapatti and *maandazi* (fried bread) to sell to the kids at the nearby school, providing herself income to buy things for the house and help with school fees. In 2000 she noticed that many police family members were farming, and she was eventually able to buy a plot from a man who was retiring. She says it is hard, but that her husband is supportive and they agree about what to do with the money they both make. She felt lucky because many of the police wives she knew had to hide their money so their husbands would not take it or 'drink it away'. She also felt fortunate to be farming on police land. The police can do what they want with the land, even though it is illegal to farm, and nobody would steal from the farm (which is not uncommon on other farms). Further, even though it is against the law to use piped water for agriculture, the farmers can do so because nobody would report them.

12.5.2 Spaces of intersection and independence

These stories highlight the larger economic or political forces that have compelled people to turn towards urban agriculture, but also the ways they personally responded, made connections and decisions, and the simple joys or frustrations they face. Many farmers are motivated by a strong sense of independence: there are few jobs in the city where you can be your own boss, and growing food means that even if you do not make any money, you can still feed your family. Many farmers had felt shamed in previous jobs, and while they know people see them as 'low class', they can hold their heads high because they are not abused by bosses or swindled out of wages.

The farms provide spaces of innovation and creativity in solving various ecological, social and economic problems. These are incredibly dynamic spaces, and farmers draw upon previous experiences in villages or other areas in the city to create farms that will best suit their livelihood needs. Because farmers have the freedom to try different strategies, they are constantly innovating solutions, contributing to an individual and a collective sense of pride. In the process, they change these spaces. Recognizing these stories is not meant to romanticize their lives: farming is difficult and often gruelling work, a point that every farmer emphasized. However, their experiences are not uniformly negative. Highlighting how these lives flow towards and intersect specifically on farms demonstrates how these spaces

provide security, insecurity, livelihoods, freedom, pride and myriad other satisfactions and dissatisfactions in the city.

12.6 The Sociality of Farms

Farms in Dar es Salaam are intensely social spaces. On any day we observed far more non-farmers than farmers moving through the space: school children, customers, passers-by, vendors, visitors and potential day labourers. Farmers are also embedded in social networks, both on and off the farms. Some farming groups have regular meetings; others hold impromptu gatherings to discuss pressing issues or emergencies. Farmers also sit together, with their friends and customers chatting, gossiping, exchanging information, talking politics, and greeting passers-by. In this section I trace some of these social connections, how they intersect with these spaces, and how they move on. This cannot be a comprehensive review of social interactions on farms; that would be impossible to document. Instead I offer a few examples of the social interconnectedness and the ways that these interactions integrate farmers and farms into the city.

I focus first on the interface between farmers and customers to show the importance of mundane interactions in daily life. I then turn to the people who move through this space on a daily basis, including school children and other local people who use the farms as a pathway to some other place. Finally, I briefly discuss the role of knowledge and emotion as they flow on and off the farms. Through these examples, I demonstrate the ephemeral, yet no less real, intersections and flows of people, information and feeling that reinforce these urban spaces.

12.6.1 Customers on the farm

Customers are obviously an important component of daily interactions. Most common are the people who stop by to purchase a bundle of vegetables for home consumption, usually the female head of household or a domestic worker. They generally stop and chat with the farmers, especially if they are regular customers, exchanging news or information. Waiting and chatting with customers is an important social time on the farms, providing an opportunity to rest from the hard labour from earlier in the day (Figs 12.1 and 12.2). The importance of this time rarely came out in interviews, but was a dominant theme in the mental maps and photo-voice images.

Farms are also nice cool places to sit in an otherwise hot, dusty and humid city. People often stop in the shade to rest on a hot day. As one farmer from Msimbazi explained:

> People will just come and sit here from nowhere, enjoy the shade, exchange news and then go away. There is no sun under here. It's like air-conditioning from God. When the sun is too hot, people come and rest here and talk to us. (Sam, Msimbazi)

Farmers talked about subjects including gossip, planting techniques and what their kids were doing. The election in October of 2010 provided ample conversation during the months of July to October. In a city where few people have access to a television and radio reception is intermittent, these informal conversations are a primary way of exchanging political news. Information flows through these spaces, and farmers are well situated outside and on pathways to see and hear reports that they use and transmit.

My research assistant and I often ate lunch at Mary's lunch stand at Drive-In. With her customers, politics was a common theme, as well as economic troubles, or new projects where someone might find a job or a new market to exploit. People shared news items and opinions readily, and disputes over different perspectives generated some intense discussions. It was in these brief encounters with customers that information and news was quickly passed through farms and outwards again to the general communities, as farmers take the information home and convey it to other areas of the city.

12.6.2 Farms as safe pathways

We visited two farms that were located on school grounds. Farmers knew many of the students

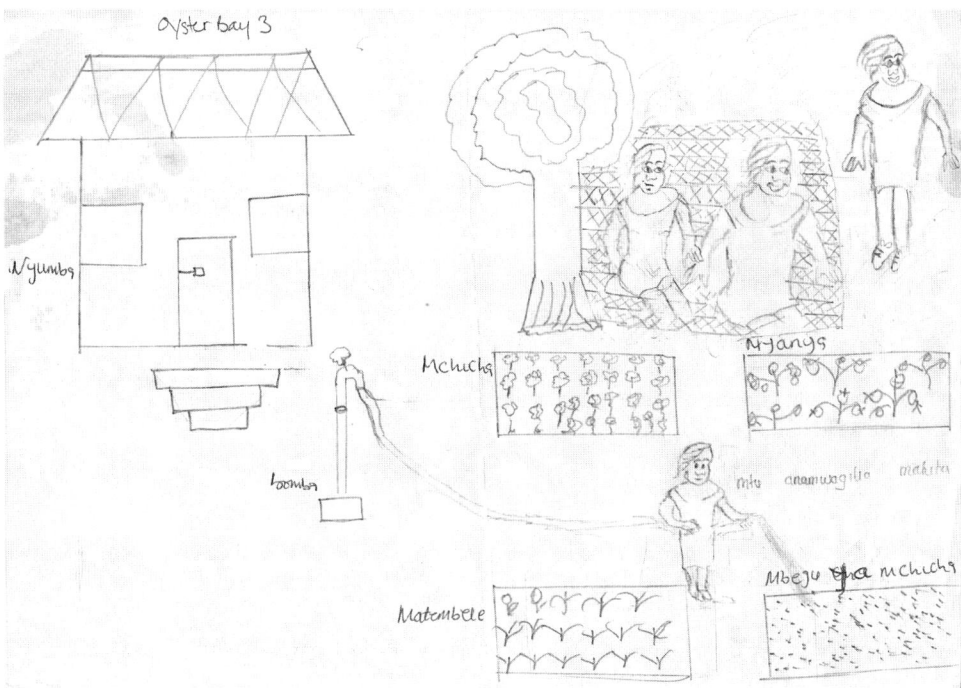

Fig. 12.1. Mental map, Oyster Bay. This map shows two farmers under a tree waiting on their mat for customers. Rebecca, the farmer who authored the map, said her favourite part of the day was talking to her fellow farmers and any customers who arrived.

Fig. 12.2. Photo-voice, Drive-In. 'I have taken this picture with one of my customers who often talked with me in the afternoons under this tree. [This relationship] contributes to the bond we have with our customers. It represents the happiness of being a farmer' (Terry).

and the teachers at the school were regular customers. Farmers did not feel a responsibility for the students, but they did feel that they were part of the authority that helped keep the students from misbehaving, rough-housing, or being too loud or disrespectful. We witnessed several occasions where students were being picked on, and the farmers would tell the aggressors to stop. Farms are a place where students can move through that feels safe, but they also represent an extension of adult authority over movement and behaviour.

Reality or not, there is a widespread perception that creating a farm and cutting down bushes removes hiding spaces for thieves. Some farms, such as Msimbazi, were known to previously harbour snakes, and eliminating the tall grasses in the valley made people feel safer in walking from the informal settlements to the roads and markets they went to on an almost daily basis.

The Ubungo farm also reflected the importance of safety in daily movement. Located at the bottom of heavily wooded hills behind the University of Dar es Salaam, it is difficult to remember that this farm is next door to the main bus terminal for the city. However, this situation makes the Ubungo farm a very important thoroughfare.

> [T]his part of the city could be a very dangerous place. People farming here help to make the city visible and remove the dangerous people that can hide in here. (James, Ubungo)

At Ubungo, we asked some of the people there how they felt about walking through the farm. Many people bought their vegetables there, but every person emphasized how much safer the area was with farmers there. This included people too young to know the area before farms existed: the narrative of safety was being passed down. We heard stories that people had been robbed, raped and even killed prior to the farms being there. There are no records kept of the changes in crime on such a small scale, but the *perception* remains that these farms provide this safe space for movement. This in turn reinforces the points of safe social flows and intersections, and therefore the importance of the presence of farms for farmers and non-farmers alike.

12.6.3 Farms as pathways of knowledge and emotion

The transmission of knowledge about the practice of farming reflects how ideas are spread and how they are grounded in specific spatial practices. Knowledge about the endeavour transforms not only these spaces, but influences, and is influenced by, knowledge as it intersects with the farm. This knowledge from the villages was brought to the city, where it influenced other people. The knowledge can also be taken home and transform other spaces in the city. Knowledge is also a component of the emotional ties that many farmers have to their plots. Farmers gain pride, a sense of belonging, and a place to teach their children lessons. All of these have the potential to have lasting impacts both on and off the farm.

Many farmers grew up in villages. While villagers usually grew staple carbohydrate crops to sell, families also grew vegetables for subsistence. Many farmers were able to bring knowledge from their villages and adapt it to the context of the city. According to one farmer:

> His parents were farmers. Farming was inside his blood. He did farming since he lived at home, so this was a very common thing for him. He thought the only thing he could do was farming, so he decided to do it just to add to his income [as a guard] because the salary was very low. (James, Drive-In)

Some farmers, however, did not learn to farm in their village. They were reliant on the knowledge that was brought and expanded by other farmers. In this case, their interactions with the farm provide them knowledge through their social ties.

> He grew up in Dar, he never thought that one day he would farm. At first when he came here and tried to farm, these women [the other farmers] were laughing at him. They would say: 'That's not how you plant the seeds. You're taking them out' because he used to dig them up again after planting. Mary used to help him plant and [apply fertilizer and pesticides], but now he's an expert. (Philip, Drive-In)

Some farmers had home gardens where they started growing some vegetables and then began to make a living from it and started larger-scale farming. Others were open-space farmers and then found areas at home in which they could grow some extra crops to supplement their household diet and incomes (Fig. 12.3). The knowledge gained at the farm was transferred to the urban home, and vice versa.

Emotions and identities flow through the practice of farming. Many farmers are proud of what they bring to the city and how farming has given them something to do that helps them build a house or support their family. Beyond these material advantages, however, are several other benefits that may seem less tangible, but are intricately woven into daily experiences.

> She says that even if she's angry, even if she's come [to the farm] with so much anger in her heart, whenever she sees all this green, her anger goes away. She relaxes when she gets to this place. (Marie, Drive-In)

> Farming has helped him become a decent person . . . Like he can get a nice woman, and a woman he wants, even though he's farming. He has the income, he has the respect to be called a man. He's proud of that . . . Some people underestimate the farmers, but he thinks that the farmers would be an example of the men who are developing right now. (Said, Ubungo)

Fig. 12.3. Photo-voice, Tazara. *From focus group*: The farmer explained that planting potato leaves on his plots had given him the idea to plant them in this unused space near his house. These are used mostly for his family, but they are also sold to neighbours, which supplements his household income.

They are happy because they feed the people of Dar es Salaam. So many people don't have to go far away to buy these things and because they plant many different kinds of vegetables, so many people are depending on them. [They are expressing a sense of pride] because they have everything here. (Phyllis and Anthony, Tazara)

People are relaxed, happy and proud as a result of these spaces and their practices within them. This is not to deny that it is hard or arduous work, but to emphasize that it is more than that. While many farmers understand that people look down on them, I found that despite this perception, farmers were proud of themselves and what they produced.

Families rarely helped on the farm, and for many of the respondents, the idea that children would help was even laughable: 'city children do not farm!' (Rachel, Msimbazi). Yet some used farming to provide their children with skills that they could fall back on if they did not find another job (Fig. 12.4). Others used the farm to teach their children lessons about finance or about hard work. They wanted to provide the children with tangible skills so that they always had something to eat. If the children did find better employment, farming, the parents argued, would provide them with other skills they would need in life.

While many of these examples may seem mundane or intangible, they are real experiences that contribute to the persistence of farms in the city. Farms are safe places where family cohesiveness is constructed and where people can rely on social connections to assist them in times of hardship. Farming creates pride and social connections that will give people the confidence that influences other components of their lives. Interactions on farms can be brief, but they are meaningful to individuals and groups as they share experiences and information. Farms provide a way to connect families and friends in places in meaningful ways that reinforce these spaces of vegetable production. These interactions are difficult at best to trace in a study such as this one, but they provide the foundation for examining the multiple ways that pathways intersect and

Fig. 12.4. Photo-voice, Oyster Bay. 'This picture shows my son picking *mchicha*. In this picture I am helping him to understand how the *mchciha* business is done and how to pick *mchicha* so that he can also do this one day in his daily life and be a good example to others. This picture shows happiness' (Ellen).

influence the lives of the people in these spaces and, in turn, reinforce the importance of these spaces in urban life.

12.7 Conclusion: Towards New Engagements with Urban Farms

This chapter has been about the movement of people and social relations within and through urban open-space farms. Too often these mundane practices, intangible interactions, and everyday materialities are taken for granted, and parts of them are analysed as if they represent a whole. But this is not possible. Urban life is too dynamic, fragmented and interconnected to truly comprehend. My approach better reflects the approach to cityness, as presented by scholars such as Jennifer Robinson (2002) and AbdouMaliq Simone (2010). It is through this understanding of a city that everyday interactions and movement can expose how larger structural issues play out on the ground, often in ways that seem intangible, but in fact are very real parts of people's lives.

This chapter begins a conversation about how 'urban work gets done' on farms (Simone, 2010, p. 2) and emphasizes the processes by which this happens. Urbanites actively negotiate and shape their spaces, take chances, and experiment to survive and indeed *live* in the city. Urbanists (planners) use buildings, roads and other infrastructure to make sense of the city; to stabilize relationships between bodies, institutions and events. They attempt to control interactions and relationships and facilitate exchange and the flow of capital. However, there is no way to fully control what people do and how they interact, whether in Dar es Salaam or London. Practices, materials and relationships all have a life beyond what urbanists attempt to control. Being open to these processes, and learning to see and understand them in their specific spatial contexts, has significant implications for how we understand cities to be made and remade. It further has the potential to reorient our approach to urban form and function in the so-called Global South and Global North.

What can be done with this approach to the city? It allows us to see how connections are built, both between people and places, through material practices and experiences. It allows us to see how people live with other people, how they engage in multiple social and economic networks for information and monetary exchange, how they spread out their investments to better ensure success, how they define themselves in relation to others, how they make claims to places, and the roles that places such as farms play in the larger story of urban development. This approach recognizes the structures that influence the flow of people and materials (e.g. economic circumstances that led people to farming), but focuses more on the agency of people as they act in material and immaterial ways to create their own sense of stability and security in the city. People I worked with were open to new possibilities, not just out of desperation, but for pride, a feeling of success, a shared sense of commonality with the people around them, and also because they needed to be.

A perspective such as this one is needed for the cities of sub-Saharan Africa. There have been endless development initiatives, workshops, meetings, discussions and programmes intended to alleviate poverty, yet poverty still seems to define these cities. People continue to arrive, putting increased pressure on development and government resources. Rather than focusing on the dystopian images of the city, the poverty and marginalization, a new focus is needed that highlights how these cities work, how people survive and even thrive in them, and why and how people continue to move there in spite of apocalyptic visions of urban life. What is needed is a realistic vision of their urban futures.

Yet it is also important to understand cities in more developed regions, too. Research on urban agriculture in the Global North tends to focus on social and ecological movements, the maintenance of identity and the desire to 'take back' urban space. Why are these motivations only ascribed to people in specific regions? How can these lessons of flow and intersection be used to better understand urban agriculture as a global spatial phenomenon? Some of the universality of urban experience can be recognized, and the way this is spatialized in specific places, in order to begin to understand the motivations, processes and development trajectories that define cities and urban form in different regional contexts.

References

Andrew, F. (2008) Vegetables in city gardens declared as silent killers. *The Guardian*, 4 July.
Armstrong, A. (1986) Colonial and neocolonial urban planning: three generations of master plans for Dar es Salaam, Tanzania. *Utafiti* 12, 43–66.
Brennan-Horley, C. (2010) Mental mapping the 'creative city'. *Journal of Maps* 6, 250–259.
Brennan-Horley, C. and Gibson, C. (2009) Where is creativity in the city? Integrating qualitative and GIS methods. *Environment and Planning A* 41, 2595–2614.
Brennan-Horley, C., Luckman, F., Gibson, C. and Willoughby-Smith, J. (2010) GIS, ethnography, and cultural research: putting maps back into ethnographic mapping. *The Information Society* 26, 92–103.
Brenner, N., Madden, D. and Wachsmuth, D. (2011) Assemblage urbanism and the challenges of critical urban theory. *City* 15, 225–240.
Briggs, J. and Mwamfupe, D. (2000) Peri-urban development in an era of structural adjustment in Africa: the city of Dar es Salaam, Tanzania. *Urban Studies* 37, 787–809.
Crang, M. (2003) Qualitative methods: touchy, feely, look-see? *Progress in Human Geography* 27, 494–504.
Darbyshire, P., MacDougall, C. and Schiller, W. (2005) Multiple methods in qualitative research with children: more insight or just more? *Qualitative Research* 5, 417–436.
Dirksmeier, P. and Helbrecht, I. (2008) Time, non-representational theory and the 'Performative Turn' – towards a new methodology in qualitative social research. *Forum: Qualitative Social Research* 9, article 55.
Dongus, S. (2001) Urban vegetable production in Dar es Salaam (Tanzania): GIS-supported analysis of spatial changes from 1992–1999. In: Drescher, A. and Mackel, R. (eds) *APT-Reports 12: Use of Resource Niches in African Ecosystems – New Research Results from Tanzania, the Chad and Namibia*. Freiburg, Germany, pp. 100–144.
Egziabher, A., Lee-Smith, D., Maxwell, D., Memon, P.A., Mougeot, L.J.A. and Sawio, C. (1994) *Cities Feeding People: An Examination of Urban Agriculture in East Africa*. International Development Research Centre, Ottawa, Canada.
Elwood, S. (2006) Beyond cooptation or resistance: urban spatial politics, community organizations, and GIS-based spatial narratives. *Annals of the Association of American Geographers* 96, 323–341.
Foeken, D. (2006) *'To Subsidise My Income:' Urban Farming in an East African Town*. Koninklijke Brill NV, Leiden, The Netherlands.
Freeman, D.B. (1991) *A City of Farmers: Informal Urban Agriculture in the Open Spaces of Nairobi*. McGill Queen's University Press, Quebec, Canada.
Gallaher, C., Kerr, J.M., Njenga, M., Karanja, N.K. and WinklerPrins, A.M.G.A. (2013) Urban agriculture, social capital, and food security in the Kibera slums of Nairobi, Kenya. *Agriculture and Human Values* 30, 389–404.
Green, A., Shuttleworth, I. and Lavery, S. (2005) Young people, job search and local labour markets: the example of Belfast. *Urban Studies* 42, 301–324.
Kato, L. (2010) Vegetables grown in Msimbazi unfit for human consumption. *Daily News*. Dar es Salaam, 12 July.
Larkin, P.J., Casterlé, B.de and Schotsmans, P. (2007) Multilingual translation issues in qualitative research: reflections on a metaphorical process. *Qualitative Health Research* 17, 468–476.
Latham, A. (2003) Research, performance, and doing human geography: some reflections on the diary-photograph, diary-interview method. *Environment and Planning A* 35, 1993–2017.
Leslie, J.A.K. (1963) *A Survey of Dar es Salaam*. Oxford University Press, New York, USA.
McFarlane, C. (2011a) The city as assemblage: dwelling and urban space. *Environment and Planning D: Society and Space* 29, 649–671.
McFarlane, C. (2011b) Encountering, describing and transforming urbanism. *City* 15, 731–739.
McIntyre, A. (2003) Through the eyes of women: photovoice and participatory research as tools for reimagining place. *Gender, Place and Culture: A Journal of Feminist Geography* 10, 47–66.
McLees, L. (2011) Negotiating access to land for urban farming in Dar es Salaam, Tanzania: histories, benefits and insecure tenure. *Journal of Modern African Studies* 49, 601–624.
McLees, L. (2012) A postcolonial approach to urban studies: interviews, mental maps and photo voices on the urban farms of Dar es Salaam, Tanzania. *The Professional Geographer* 65, 283–295.
Myers, G. and Murray, M. (2006) Introduction: situating contemporary cities in Africa. In: Murray, M. and Myers, G. (eds) *Cities in Contemporary Africa*. Palgrave Macmillan, New York, USA.
Nowell, B., Berkowitz, S., Deacon, Z. and Foster-Fishman, P. (2008) Revealing the cues within community places: stories of identity, history, and possibility. *American Journal of Community Psychology* 37, 29–46.

Packard, J. (2008) 'I'm gonna show you what it's really like out here': the power and limitation of participatory visual methods. *Visual Studies* 23, 63–77.

Page, B. (2002) Urban agriculture in Cameroon: an anti-politics machine in the making? *Geoforum* 33, 41–54.

Pavlovskaya, M. (2009) Non-quantitative GIS. In: Cope, M. and Elwood, S. (eds) *Qualitative GIS: A Mixed Methods Approach*. Sage Publications, Los Angeles, California, USA.

Rakodi, C. (2002) A livelihoods approach: conceptual issues and definitions. In: Rakodi, C. and Lloyd-Jones, T. (eds) *Urban Livelihoods: A People-centered Approach to Reducing Poverty*. Earthscan, Sterling, Virginia, USA, pp. 3–22.

Robinson, J. (2002) Global and world cities: a view from off the map. *International Journal of Urban and Regional Research* 26, 531–554.

Robinson, J. (2006) *Ordinary Cities: Between Modernity and Development*. Routledge, London, UK.

Roy, A. (2011) Slumdog cities: rethinking subaltern urbanism. *International Journal of Urban and Regional Research* 35, 223–238.

Simone, A. (2004) *For the City Yet to Come: Changing African Life in Four Cities*. Duke University Press, Durham, North Carolina, USA.

Simone, A. (2005) Introduction: urban processes and change. In: Simone, A. and Abouhani, A. (eds) *Urban Africa: Changing Contours of Survival in the City*. Zed Books, London, UK, pp. 1–26.

Simone, A. (2008) The politics of the possible: making urban life in Phnom Penh. *Singapore Journal of Tropical Geography* 29, 186–204.

Simone, A. (2010) *City Life from Jakarta to Dakar: Movements at the Crossroads*. Routledge, New York, USA.

Slater, R. (2000) Using life histories to explore change: women's urban struggles in Cape Town, South Africa. *Gender and Development* 8, 38–46.

Slater, R. (2001) Urban agriculture, gender and empowerment: an alternative view. *Development Southern Africa* 18, 635–650.

Temple, B. and Young, A. (2004) Qualitative research and translation dilemmas. *Qualitative Research* 4, 161–178.

UN-Habitat (2010) *Citywide Action Plan for Upgrading Unplanned and Unserviced Settlements in Dar es Salaam*. UN-Habitat, Nairobi, Kenya.

URT (United Republic of Tanzania) (1997) *Agriculture and Livestock Policy*. Ministry of Agriculture and Co-operative Development, Dar es Salaam, Tanzania.

URT (United Republic of Tanzania) (2000) *National Human Settlements Development Policy*. Ministry of Lands and Human Settlements Development, Dar es Salaam, Tanzania.

Young, L. and Barrett, H. (2001) Adapting visual methods: action research with Kampala street children. *Area* 33, 141–152.

13 Relying on Urban Gardens for Survival within the Building of a Modern City in Colombia

Colleen Hammelman*
University of Toronto-Scarborough, Toronto, Ontario, Canada

13.1 Introduction

Interest in urban agriculture (UA), at various scales, has been growing around the world (Mougeot, 2005; Colasanti *et al.*, 2012; Hampwaye, 2013). This interest includes a focus on the motivations and benefits of UA in relation to urban planning. This chapter adds to the discussion through investigating how displaced women in Medellín, Colombia, utilize urban gardens as a survival strategy in the face of tenuous support from local government. It contributes to literature on UA by providing a clear example of the ways in which urban planning (that is influenced by global political economic systems) impacts on a critical survival strategy: growing food. Significant research on UA has been conducted to date (Guitart *et al.*, 2012) that includes considerations of insecure land tenure (e.g. DeKay, 1997; Irvine *et al.*, 1999; Schmelzkopf, 2002; Bryld, 2003). However, Guitart *et al.* (2012) note that much of this research occurs in low-income areas of postindustrial US cities, leading the authors to ask whether similar concerns of land tenure being threatened by development occur in other countries. This chapter seeks to show that similar concerns do occur in other countries and are, importantly, driven by political economic influences on urban planning faced by cities around the world.

This research relies on interviews with 41 displaced women living in Medellín to discuss urban gardening as a survival strategy, enabling them to grow food for consumption, exchange and income. These gardens also play an important role in the Medellín government's efforts to build an image of the modern city as 'inhabited by life' (ACI Medellín, 2014; City of Medellín, 2014). The government's support for the gardens is tenuous, as it is tied to 'worlding' strategies of gaining international investment that can overlook the immediate social justice concerns of gardeners. As discussed in Chapter 1 in this volume, worlding strategies seek to make the city safe for international capital flows through projects that further integrate cities into global structures and ideologies (Roy, 2011). In particular, the gardens in Medellín are permitted under the guise of supporting displaced families, but also for preventing further geographic expansion of neighbourhoods populated by poor residents. The gardeners grow crops on the land with short-term leases, under constant fear of losing these critical resources at the whim of a changing government. This chapter contributes empirical evidence of the ways in which urban planning strategies (social urbanism in Medellín), embedded in a global political economy (via worlding), impact on the efficacy of UA as a survival strategy for low-income urban migrants.

---

*E-mail: c.hammelman@utoronto.ca

To begin, I provide a brief review of literature, addressing the relationship between UA as a survival strategy and urban development planning. Then I illustrate this relationship through the case of one neighbourhood of displaced residents in Medellín, with particular attention paid to the influence of local planning priorities carried out as worlding strategies, before concluding with broader implications for practice and research.

13.2 Urban Agriculture as a Food Insecurity Coping Strategy

UA generally involves the production of agricultural goods by urban residents (Zezza and Tasciotti, 2010). Some forms of UA include small commercial farms and community-supported agriculture, community gardens in which food or flowers are grown in 'open spaces' and managed by members of the local community (Guitart et al., 2012); backyard or household gardens (WinklerPrins, 2002; Taylor and Lovell, 2014); or some hybrid of these (Mok et al., 2013). As discussed in Chapter 1, deciding on which UA terminology to use can be difficult, as terms such as gardening, farming and agriculture all invoke different ideas about the scale and purpose of cultivation. This task becomes even more difficult when working in more than one language. The participants in this research are Spanish-speaking and generally refer to their UA spaces as *huertas*, which translates loosely to English as a smaller-scale garden. When I occasionally used the Spanish term for farm (*finca*) instead, I was quickly corrected by participants, who considered their work not to be on the scale of farming. Using the terminology of gardens, the research discussed in this chapter focuses in particular on a hybrid model of community gardens in which food is cultivated on public land by households, but lessons from this case are also evident at other scales of UA.

In the Global South, UA is frequently carried out for subsistence (Maxwell, 1999; Bryld, 2003; Lerner and Eakin, 2011) and the proportion of urban dwellers involved in agricultural activity reaches 50% in some Latin American countries (Redwood, 2009; Zezza and Tasciotti, 2010). Alongside home gardens, the reliance on community gardens to supplement food budgets of low-income families is also a growing consideration in the Global North (Maxwell, 1999; Ackerman et al., 2014). Community gardens are beneficial for providing greater accessibility and affordability of food, as well as self-reliance (Redwood, 2008; Corrigan, 2011; Mok et al., 2013). UA can support families to minimize risk of food insecurity, address cultural needs and traditions associated with agriculture and/or migrants' place of origin, organize for collective rights, and obtain both cash and non-cash capital, as families sell surplus goods on local markets (Lerner and Eakin, 2011; Taylor and Lovell, 2014; Tornaghi and Van Dyck, 2014). While UA may only provide supplemental food for a family, the corresponding savings for food budgets can be significant (Brown and Carter, 2003; Redwood, 2008). It is also regularly carried out by women in order to obtain cheap produce and grow sustainable, chemical-free food (Bryld, 2003; Buckingham, 2005; Redwood, 2008). However, while families may turn to UA in times of economic crisis, they struggle to access enough land to support the entire family and to obtain a complex set of inputs (materials, knowledge, credit) thus limiting the availability of UA as a survival strategy for those families most in need (Bryld, 2003; Frayne et al., 2014). Various challenges with UA include: insecure land tenure; risk of food contamination from chemicals present in soils and water; surface runoff from agrochemicals; and limited understanding of the carbon footprint of UA (Mok et al., 2013; Hamilton et al., 2014). Many people growing food in cities around the world do not own the land they garden, as UA is often taken up on public land (possibly with short-term leases) or private land with conflicting ownership (Brown and Carter, 2003; Redwood, 2009). This may drive gardeners to plant short-duration (less sustainable) crops or limit their investment because they fear eviction (Bryld, 2003; Redwood, 2008).

City policy makers and planners in both the Global North and Global South have taken multiple, conflicting actions relating to UA. Historically, UA has been embraced by cities in times of crisis (McClintock, 2010; Frayne et al., 2014) and dismantled as an obstacle to development and modernization as economies improve. In particular, urban planners in African cities have regularly sought to remove UA, due to concerns about public health as well as notions about building progressive, modern city images (Redwood, 2008; Morgan, 2009; Hamilton et al., 2014). In contrast, UA has a longer history of government support in Latin American cities that have undergone periods of land reform, especially in Cuba (Moskow, 1999; Galt et al., 2014; Hamilton et al., 2014). Government/city managers are generally less supportive of UA when it is perceived as having only marginal importance to the urban economy, runs counter to modernization, poses a potential health hazard, and takes up space that could be used for urban housing or other city modernization projects (Bryld, 2003).

However, these perspectives are shifting in some parts of the world – several cities in the Global South are seeing new policies in favour of UA and food-policy councils in parts of the Global North increasingly support UA (Redwood, 2008). Support for UA is also growing as part of urban sustainability agendas (see Chapter 1, this volume). However, in cities where land is at a premium, such as Medellín, visions of sustainability may come into conflict with UA practices, further threatening the work of marginalized communities. Despite the importance of UA for survival strategies among displaced people in Medellín, they experience concerns regarding uncertain land tenure in the face of changing planning priorities. This chapter contributes to the literature on UA by illustrating the complexity of land tenure and UA as a survival strategy, particularly in response to local government planning strategies for building a modern city.

13.3 Displaced Families Face Extreme Poverty, Marginalization and Food Insecurity in the Modern City

The research discussed in this chapter derives from in-depth interviews with 41 displaced women in Medellín, Colombia.[1] Medellín is located in a valley in the Andes Mountains, where it enjoys very mild weather year round, and, with a population of 2.5 million in the city, is the second-largest city in Colombia. Like many other urban areas in Latin America, its population grew rapidly in the middle of the 20th century, tripling between 1951 and 1973 (Scruggs, 2014).

The municipal government was unable to adequately respond to this rapid growth and provide sufficient housing and services to all of its residents. As a result, since the 1950s, neighbourhoods of informal settlements with limited public services, infrastructure and public space continue to sprawl up the mountainsides surrounding the central city (BID, 2009; Calderon, 2012). These neighbourhoods are largely inhabited by displaced persons violently forced out of the countryside into urban Medellín as a result of the decades-long civil war. The majority of the 200,000 displaced people living in Medellín fled violence in rural communities for the city, where they find themselves at significant risk of food insecurity in the urban, cash-based economy that does not value their agricultural skills (Fadnes and Horst, 2009). Yet internally displaced persons in Colombia are more likely than the general population to be unemployed or to be employed in the informal market because of their rural background, lack of education and absence of social networks (Aysa-Lastra, 2011). Disconnected from their home communities, families and livelihoods, displaced families often face extreme poverty, marginalization, risk of human rights abuse and food insecurity (Wyndham, 2006; Fadnes and Horst, 2009; IDMC, 2012). In 2010, 98.6% of displaced people in Colombia lived below the poverty line and 82.6% were classified as living in extreme poverty (compared to 29.1% and 8.7% respectively for the general Colombian population) (IDMC, 2012). Displaced persons in urban areas typically have less access to housing and livelihoods than their low-income counterparts (IDMC, 2014). While displaced persons can register with government agencies as displaced, entitling them to financial reparations under a 2011 Victims Law, few actively receive these reparations. Because displaced people in cities often meld into communities of urban poor and desire to stay invisible (for fear of renewed violence), there is a significant lack of understanding about how urban displacees manage their lives; accordingly, their voices are rarely heard by those in positions of power (Tibaijuka, 2010; Crisp et al., 2012).

This new reality significantly impacts displaced individuals' relationship with food. Similar to findings from other scholars on the food insecurity of displaced persons in Colombia

(Consuelo Carrillo, 2009), the majority of participants in this research ate fewer than three meals per day, skipped meals or specific food types (such as meat), and/or often relied on emergency food providers or begging to acquire sufficient food. A majority had to undertake new strategies to provide enough food for themselves and their families, such as collecting leftover or discarded food from markets, trekking over many kilometres to beg for food, finding ways to grow food within extremely limited space and time constraints, and building relationships with neighbours. These coping strategies are arduous, time-intensive, and take away from other household responsibilities.

Where space is available, displaced women grow food through urban garden projects on public land near their homes. This survival strategy has proven to be particularly effective for those with access to nearby land, but with tenuous government support. Medellín has undergone a remarkable transition from being one of the most dangerous cities in world in the 1990s to being named the International City of the Year in 2012 and hosting the United Nations' World Urban Forum in 2014. From a context of violence, ungovernable spaces and poverty, the Medellín government began an ambitious planning project in the mid-1990s, seeking to transform it and its international reputation into a 'modern city' that is safe and attractive for foreign investment, tourists and residents. These planning efforts can be understood within global urbanizing processes and neoliberal productions of space, referred to by Roy (2011) as 'worlding' processes. Worlding of cities involves the neoliberal remaking of space to be safe for capital flows. According to Roy (2011), these processes, whether the result of elite aspirations or instances of worlding from below, often serve to consolidate postcolonial sovereignty and territory in the name of building a world-class city. In this context, the city is a platform of market rule and state practices to regulate and deregulate space in accordance with neoliberal ideology. Worlding processes that bring cities further within the fold of global structures can be evident in various aesthetic, political, economic and discursive projects (Goonewardena, 2005; Roy, 2011). The Medellín government's worlding projects are most evident in its quest to be recognized as a modern international city and its transformative social urbanism agenda. This agenda has direct implications for urban gardening projects among displaced residents in the city, and these projects have been able to temporarily garner governmental support under the guise of halting informal settlement growth, thus contributing to a more aesthetically pleasing space for international visitors. Yet, this governmental support wanes when the project's value in the modern city comes into question. The remainder of this chapter illustrates the ongoing tension between worlding local government policies focused on garnering international investment and UA practices focused on improving the food security of the city's most vulnerable residents.

13.4 Utilizing Urban Gardens as a Survival Strategy

In one Medellín neighbourhood that was a focus of this research, an urban agriculture project has significantly augmented female residents' food security while also contributing to aesthetic improvements. In Chapter 1, WinklerPrins discusses the emerging role of UA in both the Global North and Global South for making the city a more 'palatable experience', particularly with regard to 'greening' cities to improve aesthetics, quality of life and sustainability. The urban gardens discussed in this chapter fill this role to the extent that they improve the quality of life for residents by addressing some food insecurity concerns. The project also improves the aesthetics of the city under the auspices of limiting the growth of informal settlements. However, the aesthetic ideal is valued more for the local government's worlding strategies. As such, there is a growing tension with other sustainability-focused projects that similarly seek to improve the international image of the city.

Most of the displaced people in the city came from agricultural towns and are more accustomed to growing their food instead of purchasing it all in stores. Some continue to foster a connection to the countryside through growing what they can in the small space available near their homes; primarily onions, herbs or flowers in pots. This is difficult, however, as there is very little land available for growing food in these densely populated neighbourhoods. Yet, in one neighbourhood on the eastern edge of the city the residents were able to secure a lease from the government to open space on a nearby hill in order to build urban gardens. This area is inhabited almost exclusively by displaced persons, who have built their homes high up the hillside, with limited access by bus to the central business district, in what is considered to be an 'invasion' or an unauthorized, unplanned neighbourhood.

In 2011, community leaders and residents came together to propose an urban gardening project. Two community leaders saw that several families were struggling to obtain sufficient food and available resources were largely located a long distance outside the neighbourhood. The community leaders identified public land nearby that was not being used. From this context and in partnership with the community, the leaders developed a proposal for an urban garden project. The community applied to local organizations (particularly one that supports victims of the war) and gained a lease to cultivate the land from local government. The garden project began with the support of a local university, government organizations and non-government organizations, and there is an ongoing relationship with some of these partners that assist the gardeners with training, fertilizer and starter plants, and visibility. The residents renew the lease with the government every year, under the stipulation that the land will not be used for anything but agriculture.

In this space, residents built 40 garden plots that are approximately 1 m × 5 m, with raised beds. The families grow a wide variety of crops, including cabbage, carrots, maize, lettuce, onions, spinach, tomatoes and herbs. The gardens are located close to the families' homes – no more than a 6-minute walk away. The women interviewed reported spending anywhere between 1 day per week, or one hour each day, to all weekend or 3–4 hours every morning in the gardens weeding, planting, watering and harvesting. The amount of time spent in the garden was influenced by the time of year and weather, the health of the women, employment status, and family care responsibilities. While some participants expressed concern that other people from the neighbourhood would steal their crops, they generally saw this as an inevitable part of having a community garden space. These gardens represent a hybrid model of community gardening with management shared between two community leaders, and residents who grow food only for their households. Each family has its own individual plot and is responsible for deciding which crops to plant, tending to those crops, and consuming, selling or exchanging the goods produced.

The crops grown in these gardens are used for consumption, exchange and sale. As found in other cases (Brown and Carter, 2003; Redwood, 2008; Corrigan, 2011; Gallaher et al., 2013; Mok et al., 2013), it provides an important barrier against food insecurity. From the 41 women interviewed for this research in Medellín, only 15 either had a plot in the garden or had a family member with a garden plot. Those who did not lease a garden plot either lived in another neighbourhood, preferred to share with a family member, or opted not to have a plot. The research participants indicated that space remained available for other families to garden in the neighbourhood who would simply need to coordinate with the community leaders managing the space.

Of the 15 women who had community garden plots, nine are the heads of their household, and three are married. Three do not have their own garden, but have a family member with a garden they can access, and of those three, one is the head of household. Importantly, all 15 women reported much less difficulty with food insecurity and much greater satisfaction with the quality of food they could obtain. For example, one woman indicated that because she has a garden, she does not have to buy vegetables and has access to better-quality vegetables and a bigger variety than before she had a garden. This is similar to research elsewhere, which found that gardens can provide an important supplement to food budgets (Gallaher et al., 2013; Hampwaye, 2013). Additionally, several of the women with a garden reported that all of their produce came from their garden, reducing their expenses on food and transportation to reach affordable food. One gardener said:

> I have food for my family because of the garden. I have had it for about five years now and can get things I couldn't get before. It was very difficult to obtain everything I needed. Having the garden means I don't have to buy vegetables and I have a lot of things I couldn't eat before... It is really important to not have to spend money on these things. (Agata,[2] Medellín, 43 years old)

The description of available produce included much more variety and satisfaction among women with a garden than those without. The sentiment that the gardens are important for their survival was shared by all of the women interviewed in this research.

While some families only use the gardens for subsistence, more than half also seek economic benefits via exchange and sale of surplus produce. If a family grows more food than they can consume, or the food is going to spoil before they can eat it, they will exchange it with neighbours, local stores or the food bank, and/or sell it at local markets. 'The things from the garden help a lot because sometimes I don't have the money to buy (the things I need) and I can take the things from the garden to the store to sell' (Olivia, Medellín, 32 years old). Several women indicated that this option is very important when they don't have enough money to buy goods such as rice, sugar or cooking oil, household items such as toothpaste or shampoo, or pay for public services such as electricity.

There are also various non-economic benefits arising from participation in the gardens. Several women expressed the importance of the gardens for connecting them to the land and the agricultural customs they had in the countryside. They also argued that the vegetables coming from the garden were better for their health and the environment's health than those found in the stores. In particular, they were pleased with the quality of vegetables that are grown organically, without genetically modified ingredients, and cultivated by their hands (see also Chapter 15, this volume). Finally, as has been reported in the literature (e.g. Chapters 5 and 8, this volume; Christie, 2004; Redwood, 2008; Gray et al., 2014; Tornaghi and Van Dyck, 2014), some of the women discussed the importance of the gardens for connecting with neighbours and providing an avenue to organize on other issues important to the community.

These gardens are a tremendous survival strategy for this community. The women rely on the gardens to meet their daily needs and are identified with the farms. Yet a majority of the women expressed concern that the government support for the gardens, and the lease to the land, might easily disappear if the city's planning priorities change. Therefore, the next section moves beyond the critical food resources provided by the gardens to discuss the links between the local government's social urbanism planning agenda and the long-term prospects of the urban gardens.

13.5 Social Urbanism and Urban Gardens in Medellín

Residents in this eastern neighbourhood in Medellín successfully tapped into the government's planning priorities to gain support and land for their urban gardens. The Medellín government's planning efforts, known as social urbanism, have gained international attention for uncharacteristically providing significant investment in the poorest and most violent neighbourhoods. The social urbanism agenda is an approach to city planning initiated by former Medellín Mayor Sergio Fajardo (2004–2007). It integrates new public spaces and public facilities with social programmes and the active involvement of local communities. It also seeks to reduce poverty and crime through state investment in social infrastructure (libraries, schools, public space). Fajardo pursued urban upgrading by enhancing mobility, the environment, housing and public space (Fajardo, 2007; Sanchez, 2010; Calderon, 2012).

Here I argue that this social urbanism agenda, in which the urban gardens are implicated, exemplifies worlding strategies as it seeks to modernize and beautify the city so as to remake its international image. While the rhetorical focus was on improving the most deprived neighbourhoods of the city, its practical motivation was making the city safe for international investment (Sanchez, 2010; Brand and Davila, 2011; Calderon, 2012; Turok, 2014). This effort sought to change the aesthetics of the city, particularly through limiting further growth of informal settlements. The negative externalities caused by problems in the poorest and most violent neighbourhoods were seen as a limitation to improving the city's international competitiveness. Fajardo was quoted in several publications emphasizing the need to transform Medellín into a vibrant city that is friendly for international business – 'once again taking its place as a business hub and tourist destination' (Hylton, 2007, p. 88) – and the city is now described by some journalists as 'media-saturated, image conscious...

[and] improving at breakneck speed' (Hylton, 2007, p. 72). Perhaps also signalling the business-focused (as opposed to solely social-focused) intentions is that funding for the social urbanism projects (that Sanchez (2010) estimates as high as 1 billion Colombian pesos or almost US$500 million) comes from a 'neoliberal constellation of corporate partners and powerful multinationals' (Scruggs, 2014, p. 3). The joint financing schemes involve contributions from the municipal and national governments, mutual benefit funds, private companies such as the major public services company EPM, and the Inter-American Development Bank (Sanchez, 2010). It is also evident in the short-term support of UA on the part of the city, so long as the gardeners use the space to prevent further growth of their informal neighbourhoods.

The social urbanism model has had success in transforming much of the city and certainly its international reputation. Since 1991, the murder rate has fallen more than 80%, residents express a renewed sense of pride and ownership of their city, the city is now the recipient of multiple awards for its resurgence and social programmes, and foreign investment is growing (Hylton, 2007; Brand and Davila, 2011; Fukuyama and Colby, 2011; Scruggs, 2014). Many believe this can be attributed in part to this new style of planning. To this end, Brand and Davila argue that some of the greatest benefits of investments in mass transit are the symbolic value of the highly visible infrastructures and aesthetic experiences that 'create a feeling of inclusion and integration into the 'modern' city, help develop local pride and promote individual self-esteem' (Brand and Davila, 2011, p. 658).

However, there remain other more mixed results leading some to perceive the transformation as a 'half-miracle' (Fukuyama and Colby, 2011). Residents in lower-income communities (including those in this study) are particularly vocal in their critique of this agenda as serving the goal of improving the city's image and middle-class interests, while significant inequality remains in the city. While there is growing wealth in the downtown areas, those living in the periphery continue to experience violence, poverty, precarious housing and limited services. As a result, there is growing intra-urban displacement occurring in the city. In the first 10 months of 2013, an estimated 5000 people were involuntarily displaced from their homes within the city (Turok, 2014).

Importantly, the social urbanism agenda is critiqued by residents, particularly the urban gardeners in my research, for not directly addressing the needs of poor residents through more traditional social services. The sentiment among several women interviewed for this research is that soccer stadiums and libraries are nice, but not as important as finding enough food to eat or being able to pay for water, electricity and transport. They distrust the government because, while it is investing significant sums of money into these projects, they struggle to obtain promised social support to displaced persons. Many displaced women indicated that the government policy is to provide displaced persons with financial assistance every 3 months, but the support rarely materializes. While the focus of the social urbanism agenda is purported to be those living in poverty in violent sectors (largely displaced neighbourhoods), it does not regularly take into account the voices of those communities. As a result, planning projects regularly risk exacerbating obstacles to survival in these communities. This divide between the everyday needs of these residents and the bigger investments of the city has led many residents to question the priorities of the government. There is also concern that the administration has overstretched its resources and is more concerned with its image than with really addressing the difficult realities of many of its residents (Brand and Davila, 2011). As such, the worlding processes evident in Medellín's social urbanism agenda have fostered impressive transformation in the city, while also reproducing contexts of poverty and marginalization.

This tension between the goals of the social urbanism agenda and the everyday needs of the city's poorer residents is particularly evident in ambivalent government support for the urban gardens. As discussed above, in order to attract foreign capital by remaking its city as modern and world-class, the government seeks also to make it an aesthetically pleasing and ordered space. This includes halting further growth of informal settlements high up the mountainsides and preserving the 'city's verdant edges' (Scruggs, 2014, p. 4). The edges of the neighbourhood rise up an iconic peak in the city that the government seeks to preserve (Barrows et al.,

2013). Understanding the importance of this, the community proposed their urban gardening project to the local government by suggesting that it would protect the existing open space from further growth of informal settlements. Instead they would plant trees and build gardens in the space, while also addressing their food security needs. The neighbourhood exists on the edge of existing housing settlements occupied by residents for many years, and thus this proposal garnered significant support from the community at the time. The utilization of UA for preserving green space is evident in other cities in both the Global North and Global South as well (Bryld, 2003). The government accepted the proposal and leased the land to the neighbourhood specifically for agricultural purposes. It was stipulated that nothing else can be built on the land. As with other examples of urban planning (Blomley, 2004), the government's social urbanism agenda seeks to build a sense of pride and ownership of the city among poorer residents. I argue that these gardens furthered this agenda by enabling residents to invest in their neighbourhood and connect to the land, while also addressing more immediate daily needs of securing food. The residents interviewed in this research were very proud of their gardens, explaining that they are very beautiful and important to them, and that they want to share their project in an effort to stimulate similar projects in other areas and cities. As one participant said: 'We want others to know about the gardens – how good the food is from them, how much they help the community, and how important it is to have them in other neighbourhoods, countries and places' (Agata, Medellín, 43 years old).

13.6 Growing Food Under the Social Urbanism Agenda

While this project seems to be a win for both the government and the neighbourhood, it is a tenuous situation that could change easily. The community is in constant fear that the government will change its planning priorities and revoke their lease to the land. Every 4 years, the municipal government changes and the people worry. Their fears are warranted. They can look to various attempts around the world to evict gardeners in the name of progress (Redwood, 2008; Saed, 2012). Additionally, many of these residents in this research reported being displaced from homes in other parts of the city and/or participating in community meetings about government plans for further development. These plans worry the residents. For example, there is a project to build and preserve a greenbelt that could expand into the land leased to the community. The greenbelt plan is a strategy for sustainability as well as urban containment that addresses issues of growing informal settlements and ecosystem loss (Chelleri and Anguelovski, 2016). Residents in nearby neighbourhoods also view the greenbelt as a strategy to attract more tourists to the city. This greenbelt plan, including a 'protection zone', proposes an area of environmental protection that includes the land the gardeners are using, but not their gardens, thus threatening their leases (Barrows et al., 2013). There are also plans to expand mass transit, build soccer stadiums and other public space projects. The residents interviewed for this research welcome investment in the neighbourhood, but they question the significant amount of money being spent on these projects and the prioritization of these projects. Instead they believe the money might be better spent on improved housing and sidewalks, and on support for community initiatives, such as the gardens.

In this context, many residents question what I describe as the neoliberal worlding of the city that pays lip service to social inclusion, yet emphasizes high-profile, aesthetically pleasing projects such as major transportation projects and a greenbelt, at the expense of addressing more immediate concerns relating to flexible transportation and obtaining food. Indeed, questions of required governmental support for UA, particularly for land, arise when the reasons for supporting these projects lie in garnering international investment. The production of space as either an urban garden in a marginalized community or a greenbelt in the name of sustainability has critical implications for who can use the space in the future. These projects in Medellín demonstrate Roy's theorization of urban planning as 'worlding from above' in which 'urban experiments are closely tied to elite aspirations and the making of world-class cities' (Roy, 2011, p. 10). Medellín's social urbanism planning agenda has successfully gained international attention through circulating around the world as an urban model. Similar to worlding practices elsewhere, with the focus on international attention-grabbing instead of local community needs, these efforts risk legitimizing 'projects of primitive accumulation, often deepening socio-spatial inequality and injustice' (Roy, 2011, p. 11). This is evident in the growing risk to the gardens from projects with a high profile and that secure international attention, such as the greenbelt.

These concerns about land tenure and divisions between marginalized resident concerns and government planning priorities are not specific to Medellín. As mentioned earlier, UA in many cities has been viewed as inconsistent with development priorities (Redwood, 2008, 2009; Morgan, 2009). In both the Global North and Global South, insecure land tenure is a significant challenge for UA (Corrigan, 2011; Guitart et al., 2012). Many people relying on UA do not own their own land. Instead they cultivate food on public land (as is the case in Medellín), lease private land, or farm illegally (Brown and Carter, 2003; Guitart et al., 2012). As little as 20% of UA in the Global South occurs on land owned by the farmer (Bryld, 2003). For those farming on public or private leases, they are often short-term leases (1–5 years) that are threatened by urban development plans, and gardeners risk losing their investments made in the land (Redwood, 2008). In some instances, vacant lots are often made available for gardeners, but only until more lucrative uses for the spaces are identified (Corrigan, 2011). This lack of land tenure (alongside the legal status of poor, migrant residents) often forces UA onto marginal lands (Redwood, 2008). Leases to public land are also threatened by historical perceptions by city planners that UA is a marginal activity counter to urban development and progress (Redwood, 2008, 2009; Morgan, 2009). While the urban gardens utilized by this group of displaced women in Medellín receive support from the local government currently, they worry that their lease to the land could be revoked at any time, thus eliminating a critical survival strategy.

13.7 Concluding Remarks

The urban gardens described in this chapter provide a critical avenue to address the food insecurity of displaced women in Medellín, while also supporting local government objectives of preserving open space and halting the growth of informal settlements. However, similar to UA in other cities, the uncertain future of land tenure for the gardens weighs on the gardeners and raises possible impediments to full investment in the gardens. Conflicting priorities for valuable urban land are evident in cities around the world and solutions are not easily found. In many instances, conflicts arise around using land for much-needed housing or for UA, two important needs for marginalized communities (Schmelzkopf, 1995; Bryld, 2003). Some scholars have offered suggestions, including securing land for the urban poor (Bryld, 2003), valuing land informally controlled by farmers, engaging advocacy groups for increasing security and access to land, and considering tenure along a spectrum of options from squatting to titling (Redwood, 2009). Where giving ownership to farmers may not be feasible, longer-term leases may be a start, as well as better understanding the value of land held outside formal systems. Additionally, where UA is perceived as marginal or of lesser value than other development or housing projects, new economic methodologies need to be developed to identify the value of the land and agricultural products for the gardeners (Redwood, 2009). To support improved land tenure for UA, more research is needed regarding the economic and non-economic benefits of UA to marginalized communities. In Medellín, these gardens and families continue to operate in tenuous circumstances that demonstrate some of the tensions between the priorities of a government seeking to remake its global image and the residents who are trying to survive every day.

Notes

[1] While men are also involved in UA in the research site, the focus of this research is on women because in these communities they are largely the heads of household responsible for obtaining food for their families. When they were in good health, the women were also largely responsible for the UA efforts.

[2] All of the quotes were transcribed and translated into English by the author. Names have been changed to preserve anonymity.

References

ACI Medellín (2014) Medellín wants to be a center of innovative development in Latin America. Agency for Cooperation and Investment of Medellin and the Metropolitan Area (ACI). Alcadia de Medellín. Available at: http://www.medellin.gov.co (accessed 21 October 2014).

Ackerman, K., Conrad, M., Culligan, P., Plunz, R., Sutto, M.P. and Whittinghill, L. (2014) Sustainable food systems for future cities: the potential of urban agriculture. *The Economic and Social Review* 45(2), 189–206.

Aysa-Lastra, M. (2011) Integration of internally displaced persons in urban labour markets: a case study of the IDP population in Soacha, Colombia. *Journal of Refugee Studies* 24(2), 277–303.

Barrows, L., Bu, L., Calvin, E., Krassner, A., Quinn, N., Richardson, J., Sollenberger, G., Irazábal, C. and Buchholz, N. (2013) *Growth Management in Medellín, Colombia*. Columbia University, New York, USA.

BID (2009) *Medellin – Transformacion de una ciudad*. BID – Banco Interoamericano de Desarrollo, Alcaldia de Medellin 2008–2011, Medellin, Colombia.

Blomley, N. (2004) Un-real estate: proprietary space and public gardening. *Antipode* 36, 614–641.

Brand, P. and Davila, J.D. (2011) Mobility innovation at the urban margins. *City: Analysis of Urban Trends, Culture, Theory, Policy, Action* 15(6), 647–661.

Brown, K. and Carter, A. (2003) *Urban Agriculture and Community Food Security in the United States: Farming from the City Center to the Urban Fringe*. Community Food Security Coalition's North American Urban Agriculture Committee, Venice, California, USA.

Bryld, E. (2003) Potentials, problems, and policy implications for urban agriculture in developing countries. *Agriculture and Human Values* 20(1), 79–86.

Buckingham, S. (2005) Women (re)construct the plot: the regen(d)eration of urban food growing. *Area* 37, 171–179.

Calderon, C. (2012) Social urbanism – participatory urban upgrading in Medellín, Colombia. In: Lawrence, R.J., Turgut, H. and Kellett, P. (eds) *Requalifying the Built Environment: Challenges and Responses*. Hogrefe Publishing, Gottingen, Germany.

Chelleri, D.L. and Anguelovski, I. (2016) Are urban sustainability and resilience legitimizing social un-justice? Aesop Young Academics Network blog, 6 April. Available at: https://aesopyoungacademics.wordpress.com/2016/04/06/are-urban-sustainability-and-resilience-legitimizing-social-un-justice (accessed 25 November 2016).

Christie, M.E. (2004) The cultural geography of gardens. *The Geographical Review* 94(3), iii–iv.

City of Medellín (2014) *Medellín, City Inhabited by Life*. City of Medellín, Medellín, Colombia.

Colasanti, K.J., Hamm, M.W. and Litjens, C.M. (2012) The city as an 'agricultural powerhouse'? Perspectives on expanding urban agriculture from Detroit, Michigan. *Urban Geography* 33(3), 348–369.

Consuelo Carrillo, A. (2009) Internal displacement in Colombia: humanitarian, economic and social consequences in urban settings and current challenges. *International Review of the Red Cross* 91, 527–546.

Corrigan, M.P. (2011) Growing what you eat: developing community gardens in Baltimore, Maryland. *Applied Geography* 31, 1232–1241.

Crisp, J., Morris, T. and Refstie, H. (2012) *Displacement in Urban Areas: New challenges, New Partnerships*. Office of the United Nations High Commissioner for Refugees, Geneva, Switzerland.

DeKay, M. (1997) The implications of community gardening for land use and density. *Journal of Architectural & Planning Research* 14, 126–149.

Fadnes, C. and Horst, E. (2009) Responses to internal displacement in Colombia: guided by what principles? *Refuge* 26(1), 111–120.

Fajardo, S. (2007) *Medellin, From Fear to Hope* (conference). J.F. Kennedy School of Government, Harvard University, Boston, Massachusetts, USA.

Frayne, B., Crush, J. and McLachlan, M. (2014) Urbanization, nutrition and development in Southern African cities. *Food Security* 6(1), 101–112.

Fukuyama, F. and Colby, S. (2011) Half a miracle: Medellín's rebirth is nothing short of astonishing. But have the drug lords really been vanquished? Foreign Policy, 25 April. Available at: http://foreignpolicy.com/2011/04/25/half-a-miracle (accessed 15 November 2016).

Gallaher, C.M., Kerr, J., Njenga, M., Karanja, N. and WinklerPrins, A.M.G.A. (2013) Urban agriculture, social capital, and food security in the Kibera slums of Nairobi, Kenya. *Agriculture and Human Values* 30, 389–404.

Galt, R.E, Gray, L.C. and Hurley, P. (2014) Subversive and interstitial food spaces: transforming selves, societies, and society–environment relations through urban agriculture and foraging. *Local Environment* 19.2, 133–146.

Goonewardena, K. (2005) The urban sensorium: space, ideology and the aestheticization of politics. *Antipode* 37(1), 46–71.

Gray, L., Guzman, P., Glowa, K.M. and Drevno, A.G. (2014) Can home gardens scale up into movements for social change? The role of home gardens in providing food security and community change in San Jose, California. *Local Environment* 19, 187–203.

Guitart, D., Pickering, C. and Byrne, J. (2012) Past results and future directions in urban community gardens research. *Urban Forestry and Urban Greening* 11, 364–373.

Hamilton, A., Burry, K., Mok, H.-F., Barker, S.F., Grove, J.R. and Williamson, V. (2014) Give peas a chance? Urban agriculture in developing countries. A review. *Agronomy and Sustainable Development* 34, 45–73.

Hampwaye, G. (2013) Benefits of urban agriculture: reality or illusion? *Geoforum* 49, 87–88.

Hylton, F. (2007) Medellín's makeover. *New Left Review* 44, 71–89.

IDMC (2012) Internal Displacement Global Overview 2011: People internally displaced by conflict and violence. Internal Displacement Monitoring Centre. Available at: http://www.internal-displacement.org/publications/2012/internal-displacement-global-overview-2011-people-internally-displaced-by-conflict-and-violence (accessed 25 November 2016).

IDMC (2014) Global Overview 2014: People Internally Displaced by Conflict and Violence. Internal Displacement Monitoring Centre. Available at: http://www.internal-displacement.org/assets/publications/2014/201405-global-overview-2014-en.pdf (accessed 23 October 2015).

Irvine, S., Johnson, L. and Peters, K. (1999) Community gardens and sustainable land use planning: a case-study of the Alex Wilson Community Garden. *Local Environment* 4, 33–46.

Lerner, A. and Eakin, H. (2011) An obsolete dichotomy? Rethinking the rural–urban interface in terms of food security and production in the global south. *The Geographic Journal* 177(4), 311–320.

Maxwell, D. (1999) The political economy of urban food security in sub-Saharan Africa. *World Development* 27(11), 1939–1953.

McClintock, N. (2010) Why farm the city? Theorizing urban agriculture through a lens of metabolic rift. *Cambridge Journal of Regions, Economy and Society* 3, 191–207.

Mok, H.-F., Williamson, V.G., Grove, J.R., Burry, K., Barker, S.F. and Hamilton, A.J. (2013) Strawberry fields forever? A review of urban agriculture in developed countries. *Agronomy and Sustainable Development* 34, 21–43.

Morgan, K. (2009) Feeding the city: the challenge of urban food planning. *International Planning Studies* 14, 341–348.

Moskow, A. (1999) Havana's self-provision gardens. *Environment and Urbanization* 11, 127–134.

Mougeot, L. (ed.) (2005) *Agropolis: The Social, Political and Environmental Dimensions of Urban Agriculture*. Earthscan, London, UK and IDRC, Ottawa, Canada.

Redwood, M. (2008) *Agriculture in Urban Planning: Generating Livelihoods and Food Security*. Earthscan, London, UK.

Redwood, M. (2009) Tenure and land markets for urban agriculture. *Open House International* 34(2), 8–14.

Roy, A. (2011) Urbanisms, worlding practices and the theory of planning. *Planning Theory* 10(6), 6–15.

Saed (2012) Urban farming: the right to what sort of city? *Capitalism Nature Socialism* 23, 1–9.

Sanchez, A. (2010) Social urbanism: the metamorphosis of Medellín. *Barcelona Metropolis*, Winter, 1–5.

Schmelzkopf, K. (1995) Urban community gardens as a contested space. *Geographical Review* 85(3), 364–372.

Schmelzkopf, K. (2002) Incommensurability, land use, and the right to space: community gardens in New York City. *Urban Geography* 23, 323–343.

Scruggs, G. (2014) Latin America's new superstar: how gritty, crime-ridden Medellín became a model for 21st-century urbanism. Next City, 31 March. Available at: https://nextcity.org/features/view/medellins-eternal-spring-social-urbanism-transforms-latin-america (accessed 15 November 2016).

Taylor, J.R. and Lovell, S.T. (2014) Urban home food gardens in the Global North: research traditions and future directions. *Agriculture and Human Values* 31, 285–305.

Tibaijuka, A. (2010) Adapting to urban displacement. *Forced Migration Review* 34, 4.

Tornaghi, C. and Van Dyck, B. (2014) Research-informed gardening activism: steering the public food and land agenda. *Local Environment: The International Journal of Justice and Sustainability* 20(10), 1247–1264.

Turok, I. (2014) Medellín's 'social urbanism' a model for city transformation. Mail & Guardian. Available at: http://mg.co.za/article/2014-05-15-citys-social-urbanism-offers-a-model (accessed 24 September 2014).

WinklerPrins, A.M.G.A. (2002) House-lot gardens in Santarém-Para, Brazil: linking rural with urban. *Urban Ecosystems* 6(1), 43–65.

Wyndham, J. (2006) A developing trend: laws and policies on internal displacement. *Human Rights Brief* 14, 7–12.

Zezza, A. and Tasciotti, L. (2010) Urban agriculture, poverty, and food security: empirical evidence from a sample of developing countries. *Food Policy* 35(4), 265–273.

14 Regreening Kibera: How Urban Agriculture Changed the Physical and Social Environment of a Large Slum in Kenya

Courtney M. Gallaher*
Northern Illinois University, DeKalb, Illinois, USA

14.1 Introduction

Our world is becoming a 'Planet of Slums' (term coined by Davis, 2006) as a result of rapid population growth and unplanned urbanization. Globally, more than a billion people now live in slums. In sub-Saharan Africa, slums are growing faster than its cities, with the majority of population growth occurring in densely packed, informal settlements that are associated with a host of social, economic and environmental problems (Davis, 2006). The inhabitants of these slums must contend on a daily basis with a range of significant environmental justice issues, including inadequate housing, sanitation services and access to water, and exposure to a range of pollutants. Additionally, due to the density of the slums, residents have little exposure to nature (e.g. trees, patches of grass, wild birds or animals) in the way that residents of other areas of these cities do. Particularly for residents who have migrated to the city from rural areas, this represents a significant rift from the ways they are accustomed to interacting with their environments.

Nairobi, Kenya, is a highly developed city and is considered to be the business hub of East Africa. It has also been described as one of the most unequal cities in the world, based on its Gini income inequality coefficient, with many of the richest and poorest people in Africa living in very close proximity (UN-Habitat, 2011). Kibera, one of the largest slums in sub-Saharan Africa, is situated amid luscious green, upper-class neighbourhoods; however, upon entering the slum from the surrounding environs, one is struck by the stark contrast of Kibera's physical environment to the open green spaces outside the slum. The slum sits adjacent to a private golf course and neighbourhoods where tall trees create a verdant canopy over the roads. Inside Kibera, open sewage trenches run next to small iron-sheeted shops and houses, dusty footpaths wind their way between narrowly spaced mud houses and pit latrines, and piles of trash sit near roadsides or behind homes. This densely populated informal settlement bears little resemblance to its past as a forest,[1] when it was initially allocated as a housing settlement for Nubian soldiers who served in the Kenyan army during the First World War. Today, the former lush green forest has been replaced by dense, unplanned settlements and trees are nearly absent from the slum.

---

*E-mail: cgallaher@niu.edu

Over the last 10 years, patches of green have started to reappear throughout Kibera, as community members have started small urban farms or gardens near many of the houses. Most recently, thousands of residents have begun sack gardening, which is a form of urban agriculture that involves planting crops into the top and sides of large sacks filled with soil. With very little open space in Kibera, typical forms of urban agriculture are not attainable for most households, but sack gardening allows people to grow a larger number of plants into relatively small spaces, by making use of the vertical space occupied by the sacks (Fig. 14.1). As urban residents are increasingly isolated from nature, trees and gardens provide them with a link to nature. Urban green spaces have been shown in other contexts to have positive impacts on numerous aspects of community well-being, but have not been examined in the context of urban agriculture in the Global South (but see Chapters 11, 12 and 19, this volume). This chapter considers the effects this greening of the landscape that sack gardening has had on people's perceptions of the environment by examining three key issues:

1. what environmental issues residents of Kibera perceive to be most serious;
2. who residents believe is responsible for addressing these environmental problems;
3. to what extent they believe sack gardening is changing the environment of Kibera.

14.2 Urbanization and Separation from Nature

Industrialization and rapid urbanization over the past century and a half has fundamentally changed the way in which humans interact with nature. While urbanization in the first half of the 20th century was primarily concentrated in developed countries, most urbanization is now taking place in Africa and Asia. Although it can be associated with substantial economic benefits, rapid and unplanned urbanization in many parts of Africa has resulted in accelerated environmental degradation and increases in food insecurity and poverty. In particular, significant negative environmental effects are associated with the rapid growth of informal settlements, like Kibera, throughout sub-Saharan Africa. According to estimates by UN-Habitat, approximately 60–80% of Kenya's urban population now resides in one of the country's many informal settlements (UN-Habitat, 2009).

Rapid urbanization has resulted in a separation of humans from nature, or a metabolic rift. McClintock (2010) draws on Marxist theories about capitalism and nature to explain the relationship between different forms of metabolic rift and the practice of urban agriculture in the Global North (GN) and the Global South (GS). He argues that many in the Global South experience this metabolic rift as *social rift*, which refers to the commodification of both land and labour. In these regions, rural populations, previously dependent on agriculture, have been driven to megacities and slums due to a variety of institutional, economic and environmental pressures. Low wages in urban areas combined with dispossession of their rural landholdings often drive these urban migrants to practise urban

Fig. 14.1. Farmers in Kibera grow food in sack gardens in the limited open spaces available in the slum. Sack gardening maximizes yield by making use of the vertical space occupied by the sacks. Rows of green sacks are now frequently seen between houses or along the roadside. (Photo by author, 2010.)

agriculture. Many individuals experience a more personal perception of themselves as separate from nature, which McClintock (2010) terms *individual rift*. It is this individual rift that has driven people to find ways to reconnect with nature, which McClintock argues has driven much of the urban agriculture movement in the Global North. Re-engaging with food production and consumption processes helps urban gardeners to overcome individual metabolic rift.

The disconnect between humans and nature, as captured by theories of metabolic rift, is evident in other studies on urbanization and people's relationship with nature. Turner *et al.* (2004) measured biological diversity in urban areas on a global scale and found that urban residents live in areas of significantly impoverished biodiversity, meaning the baseline for ecological health in urban areas is significantly poorer than in urban areas. They suggested that people are losing the opportunity to interact with and appreciate nature in urban areas. Vining *et al.* (2008) found that individual concern for the environment is correlated with a sense of connectedness to nature. As a number of authors have noted, for many people gardening is a means of reconnecting with nature in urban areas that otherwise feel devoid of nature or 'wilderness' (e.g. Clayton, 2007; McClintock, 2010; Cheng and Monroe, 2012; Freeman *et al.*, 2012).

14.3 Urban Green Spaces

14.3.1 Benefits of urban green spaces

Green spaces provide an important link to nature for residents of urban areas. Types of green spaces vary, but often include parks, urban forest cover in residential areas and along roadways, green roofs and community gardens. The benefits of urban green spaces on the physical environment are wide ranging. Urban trees have been shown to filter dust from the air and provide some protection from UV radiation (Tyrväinen *et al.*, 2005), improving overall air quality. Urban forests also help improve the hydrology of an area by reducing surface runoff and dampening the peak flow of streams following a storm event (Tyrväinen *et al.*, 2005). Urban forests and urban gardens have been demonstrated to help preserve or enhance biodiversity within cities (Goddard *et al.*, 2010).

Benefits of urban green spaces extend beyond their contributions to the health of the physical environment to the health benefits of the residents living near them. Numerous studies suggest that living or working in proximity to an urban green space can result in improvements to people's physical health. A study in Japan found that residents who had access to plentiful green space had a lower mortality risk than those who did not (Takano *et al.*, 2002). A large epidemiological study in the Netherlands demonstrated a positive relationship between the percentage of green space in people's living environment and the general health of the residents, especially those of lower socio-economic status (Maas *et al.*, 2006). A similar study in the US found residents of neighbourhoods with abundant green space to be in better health overall, especially the elderly and people of low socio-economic status (De Vries *et al.*, 2003). Few studies have examined gendered differences in the impacts of green spaces, but existing studies have come to mixed conclusions. A study in the UK found that a greater percentage of green space in neighbourhoods was related to improved health outcomes (cardiovascular disease mortality, respiratory disease mortality and self-reported limiting long-term illness) for men, but green spaces had no effect on women's health outcomes (Richardson and Mitchell, 2010).

Urban green spaces also provide numerous social benefits. In low-income, urban areas of the US, the presence of well-used green spaces has been linked to stronger ties to neighbours, a greater sense of safety (Kweon *et al.*, 1998), fewer incidences of graffiti and other incivilities (Kuo *et al.*, 1998), and fewer crimes (Kuo and Sullivan, 2001).

14.3.2 Environmental justice and green spaces

As areas become increasingly urbanized, access to green space typically declines. This particularly affects people from low socio-economic backgrounds, who lack the resources to move outside of cities to areas with better access to green spaces (Maas *et al.*, 2006). Green spaces can be thought of as both a public and private good, depending on who is planting the trees or creating the green space. But the urban poor are much less likely to be able to produce and maintain their own green spaces, so they are fully dependent on public-sector investment in the maintenance of trees and parks to be able to access green spaces (Heynen et al., 2006).

Environmental justice movements have historically focused on unequal exposure to environmental contaminants, but inequitable access to environmental resources is beginning to gain attention. Because reduced access to green spaces leads to a reduced quality of life, the spatial distribution of and access to green spaces has begun to be recognized as an important issue within environmental justice. A study of urban forest cover in Milwaukee found that the city's green spaces were distributed unevenly, based on analyses of household income, housing market characteristics, racial and other ethnic factors. While households of higher socio-economic status had easy access to urban forest cover, those of lower socio-economic status were limited to accessing forest cover along roads and in city parks (Heynen et al., 2006). In another study, a GIS analysis of public green spaces in nine towns in South Africa found that per capita, green space was highest in the affluent, white areas of town and lowest in the largely black, low-cost housing settlements (McConnachie and Shackleton, 2010).

Community gardens have begun to serve as green spaces in communities that otherwise lack access to urban forests or parks. In many parts of the GN, especially in North America, community gardens and, more broadly, urban agriculture movements, can be viewed as a response to urban decay. For example, in the 1970s in New York City, many communities took it upon themselves to revitalize areas that were crime- and trash-ridden by cleaning them up and planting community gardens (Tidball and Krasny, 2007). In Detroit, a regreening of the city can be seen, as a vibrant urban agriculture movement has emerged using large areas of abandoned or vacant urban land that has been taken over by community or backyard gardens (Colasanti et al., 2012). Numerous cities throughout North America have experienced various degrees of regreening, as community gardens have grown in popularity (e.g. Twiss et al., 2003; Baker, 2004; Corrigan, 2011; Taylor and Lovell, 2012). These community gardens provide more than sources of food for the local community; they are creating public urban green spaces in areas where residents otherwise lack access to such spaces. This form of urban community greening relies on the active participation of urban residents, who strive to build healthier sustainable environments. While much has been documented about the social benefits of urban green spaces and the community garden movement in North America, very little research has been done on inequity in access to green spaces or the importance of urban gardens as green spaces in the sub-Saharan African context. Yet residents of urban areas in Africa often face some of the greatest environmental injustices.

In this chapter, I focus on green spaces because of the potential of urban gardens to create positive change in the community, but acknowledge that inequity in access to green space is only one of a myriad of environmental injustices faced by residents of the Kibera slums. Faced with these problems, communities may feel disempowered by the combination of so many environmental and social injustices. Sack gardening, in creating both physical green spaces and social networks of gardeners, has the potential to change perceptions of the physical and social environment and address some of the environmental injustices experienced by residents of Kibera.

14.4 Study Area and Methods

14.4.1 Study area

Kibera is East Africa's largest slum, with more than 200,000 residents occupying about 2.5 km^2, making it one of the more densely populated urban settlements in the world (Desgroppes and Taupin, 2011). It is located 7 km from downtown Nairobi, and is situated among some of the most attractive neighbourhoods in Nairobi. Kibera's history dates to the First World War, when the Kenyan colonial government designated a forested area that is now Kibera as a military reserve and site of temporary residence for Sudanese (Nubian) soldiers who had served as part of the King's African Rifles (KAR) and were unable to return to Sudan after their service (Balaton-Chrimes, 2011). The British colonials made no effort to repatriate or resettle the Nubians to Sudan, nor did they give them official title deeds for the land in Kibera. However, residents of Kibera were denied official land tenure (Parsons, 1997). After the Second World War, severe housing shortages in Nairobi caused a large influx of East Africans into Kibera, and Nubians began renting out parts of their farms or houses to these migrants from other parts of Africa. During the decades that followed, Kibera was largely ignored by the government, with minimal investment in infrastructure within the slum, creating major environmental justice issues for residents who still lack access to any formal waste management systems or reliable sources of water. What started as a forest preserve decades ago has now become a densely packed informal settlement with almost no tree cover or green on the landscape.

14.4.2 Data collection and analysis

This research was part of a larger, mixed-methods study examining the impact of sack gardens on household food security, urban livelihood strategies and exposure to environmental contaminants via urban farming (Gallaher et al., 2013a, b, 2015). The overall project comprised open-ended qualitative interviews with individual farmers, a large household survey with farmers and non-farmers, collection of soil, plant and water samples and, finally, focus groups with a subset of participants. Data for this portion of the project is derived primarily from the focus group discussions.

To gather more information about how farmers and non-farmers understand environmental problems in Kibera, and their views on how sack farming had impacted the physical and social environment of Kibera, I conducted seven focus group sessions with farmers and non-farmers, grouped accordingly: two with male non-farmers, two with female non-farmers, two with female farmers and one with male farmers. Focus group discussions were focused around three key themes:

1. key environmental problems in Kibera;
2. who is responsible for dealing with environmental problems in the slum;
3. what impact has sack gardening had on the environment in Kibera.

Each focus group session consisted of 5–12 people and were held at the Kibera Girls Soccer Academy School in Kibera, because it is well known and centrally located. Focus group sessions were audio-recorded, transcribed in Kiswahili, and then translated into English with the help of a research assistant. Transcripts were analysed using thematic analysis, to look for key themes that appeared in the data (Guest et al., 2011). QSR NVivo 9 software was used to code the transcripts and facilitate aggregation of data along major themes.

14.5 Findings

14.5.1 Key environmental issues in Kibera

I initially conceptualized the environment as purely the physical environment but found that residents moved fluidly in their discussions between the physical and social environments of Kibera. Given the high population density in Kibera, the built and social environment strongly influences the way in which people experience the limited aspects of the physical environment that are accessible in Kibera. This mixing of the physical and social environments reflect an urbanization of nature which has fundamentally altered the way in which people experience their environment compared to rural counterparts. Similar to the Global North, separation from nature has driven a resurgence in urban agriculture as a means of reconnecting with nature. In Kibera, the lines between the natural and socially constructed environment have blurred, leaving open the possibility for urban agriculture as a way to heal this metabolic rift (McClintock, 2010).

Participants in all focus groups easily identified a large number of environmental problems in the Kibera slums (Table 14.1). The environmental problems described by our focus group participants were consistent with those that have previously been identified in the literature (e.g. Njeru, 2006; Omambia, 2010; Hardoy et al., 2013). Many of the physical environmental problems that were of concern to participants related to overcrowding and poor physical infrastructure or housing. Poorly constructed mud homes were easily vandalized, fell apart

Table 14.1. A summary of the key environmental problems in Kibera identified by focus group participants. They are divided into problems primarily affecting the physical environment, and those affecting the social environment of the slum.

Physical environment	Social environment
Air quality – smoke from burning trash, cooking with charcoal in confined spaces, smells from rotting detritus near roads	**Corruption** – police must be bribed to investigate crimes, perpetrators of crimes can bribe their way out, jobs are given only to those with connections
Animals – problems with rodents in homes, improper disposal of livestock manure, feral dogs and cats	**Crime** – very high crime rates (petty theft and violent crime) in Kibera, lack of police response
Electrical wires – lack of formal electrical grid has led to amateurly rigged wiring to homes, live wires have fallen and killed children and adults	**Drug abuse** – concerns about drug addiction, especially among young men, lack of police response, gang violence related to drugs
Poor-quality housing – homes are poorly constructed and not maintained by landlords, and are often damaged during storms	**Food quality** – inadequate access to safe and high-quality food, locally sold food is often expired or cooked with dirty water or ingredients, purchased foods often not what they seem (e.g. beef may be dog meat, milk has been thinned with flour and water)
Overcrowding – high-density housing, lack of space for waste disposal, concerns over spread of diseases	
Toilets – serious shortage of toilets for residents, high cost of using toilets, haphazard disposal of human waste creates public health problems	**Sexual violence** – participants repeatedly described concerns about sexual violence against women and girls, lack of police response, retaliation for reporting it to police
Waste disposal – lack of any formal solid or liquid waste disposal programme, waste often swept into open drainage ditches or thrown into other people's yards, attracts flies and animals, spreads diseases	**Reputation of Kibera** – residents of the slum are discriminated against when trying to conduct business outside of Kibera, labelled as criminals and low-lifes
Water – frequent water shortages, high cost of water, water is frequently polluted when water pipes passing through sewage ditches break	**Unemployment** – high unemployment, lack of education and opportunities for youth leads to criminal activities
	Underground economy – residents constantly question quality of goods they purchase, drugs from pharmacies are often fake, business transactions frequently illegal

and needed to be repaired after a major rain. Participants stated that rodents moved easily in and out of cracks in the house walls, eating food in the house and spreading diseases. With no formal or legal infrastructure to provide electricity to slum residents, focus group participants described an illegal network of wires that had been patched together. This informal electrical grid was a major health hazard, especially if live wires fell to the ground, electrocuting people who walked on them or children who picked them up to play with.

Other major concerns about the physical environment related to the lack of formal sanitation services. Pit latrines in Kibera are typically shared by 50–150 people and people must pay a fee to access the latrine (Mutisya and Yarime, 2011). Without proper access to a latrine, focus group participants described being forced to collect and dump their excrement in open drainage ditches that pass between the houses and run along the roads. These drainage ditches eventually drain into the reservoir near Nairobi Dam on the southeast side of the slum. Another solution to the toilet problem involved disposing of excrement in plastic bags (Njeru, 2006). As one focus group participant described:

> Sometimes you may find that some of the houses do not have toilets. You find that some of them opt to use plastic bags and later on dispose of them, but they just throw them anywhere. It might even fall on someone.

This solution to disposing of human waste is widely recognized and has previously been described in the literature as 'flying toilets' or 'scud missiles' (Njeru, 2006; Mutisya and Yarime, 2011). Lack of sanitation services also means that residents have no way to dispose of solid waste. Less than 1% of Kibera's residents are served by a public garbage collection system. Instead most households dispose of their waste by dumping it in their own neighbourhoods, or burning or burying it in their own compound. These ditches follow the same narrow footpaths between houses that people walk on daily, and which children play on, and are used as dumping points for emptying waste water from latrines, and for households to dispose of grey water from bathing in their house. The drains are inadequate in size and are poorly maintained and operated, resulting in extensive environmental pollution, health risks and danger to the inhabitants (UN-Habitat, 2014). Focus group participants described throwing their trash into open dumpsites or sweeping it into drainages. The trash would then flow downstream, clogging open waterways and creating large, informal dumpsites throughout the slum (Fig. 14.2).

Poor air quality, due to particulate matter released by charcoal cooking stoves and rotting smells from the drainage ditches were a strongly voiced concern during the focus group sessions. Women in particular were concerned with indoor air quality when they cooked over charcoal stoves inside their homes. They or their children frequently developed asthma as a result of inhaling the charcoal smoke.

The last and most frequently discussed concern about the physical environment related to inadequate access to water, and the poor quality of available water in Kibera. In all focus group sessions, participants had major concerns about both access to and quality of water. Because there is no formal water supply in Kibera, residents must obtain water from water vendors who have illegally connected to the city's water supply with plastic PVC pipes that run overland

Fig. 14.2. Drainage ditches run between most homes in Kibera and flow towards the Nairobi Dam, just outside the slum. Because residents have no way to formally dispose of solid waste, these drainage ditches have become informal dumpsites, where residents dispose of trash and then wait for the water to carry it away. (Photo by author, 2011.)

throughout Kibera. These pipes often crack and dirt and raw sewage contaminates the water. As one female non-farmer described:

> In my area we experience all these problems but the most disturbing one is the water situation. Most of the water pipes pass through dirty trenches or sewers which is the same water we use domestically.

Participants also complained about the frequent water shortages that occurred when the city shut off water to the main pipes supplying Kibera. Water shortages could last anywhere from a few hours to weeks, so residents were forced to stockpile large quantities of water in jerry cans in their homes. During water shortages, the price of water increased substantially, which was a problem for all residents, but made it prohibitively expensive for farmers to purchase water for their sack gardens.

Although I had planned only to ask about perceptions of problems with the physical environment, residents consistently wove discussions about the physical environment in and out of observations about the social environment in Kibera. They aptly perceived that the physical and social environments are linked, with one influencing the other. Participants voiced a long list of concerns about the social environment of Kibera (Table 14.1) but those most frequently observed were concerns about crime (petty theft and violent crime), sexual violence against women and children, high unemployment, and food quality.

14.5.2 Responsibility for the environment

From the seven focus group sessions, it was abundantly evident that the majority of residents of Kibera felt quite powerless to change anything about the local environment and they viewed many of these physical and social environmental issues as problems without solutions. There were, however, some interesting distinctions based on gender and participation in sack gardening. All focus group participants described similar types of environmental problems in the Kibera slums. However, when asked about responsibility for these problems, the non-farmers were much more likely to describe ways in which they felt powerless to change the situation. Among women non-farmers, this powerlessness was most typically framed in the context of landlord/renter relationships, where they felt unable to approach local authorities about environmental problems for fear of being evicted from their current residence. When asked whether they ever approached their landlords about issues with overflowing toilets or houses that were falling apart, one woman said: 'There are lots of people here, so you will be sidelined. If you go to the landlord, he will scream and tell you to look for another place to live.' There was general consensus among the women non-farmers that fear of losing their homes and having no place for their children to live often prevented them from trying to fight for their rights to have regular access to clean water, solid waste sanitation, basic home repairs, or to have the outhouse latrines cleaned or repaired.

Men non-farmers who participated in focus groups also discussed feeling powerless to address environmental problems in Kibera. In contrast to the women participants, the men's frustrations were primarily related to inadequate political representation and perceived corruption among elected officials, the police and non-profit organizations working in the area. They felt that initiatives to clean up the environment of Kibera were destroyed by local leaders who used them for their own profits instead. For example, the Nairobi City Council briefly operated a programme in Kibera called *Kazi kwa Vijana*, a youth service corps designed to hire young men to participate in cleaning up the slums. The male participants, however, described nepotism in the selection of youth paid to work for the programme and, after about a year, the programme stopped functioning. As one male participant said:

> With the leaders we have these days, it has become such that whatever project comes our way, the chairman or chief must be present and then they hand out the jobs created to their relatives or people they know personally ... All I'm stating is that our leaders contribute to this problem a lot. You have to bribe your way into a job these days.

The responses of the men and women focus group participants were very consistent with previous research on environmental risk perception, which has demonstrated that women often frame environmental problems in terms of threats to their homes, children and extended family, while men tend to perceive environmental risk in the context of their working lives (Flynn et al., 1994; Gufstafson, 1998). It also illustrates a point made by Brulle (2000), who argues that environmental problems are inherently linked to how human society is organized. Those who are the most socially vulnerable also bear the brunt of environmental problems, yet are the least capable of enacting change.

Male and female farmers who participated in the focus group discussions also expressed similar frustrations with their powerlessness to address widespread environmental problems in the slum for many of the same reasons cited by the non-farmers: fear of eviction, corruption and neglect by the city council and police. Interestingly, although the questions were framed in the same manner as with the non-farmers, the farmers were much more likely to discuss personal responsibility towards the environment and the desire to find ways to improve the environment in Kibera, saying things like 'if we are clean, the environment will be clean', or 'cleanliness starts with yourself'. Both men and women farmers described efforts to clean up the areas near their sack gardens, including sweeping away trash on a daily basis, erecting makeshift fences to protect their gardens, and conversing with neighbours about not haphazardly throwing trash.

Although these are small actions in the context of the major environmental justice issues in Kibera, sack gardening, nonetheless, has empowered farmers to think differently about the problems they face. This point will be explored further in the following section, which examines the impacts of sack gardening on the physical and social environment of Kibera.

14.5.3 Changes to the physical environment of Kibera

The final goal of these focus group discussions was to investigate the extent to which focus group participants felt that urban agriculture, in the form of sack gardening, had impacted the environment of Kibera. Both farmers and non-farmers felt that sack gardening had significantly improved the physical environment of the Kibera slums by changing its overall appearance. As described, a lack of municipal solid waste management in Kibera has resulted in large quantities of trash scattered throughout living areas in the slum. A common theme shared by focus group participants was that sack gardening has encouraged people to remove trash from the areas surrounding the sacks in an effort to prevent contamination of the vegetable, making the environment near the gardens more attractive. One female farmer explained about sack gardening, 'When you look back, you realize that there is no more garbage in the area because of this type of cultivation.' The focus group of male gardeners affirmed that gardening brings beauty to the landscape of Kibera and the changes are noticeable even from a distance within the slum. Places that were once dirty have been cleaned up and now contain the green of sack gardens.

Another common theme from the focus group discussions related to the perceptions of improved air quality. Both male and female participants believed that sack gardening had significantly improved air quality in Kibera. Several focus groups made the analogy between trees producing oxygen and their sack gardens helping to purify the air in Kibera, and that the air surrounding the sack gardens feels cleaner to breathe. One of the male farmers summarized the change by saying:

> You see the way it looks here [in Kibera]? There is no fresh air, there are no trees. Maybe the gardening project that was started has made it better. Now when someone comes to visit my house, they claim the air here has changed a bit. Normally the air we breathe in is totally polluted.

As these points illustrate, the introduction of sack gardens into the landscape of Kibera has provided a small way for gardeners to reconnect with nature. The mere presence of the sack gardens has altered their perceptions of the physical environment, such that they feel the air is cleaner near the sack gardens and there is now beauty in the landscape. Many of the sack farmers are migrants from rural areas who have come to live in Kibera. Sack gardening is a reminder of their connection to the natural world and appears to have created a greater sense of appreciation for the natural world.

14.5.4 Changes to the social environment of Kibera

In addition to the perceived impacts on the physical environment, focus group participants noted many positive changes in the social environment of Kibera as a result of sack gardening. All farmers described the benefits of gardening in terms of improved access to fresh food but they were equally vocal about the positive effects of sack gardening on their communities. Focus group participants explained the many ways in which sack gardening increased cooperation between farmers and their neighbours, including through sharing of labour for planting and weeding, and sharing harvested food from the garden.

In the GS, numerous studies have found community gardens build social capital (e.g. Glover et al., 2005; Kingsley and Townsend, 2006; Alaimo et al., 2010). The social impacts described by these farmers go beyond just improving social capital. Many of the farmers in Kibera, especially women, emphasized that sack gardening had given them a stronger sense of purpose in life. In the context of high levels of unemployment in Kibera, gardening gave these women an outlet for them to work and provide for their families, as well as interact more with their friends and neighbours. They also found a sense of purpose in being able to share their knowledge about gardening with others. One of the female farmers said simply: 'In general, sack gardening has made me feel like I am of use to society.' This relationship between gardening and empowerment is similar to that documented by Slater (2001), who found that the act of gardening was empowering for women in South Africa because it provided an outlet to cope with trauma, gave them a sense of putting down roots in an urban community, and helped fulfil their societal roles as wives and mothers. See also Chapter 12, this volume, for a case from Tanzania. Empowerment is a strong benefit of community gardens in the Global North: see Chapters 5 and 8, this volume.

Changes in the physical environment related to sack gardening also led to social changes in Kibera. Because areas that were previously vacant were now being cultivated and attended to, some participants commented that these areas were now less dangerous. For example, farmers in one neighbourhood in Kibera often chose to keep sack gardens near the river where water was easier to transport. These focus group participants said that not only did the sack gardens help to beautify the landscape near the river, but crime in that area had gone down because it was no longer deserted.

Finally, Travaline and Hunold (2010) argue that urban gardening helps to build community and transforms residents into ecological citizens. This idea of creating ecological citizens was observed to a limited extent in Kibera. Some of the sack gardeners in Kibera had very limited prior knowledge of farming, and children growing up in Kibera have almost no exposure to agriculture. Gardeners were proud to be able to educate their children about farming by teaching them about sack gardening. One of the female farmers happily explained:

> Sack gardening has become a practical lesson for my children. It has helped them know the different types of crops grown in farms. At least now if I decide to take them upcountry [to our ancestral home] I will not be embarrassed that they do not know things.

Overall, both farmers and non-farmer focus group participants felt that sack gardening had positively changed the social environment of Kibera. One female farmer summarized the sentiments of the focus group participants by saying: 'At the end of the day the whole community is thinking as one and helping each other, another benefit that sack farming has brought to our community.'

14.6 Conclusions

Residents' perceptions of key environmental problems in Kibera were consistent with the literature documenting major environmental problems in slums globally, and specifically in Kibera. Participants in this study identified more strongly with problems in the social environment rather than the physical environment, but identified a range of social and physical environmental problems in Kibera (Table 14.1). Discussions about who holds responsibility for dealing with these environmental problems underlined the ways in which many residents of Kibera feel powerless to change their physical and social environments due to problems of corruption and violence in the slums. My research illustrates how sack gardening has helped, in a small way, to empower residents to improve their physical and social environments. Many of the social benefits that residents described may be more important than the physical changes that sack gardening has brought to Kibera. This is consistent with the literature on green spaces, which emphasizes the positive social and psychological benefits of green spaces, particularly in disadvantaged communities (Kuo et al., 1998; Kweon et al., 1998; Kuo and Sullivan, 2001). While Kibera is unlikely to ever return to its original forested state, this regreening of the slum via sack gardening has provided an important way for people to reconnect with nature in a highly urbanized landscape, and heal the individual metabolic rift they experience as urban slum dwellers.

Research on urban agriculture in the Global South has traditionally been descriptive and applied, documenting how people are involved in urban agriculture or focusing on the improvements in household food security associated with urban farming (e.g. Maxwell, 1995; Binns and Lynch, 1998; Tevera, 1996; Zezza and Tasciotti, 2010; Chapter 2, this volume). Our broader study of sack gardening affirmed that this type of urban agriculture does contribute to improvements in household food security for residents of the Kibera slums (Gallaher et al., 2013a). Importantly, evidence from these focus group discussions contributes to a body of literature, demonstrating that the value of urban gardening extends beyond contributions to household food security by building social capital and empowering local communities. While the social benefits of urban agriculture have been well documented in the Global North (e.g. Teig et al., 2009; Alaimo et al., 2010; Firth et al., 2011; Gray et al., 2014), these types of studies have been much more limited in the Global South. Given the potential of urban agriculture to address both pragmatic issues of household food security and broader issues of environmental justice, the facilitation of urban agricultural activities is particularly important for disadvantaged populations, such as those living in the Kibera slums. Space constraints often prevent the practice of urban agriculture in informal settlements, but our research suggests that even small-scale interventions, like sack gardening, provide a range of benefits to inhabitants beyond improving food access and should be promoted in other regions of the world.

Acknowledgements

This research was supported with funds from the Geography and Spatial Sciences Program at the US National Science Foundation, award BCS-1030325, as well as the Society of Women Geographers Pruitt Dissertation Research Fellowship. I would like to express my sincere thanks to the households who participated in this research project. I am also immensely grateful to my team of collaborators, including Antoinette WinklerPrins, Nancy Karanja, Mary Njenga, Dennis Mwaniki, Catherine Wangui, George Aloo, Joel Boboti, Baraka Mwau, Jack Odero and Jamie Clearfield, who contributed to this project in a variety of ways.

Note

[1] The term Kibera comes from a Nubian word, *Kibra*, meaning forest (Balaton-Chrimes, 2011).

References

Alaimo, K., Reischl, T.M. and Allen, J.O. (2010) Community gardening, neighborhood meetings, and social capital. *Journal of Community Psychology* 38, 497–514. DOI: 10.1002/jcop.20378.

Baker, L.E. (2004) Tending cultural landscapes and food citizenship in Toronto's community gardens. *The Geographical Review* 94(3), 305–325.

Balaton-Chrimes, S. (2011) The Nubians of Kenya and the emancipatory potential of collective recognition. *Australasian Review of African Studies* 32(1), 12–31.

Binns, T. and Lynch, K. (1998) Feeding Africa's growing cities into the 21st century: the potential of urban agriculture. *Journal of International Development* 10(6), 777–793.

Brulle, R.J. (2000) *Agency, Democracy, and Nature: The US Environmental Movement from a Critical Theory Perspective*. MIT Press, Cambridge, Massachusetts, USA.

Cheng, J.C.H. and Monroe, M.C. (2012) Connection to nature: children's affective attitude toward nature. *Environment and Behavior* 44(1), 31–49.

Clayton, S. (2007) Domesticated nature: motivations for gardening and perceptions of environmental impact. *Journal of Environmental Psychology* 27(3), 215–224.

Colasanti, K.J., Hamm, M.W. and Litjens, C.M. (2012) The city as an 'agricultural powerhouse'? Perspectives on expanding urban agriculture from Detroit, Michigan. *Urban Geography* 33(3), 348–369.

Corrigan, M.P. (2011) Growing what you eat: developing community gardens in Baltimore, Maryland. *Applied Geography* 31(4), 1232–1241.

Davis, M. (2006) *Planet of Slums*. Verso, London, UK.

De Vries, S., Verheij, R.A., Groenewegen, P.P. and Spreeuwenberg, P. (2003) Natural environments – healthy environments? An exploratory analysis of the relationship between greenspace and health. *Environment and Planning A* 35(10), 1717–1732.

Desgroppes, A. and Taupin, S. (2011) Kibera: the biggest slum in Africa? *Les Cahiers de l'Afrique de l'Est* 44, 23–34.

Firth, C., Maye, D. and Pearson, D. (2011) Developing 'community' in community gardens. *Local Environment* 16(6), 555–568.

Flynn, J., Slovic, P. and Mertz, C.K. (1994) Gender, race, and perception of environmental health risks. *Risk Analysis* 14, 1101–1108.

Freeman, C., Dickinson, K.J., Porter, S. and van Heezik, Y. (2012) 'My garden is an expression of me:' Exploring householders' relationships with their gardens. *Journal of Environmental Psychology* 32(2), 135–143.

Gallaher, C.M., Kerr, J.M., Njenga, M., Karanja, N.K. and WinklerPrins, A.M.G.A. (2013a) Urban agriculture, social capital, and food security in the Kibera slums of Nairobi, Kenya. *Agriculture and Human Values* 30(3), 389–404.

Gallaher, C.M., Mwaniki, D., Njenga, M., Karanja, N. and WinklerPrins, A.M.G.A. (2013b) Real or perceived: the environmental health risks of urban sack gardening in Kibera slums of Nairobi, Kenya. *Ecohealth* 10(1) 9–20.

Gallaher, C.M., Njenga, M., Karanja, N.K. and WinklerPrins, A.M.G.A. (2015) Creating space: sack gardening as a livelihood strategy in the Kibera slums of Nairobi, Kenya. *Journal of Agriculture, Food Systems and Community Development* 5(2), 155–173.

Glover, T.D., Parry, D.C. and Shinew, K.J. (2005) Building relationships, accessing resources: mobilizing social capital in community garden contexts. *Journal of Leisure Research* 37(4), 450.

Goddard, M.A., Dougill, A.J. and Benton, T.G. (2010) Scaling up from gardens: biodiversity conservation in urban environments. *Trends in Ecology and Evolution* 25(2), 90–98.

Gray, L., Guzman, P., Glowa, K.M. and Drevno, A.G. (2014) Can home gardens scale up into movements for social change? The role of home gardens in providing food security and community change in San Jose, California. *Local Environment* 19(2), 187–203.

Guest, G., MacQueen, K.M. and Namey, E.E. (2011) *Applied Thematic Analysis*. Sage Publications, Thousand Oaks, California, USA.

Gufstafson, P.E. (1998) Gender differences in risk perception: theoretical and methodological perspectives. *Risk Analysis* 18, 805–811.

Hardoy, J.E., Mitlin, D. and Satterthwaite, D. (2013) *Environmental Problems in an Urbanizing World: Finding Solutions in Cities in Africa, Asia and Latin America*. Routledge, London, UK.

Heynen, N., Perkins, H.A. and Roy, P. (2006) The political ecology of uneven urban green space: the impact of political economy on race and ethnicity in producing environmental inequality in Milwaukee. *Urban Affairs Review* 42(1), 3–25.

Kingsley, J.Y. and Townsend, M. (2006) 'Dig in' to social capital: community gardens as mechanisms for growing urban social connectedness. *Urban Policy and Research* 24(4), 525–537.

Kuo, F.E. and Sullivan, W.C. (2001) Environment and crime in the inner city: does vegetation reduce crime? *Environment and Behavior* 33(3), 343–367.

Kuo, F.E., Sullivan, W.C., Coley, R.L. and Brunson, L. (1998) Fertile ground for community: inner-city neighborhood common spaces. *American Journal of Community Psychology* 26(6), 823–851.

Kweon, B.S., Sullivan, W.C. and Wiley, A. (1998) Green common spaces and the social integration on inner-city older adults. *Environment and Behavior* 30(6), 832–858.

Maas, J., Verheij, R.A., Groenewegen, P.P., De Vries, S. and Spreeuwenberg, P. (2006) Green space, urbanity, and health: how strong is the relation? *Journal of Epidemiology and Community Health* 60(7), 587–592.

Maxwell, D.G. (1995) Alternative food security strategy: a household analysis of urban agriculture in Kampala. *World Development* 23(10), 1669–1681.

McClintock, N. (2010) Why farm the city? Theorizing urban agriculture through a lens of metabolic rift. *Cambridge Journal of Regions, Economy and Society* 3(2), 191–207.

McConnachie, M.M. and Shackleton, C.M. (2010) Public green space inequality in small towns in South Africa. *Habitat International* 34(2), 244–248.

Mutisya, E. and Yarime, M. (2011) Understanding the grassroots dynamics of slums in Nairobi: the dilemma of Kibera informal settlements. *International Transaction Journal of Engineering, Management, and Applied Sciences and Technologies* 2(2), 197–213.

Njeru, J. (2006) The urban political ecology of plastic bag waste problem in Nairobi, Kenya. *Geoforum* 37(6), 1046–1058.

Omambia, A.N. (2010) Sanitation in urban slums: perception, attitude and behavior: the case of Kibera, Nairobi, Kenya. *Journal of Environmental Science and Engineering* 4(3), 70–80.

Parsons, T. (1997) 'Kibra is our blood': the Sudanese military legacy in Nairobi's Kibera location, 1902–1968. *The International Journal of African Historical Studies* 30(1), 87–122.

Richardson, E.A. and Mitchell, R. (2010) Gender differences in relationships between urban green space and health in the United Kingdom. *Social Science and Medicine* 71(3), 568–575.

Slater, R.J. (2001) Urban agriculture, gender and empowerment: an alternative view. *Development Southern Africa* 18(5), 635–650.

Takano, T., Nakamura, K. and Watanabe, M. (2002) Urban residential environments and senior citizens' longevity in megacity areas: the importance of walkable greenspaces. *Journal of Epidemiology of Community Health* 56, 913–918.

Taylor, J.R. and Lovell, S.T. (2012) Mapping public and private spaces of urban agriculture in Chicago through the analysis of high-resolution aerial images in Google Earth. *Landscape and Urban Planning* 108(1), 57–70.

Teig, E., Amulya, J., Bardwell, L., Buchenau, M., Marshall, J.A. and Litt, J.S. (2009) Collective efficacy in Denver, Colorado: strengthening neighborhoods and health through community gardens. *Health and Place* 15(4), 1115–1122.

Tevera, D.S. (1996) Urban agriculture in Africa: a comparative analysis of findings from Zimbabwe, Kenya and Zambia. *African Urban Quarterly* 11(2–3), 181–187.

Tidball, K.G. and Krasny, M.E. (2007) From risk to resilience: what role for community greening and civic ecology in cities. In: Wals, A. (ed.) *Social Learning Towards a More Sustainable World*, Wageningen Academic Publishers, Wageningen, The Netherlands, pp. 149–164.

Travaline, K. and Hunold, C. (2010) Urban agriculture and ecological citizenship in Philadelphia. *Local Environment* 15(6), 581–590.

Turner, W.R., Nakamura, T. and Dinetti, M. (2004) Global urbanization and the separation of humans from nature. *Bioscience* 54(6), 585–590.

Twiss, J., Dickinson, J., Duma, S., Kleinman, T., Paulsen, H. and Rilveria, L. (2003) Community gardens: lessons learned from California healthy cities and communities. *Journal of Information* 93(9), 1435–1438.

Tyrväinen, L., Pauleit, S., Seeland, K. and de Vries, S. (2005) Benefits and uses of urban forests and trees. In: Konijnendijk, C., Nilsson, K., Randrup, T. and Schipperijn, J. (eds) *Urban Forests and Trees*. Springer, Berlin, Germany, pp. 81–114.

UN-Habitat (2009) *UN-Habitat and the Kenya Slum Upgrading Programme*. UN-Habitat, Nairobi, Kenya. Available at: http://unhabitat.org/books/un-habitat-and-the-kenya-slum-upgrading-programme-strategy-document (accessed 23 November 2016).

UN-Habitat (2011) *State of the World's Cities 2010/2011: Bridging the Urban Divide*. Earthscan, London, UK.

UN-Habitat (2014) *Kibera: Integrated Water Sanitation and Waste Management Project*. UN-Habitat, Nairobi, Kenya. Available at: http://unhabitat.org/books/kibera-integrated-water-sanitation-and-waste-management-project (accessed 3 March 2015).

Vining, J., Merrick, M.S. and Price, E.A. (2008) The distinction between humans and nature: human perceptions of connectedness to nature and elements of the natural and unnatural. *Human Ecology Review* 15(1), 1.

Zezza, A. and Tasciotti, L. (2010) Urban agriculture, poverty, and food security: empirical evidence from a sample of developing countries. *Food Policy* 35(4), 265–273.

15 Farm Fresh in the City: Urban Grassroots Food Distribution Networks in Finland

Sophia E. Hagolani-Albov[1]* and Sarah J. Halvorson[2]

[1]*University of Helsinki, Helsinki, Finland;* [2]*University of Montana, Missoula, Montana, USA*

It was March, there were a few inches of snow on the ground, and the air was bitingly cold. I was invited by the founder of the REKO Circles [food distribution networks] to accompany him to the weekly pick up scheduled to take place in the midafternoon, which at that latitude was right before nightfall. We arrived shortly after the start of the event at a parking lot that was in a forgotten corner of Pietarsaari, Finland. Cars were parked every few spaces and there was a group of people clustered around each car. The temperature hovered around freezing and products were exchanged quickly and efficiently through open trunks or out of backseats. As I watched the scene unfold in front of me, I was amazed to realize that 30 minutes ago this had been an empty parking lot and in another 30 minutes all the producers and consumers would be gone. The parking lot would be cold and silent again; the only hint of this 'instant' market would be the trampled snow.

(Field notes, 20 March 2014)

15.1 Introduction

This chapter probes two questions: What is the role of grassroots food distribution networks in Finland, and to what extent are these networks creating farmer–consumer linkages that support a robust urban foodshed? To address these questions, this chapter entails a discussion and comparison of two such networks in Finland: Community Supported Agriculture (CSA) as represented by the *Herttoniemen Ruokaosuuskunta* (Herttoniemi Food Cooperative); and REKO Circles or *Rejäl Konsumtion*, which is Swedish for the principle of 'fair consumption' that underlies this unique food distribution network. These food distribution systems operate in peri-urban and urban areas and are reflective of creative farmer-to-consumer interactions operating outside the dominant Finnish oligopolized grocery store-driven food chain. The following paragraphs uncover various facets and nuances of these systems that are rooted in a desire for farm fresh food as well as the support of Finnish farmers.

Urban agriculture (UA) that thrives at the edge of the Arctic Circle is conceptually rather unexpected. Nevertheless, empirically urban agriculture even at far northern latitudes, such as in Finland, as well as elsewhere, serves to supplement and transform foodsheds. The idea of 'foodshed' is a dynamic conceptual framing device for this chapter. Foodshed is a term which was brought into popular use in the 1990s and provides a conceptual framework to 'facilitate critical thought about where our food is coming from and how it is getting to us' (Kloppenburg *et al.*, 1996, p. 33). The foodshed concept is

---

*Corresponding author; e-mail: sophia.hagolani-albov@helsinki.fi

described, 'Not as a doctrine to be followed, but a set of principles to be explored' (Kloppenburg et al., 1996, p. 33). The vision of a foodshed, as opposed to a food economy, integrates concepts and values which extend beyond the economic drivers associated with a globalized food system (Kloppenburg and Lezberg, 1996). The observed and perceived transformative forces presented by urban food production in and around northern cities of the Global North have been evaluated in the academic literature. The aim of this chapter is to bring into focus urban-based engagement with food production in Helsinki, Finland's capital city, at 60.1°N latitude through a comparison of the aforementioned Herttoniemi CSA and REKO Circle systems.

15.2 Background to Urban Agriculture

The practice of urban agriculture is not a new phenomenon. However, it has been receiving increased academic and media interest, as it is often tied to trends in sustainable development and a recent 'sustainability-environmental turn' (Moore, 2006; Tornaghi, 2014, p. 555). Sustainability is not the sole reason why people choose to engage in urban agriculture. Acts of urban agricultural practice can be interpreted as part of a process to support and further urban development and green planning initiatives (Colasanti et al., 2012; McLain et al., 2014). The interplay between planning initiatives, urban gardens and politics has been underscored (McClintock, 2014). In addition, urban agriculture has been framed as a livelihood strategy and a means to the development of social capital (Gallaher, 2012; Gallaher et al., 2013a). Importantly, gender roles and family structure can be examined through the lens of urban agricultural practices (WinklerPrins, 2002; Murrieta and WinklerPrins, 2003; WinklerPrins and de Souza, 2005; Chapters 2, 3, 11, 12, 13 and 14, this volume). Home gardens can be a way to link urban dwellers with their countryside heritage or carry the traditions of the country into city life (WinklerPrins, 2002). The garden is not simply a garden; it is an extension of the living space and the social interactions of family, household, neighbourhood and community (WinklerPrins and de Souza, 2009). The urban garden as a stage for social relations has been expressed in the conceptualization of the home garden as a female-constructed place and a coping mechanism integral to survival in urbanized spaces (Murrieta and WinklerPrins, 2003; WinklerPrins and de Souza, 2005).

The right to the city as expressed through concepts of food justice and alleviation of food insecurity are also common themes in studies focused on UA. The social spaces of urban agricultural practice have been characterized as politicized spatial platforms for gaining political agency within the context of the city (White, 2015; Shillington, 2013; Chapters 5, 8 and 13, this volume). The theme of community building as a goal or by-product of urban agricultural enterprises is discussed at length in case studies undertaken in both the Global North and Global South (Gallaher et al., 2013a, b; White, 2015). Further, delineations between the Global North and the Global South beyond the community-building dimension have been put forth in the literature which surrounds UA (see Chapter 1, this volume).

The urban grassroots food distribution networks in Helsinki and other cities in Finland presented below serve as vibrant examples of the (re)localization movement and the (re)connection of food, people and place (see Chapter 1, this volume). Both the innovation and institutionalized segments of the portfolio of urban agricultural activity in Finland are working to (re)localize the foodshed in ways that reflect a deeply rooted Finnish food heritage.

15.3 Policy and Programmes Impacting Urban Agriculture in Finland

The European Union's (EU) Common Agricultural Policy (CAP) defines an agricultural area as 'any area taken up by arable land, permanent grassland or permanent crops' (European Commission, 2014, p. 2). The CAP is fundamentally tied to agriculture in the context of rural development. However, the CAP does not currently provide specific mechanisms to develop policy concerning urban agriculture and urban food-growing initiatives. Given that the CAP is the primary directive for Finnish agricultural policy, the exclusion of urban agriculture in the CAP creates a subsequent exclusion of urban agriculture at the Ministry of Agriculture and Forestry and other entities which operate at the ministerial level in Finland. As a result, questions of urban agricultural policy are handled at the municipal level.

This situation is poised to potentially change in a future iteration of the CAP, as urban agriculture gains traction both popularly and in the research agenda. Currently, a research effort supported by the European Cooperation in Science and Technology (known by its acronym COST) is underway to assess urban food regimes and concomitant planning systems. COST is an intergovernmental framework allowing for coordination of national research agendas at a European level (COST, 2015). One of the main goals of the COST Action is to develop a European understanding and agenda on urban agriculture. This coordination of national research agendas helps to ensure that research efforts are productive while not duplicative and helps to identify gaps in the overarching research agenda. COST does not set research priorities or directly fund research efforts (COST, 2015). Finland is eligible to participate in this COST Action on Urban Agriculture; however, to date no Finnish research teams have come forward with proposals to participate in this urban agriculture-focused opportunity.

The omission of urban agriculture at the ministerial level is not reflected at the municipal level. The city government of Helsinki is active and supportive of urban agricultural endeavours and promoting the cultural heritage which surrounds food and agriculture in Finland. This support is evident in the active identification of interstitial areas appropriate for agricultural experiments and the inclusion of spaces for allotment gardening, both historically and currently, in the city plan (Rinne, 2014).

The city of Helsinki ideologically supports urban agriculture and has included it as a facet of the Helsinki Culinary Culture Strategy. This strategy represents a set of efforts developed and implemented by the City of Helsinki municipal authorities to bolster and stimulate the food culture in Helsinki (City of Helsinki, 2009). The city directly supports urban agriculture through the long-term provision of dedicated spaces in the city, thereby overcoming a major limiting factor to urban agricultural endeavours in both the Global North and the Global South historically and in the present day (Moore, 2006; Gallaher, 2012). In the case of the allotment gardens, the contracts with the city have been signed through to 2026, but the city as an entity does not participate in the management of these allotment gardens (City of Helsinki, 2014). Interestingly, in 2012, members of the *kaupunginvaltuutetut* (City Council members) set up their own urban gardening crew and City Hall opened a small balcony garden to symbolize support for urban agricultural endeavours (City of Helsinki, n.d.).

15.4 Methods

Our approach to this research has been qualitative in nature, with the goal of combining several different kinds of qualitative data, including semi-structure interviews, participant observation and participatory mapping. The use of interviews helped to identify the factors in the growth of the urban agricultural sector which are relevant and important to the interviewees (Hay, 2005). The themes which emerged from the multiple series of interviews address:

1. the perceptions of organic, local and Finnish cultural heritage values as expressed through participation in grassroots food distribution networks;
2. views on urban agriculture and its contributions to cultural heritage, food quality and sustainability;
3. attitudes about the role of geographic proximity and trust in food procurement decisions.

15.5 Urban Agriculture Sites and Spaces

There is an abundance of space for urban agricultural production within the urban spaces of Finland, especially Helsinki, which is both the largest city and part of the most populous metropolitan area in the country. There are two main categories of urban agricultural endeavours in Helsinki: those which are institutionalized at the municipal level and those which have arisen as a result of grassroots organization efforts. The *Viljelyspalsta* (allotment gardens) and *Siirtolapuutarhat* (cottage allotments) are both considered to be institutionalized as they have longstanding connections to the city government and are integrated into the city plan. There are also box and sack gardens in interstitial spaces in the city, in addition to home gardens, which are not institutionalized, but fall beyond the scope of this chapter. I will focus specifically on the non-institutionalized consumer-driven grassroots organizations which are represented by the Herttoniemi CSA and the REKO Circles. These endeavours are characterized as consumer-driven because they were started by consumers as an avenue to increase farmer–consumer interaction as opposed to purely profit-driven motives.

15.6 Herttoniemi Community Supported Agriculture (CSA)

The Herttoniemi CSA is a food cooperative formed in 2010 by members of the Herttoniemi food circle. The name in Finnish is *Ruokaosuuskunta*, which translates as 'food cooperative'. The Herttoniemi part of the name is tied to place; that is, the part of the city where the food circle and subsequent CSA were founded. Herttoniemi is a suburb of Helsinki, made up of four neighbourhoods, located east of the downtown area and easily reachable by metro. The idea for the CSA came from a single member. The actual work to set up the CSA project was completed by several members following a well-attended brainstorming meeting. The CSA currently has approximately 200 members, and the CSA fields are located in Vantaa, one of the three cities making up the Helsinki urban centre, or 'greater city' (Dijkstra and Poelman, 2012). The land where the fields are located is privately owned and leased on a 10-year term to the CSA. The fields are quite close to the international airport and under flight paths, with houses visible along one of the field edges. The area is considered peri-urban or suburban, but is in close proximity to the Vantaa urban centre.

Many of the members of the CSA have limited interaction with the physical fields and interact only at the pickup locations. There are four pickup locations in the city of Helsinki. Two are in library areas, one at a coffee shop, and one at a recently redeveloped downtown retail and industrial complex. All pickup locations are easily reached via public transportation. The members of the CSA range from young professionals to retired people; the unifying thread between the members is an interest in local food. The members must pay for the season share before the start of the season and are required to donate 10 hours of work each season or to pay a fine for any volunteer hours not completed. The members of the Herttoniemi CSA indicated several reasons they have chosen to be involved with this organization, ranging from political motivations to environmental concerns. These motivations will be explored later in the chapter through a collection of narratives from the members of the CSA.

In the CSA, food is grown by hired farmers for distribution to the members of the CSA. The farmers are hired by the members of the CSA; the land being farmed is under lease to the CSA and does not belong to these farmers. Two of the farmers of the CSA highlighted the land-lease arrangement and its implications for their own personal relationship to the land as one of the reasons this CSA concept really worked for them. Both farmers indicated that if the land was solely owned by them, then the CSA might not work as well, because they might have interests that are different from the interests of the members. Their priorities to meet open market demands could potentially override the members' produce preferences. The CSA members pay for their share in advance of each growing season, and during the growing season they receive their share of the vegetables harvested that week. The members do not get to pick which vegetables they receive, whereas in a food circle or a REKO Circle, the consumer only orders and purchases what he or she wants. This CSA was the first of its kind in Finland and has often been used as a model for other CSAs across the country. There are approximately 15 CSA experiments in Finland at the time of this writing, all founded within the last 5 years.

Figs 15.1 and 15.2. These photos illustrate how the winter REKO exchange in Jakobstad does not look like a typical farmers' market: there is no emphasis on convincing the consumer to buy the products, as the selling has taken place prior to meeting up. These photos were taken at the beginning of the pickup, which lasts approximately an hour. (Photographs by Sophia Hagolani-Albov.)

15.7 REKO Circles

The other grassroots movement represented by REKO Circles is organized primarily through social media and focuses on connecting producers directly to consumers. The REKO Circles were the brainchild of a Swedish-speaking Finnish organic beef farmer, Thomas Snellman, in a small city in western Finland. He was inspired by the Associations for the Preservation of Peasant Farming (AMAP) system during a visit to France in November 2012. The AMAP system follows what in the US would be known as a 'box system', consisting of consumers buying several weeks' worth of produce boxes in advance, usually a complete growing season (Curtet and Girard, 2012). This is akin to a CSA, but instead of all the produce originating from a single diversified farm, it brings together several local and organic producers to contribute toward the assembly of each box. Significantly, the REKO founder deviated from the AMAP system by designing his circles without any long-term contract. The act of participation in the REKO Circles is free for both the producers and the consumers. Money only changes hands when a specific product is bought, and this exchange happens directly between the producer and consumer. There is no professional administration for the food circles. The communication between the producers and consumers occurs almost exclusively through Facebook, as described by one of the participating members, as follows.

> The idea of REKO [is] that there will be no management – everything takes care of itself through the Facebook page. Producers put themselves, their products [on the Facebook page] before each delivery and customer orders [through Facebook] during the producer's post. (Interviewee E-8)

The reliance on social media means that the REKO model in its current form is only realistic in regions where there is consistent access to the internet and use of social media. In 2014 in Finland, approximately 86% of the population between the ages of 16 and 89 had used the internet in the last 3 months and 64% are on the internet multiple times per day. In addition, 51% of the population has used social media in the last 3 months (OSF, 2014). This emphasis on the use of social media for REKO-specific communication inherently makes the REKO Circle model challenging for less wired areas of the world, be they in the Global South or the Global North, as not all parts of the world are similarly connected via the internet or engaged with social media (World Bank, 2016).

In addition to the internet and social media, REKO Circles rely on trust in the relationship between the producer and consumer. In fact, from the consumer side, the foremost articulated benefit of participating in a REKO Circle was reinserting the concept of trust back into the food purchasing process, as described by a REKO consumer:

> REKO is an answer to consumers concern about food safety – as you get to know the producer you build trust, you learn where it comes from, you can ask about growing etc. Thus it also strengthens the value of the food. (Interviewee E-10)

From the producer side, there is the benefit of knowing how much of each product will be sold to consumers who are committed to the exchange before arriving at the selling point. This strategy is ideal for smaller producers to access a market and minimize the risk because all the products are presold. For example:

> For me as a producer this is actually the best part of the concept. I don't have to guess how [many] potatoes I need to bring to the market and I don't risk having to bring half of [them] back home. The consumer on the other hand doesn't risk going to the market and returning empty-handed because the potato farmer had underestimated the demand. (Interviewee E-9)

The only requirement to start a REKO Circle is a group of enthusiastic consumers who are ready to gather producers to sell directly to future members of the REKO Facebook group. This Facebook group is set up, and all consumers and producers join the group. The group is closed, but anyone is allowed to submit a request to join. These requests are then approved by a volunteer administrator. Producers advertise what they have for sale that week through a posting on the Facebook page. Consumers are able to respond directly to each post from producers when they would like to make a purchase, indicating how much they would like of the item offered for sale. There is a predetermined meeting time, as illustrated by the field notes at the beginning of this chapter, where all consumers and producers come together to exchange the money and products as they have prearranged through the Facebook group. The pickup location is usually in a parking lot and does not look like a typical farmers' market: there are no stalls or booths with displays, as there is no need to entice the customer. The pickup time is very limited, half hour to an hour, as all products are already presold. This process is summarized by a REKO group organizer as follows.

> At REKO . . . [the] customer orders products [in advance] via a Facebook page. This means that the producer knows exactly how much [of] the goods to [take] to the time of delivery, which means there is no waste. Delivery lasts for only 1 hour, which is advantageous for producers who do not need to stand a whole day, as is the case in an ordinary market. (Interviewee E-8)

The REKO Circles are answering a need that is not filled through traditional channels of distribution. To get to the heart of its revolutionary nature, one must understand the state of the supply chain in Finland. While there is no shortage of physical places to obtain groceries in Finland, there is limited choice of grocery store brands. As we elaborate below, there are two main supermarket chains which hold a 78.8% market share of the grocery business in Finland (Kesko, 2015). This market dominance limits local and small-scale production being selected for sale by these grocery giants. One of the farmers described his frustration with the commercial grocery chains.

> They even say that the size of farms has been controlled by the two grocery companies. Farms have to have the amount of animals [or] size of crop that one of the two major companies is willing to buy. (Interviewee A-15)

Only very recently has the Finnish grocery sector been infiltrated by grocery stores which are outside the two dominant chains. However, both facets of the supermarket system, Finnish-owned and foreign owned, do not tend to have large selections of organic food and when there is organic it is often imported from other parts of the EU and beyond. Even when there is a food product which is produced in Finland, a store's supply will often come from abroad, or less expensive foreign-produced fruits, vegetables, meat and processed foods can be found alongside the Finnish-produced food.

> Many want to make their food from scratch, preferably organic and since the supermarkets have been slow in responding, REKO is filling that need embedded in this cultural shift. (Interviewee E-14)

The idea that Finnish consumers want to know where their food comes from was also expressed numerous times by interviewees.

The two key points to the success of the REKO are the added value of local and 'known' food and the concept of interpersonal trust as an important consideration in food decisions. Trust is a theme often discussed in the context of Finnish consumer food decisions. This focus on trust could be a lingering result of the upswing in food safety issues which accompanied Finland's entry into the EU. Finland, long regarded as a peripheral country, experienced previously unknown issues in food safety in the early 2000s. These events perhaps have served to make consumers wary of unknown or non-Finnish food sources (Raento, 2010). The dissatisfaction with the current grocery system was reiterated by interviewees. Interview data and field observations confirm the strong desire for Finnish consumers to be able to interact with the producers; as one interviewee expressed, 'REKO is a way to put a face behind the products' (Interviewee E-8).

The REKO Circle as a means of mobilizing around food distribution has diffused relatively quickly through the Swedish-speaking part of Finland. The first delivery occurred in Jakobstad (Pietarsaari in Finnish), located on the coast of western Finland, on 6 June 2013. At the time of writing, and as reported by the founder of the REKO model, approximately 120 Circles have been established throughout Finland and more are being regularly founded every month. The largest concentration of Circles is in the western part of Finland, where the movement was founded. However, Circles are taking root in Finnish-speaking areas and have extended all the way across the country to the eastern edge, thereby encouraging 'a new way to shop', as one study participant put it, in other parts of the country. The simplicity with which the Circles are set up has facilitated their diffusion across Finland. The REKO concept has even been recognized by the Finnish government, and active leaders within the REKO Circles are being asked to share their expertise in local food marketing with the government. The following is a description of the REKO trajectory, which speaks of the activism of leaders, the ease with which new circles can be developed, the excitement over the access to local food, and the interest from the government in the activities of the REKO Circles.

> REKO is growing very [quickly] all the time. This week we will have at least three new circles. Last Thursday I was ... presenting the REKO concept to 20 project workers from [the] southwest. Not one of them had heard about REKO before, but they were very enthusiastic about this, and this morning a lady called me and said they have already started the preparation at two places. (Interviewee E-13)

15.8 Implications of CSAs and REKO Circles for Finnish Farmers

The development of these grassroots food distribution networks from the bottom up is a deviation from the typical flow of agricultural development in Finland. In numerous discussions with interviewees about the policies surrounding agricultural regulation, there was a widely held perception that while there were some grassroots efforts within the organic agricultural sector, the majority of the rules and regulations in organic and conventional agriculture were developed from the top – EU and ministerial level – and handed down to the farmers. This top-down policy flow was articulated by 74% of the farmers interviewed ($n=27$). One farmer expanded on this thought:

> It comes from the top down, farmers don't have much to say to the policy ... It is the same in conventional farming. And it is the same in every European country, but I do not think all European countries obey them as much as Finland does. (Interviewee A-45)

In spite of the feeling that Finland's farmers are prepared to obey the rules, 55% of the 18 farmers who were asked this question indicated that they had little or no voice to influence this policy or the overall functioning of the agricultural sector. These farmers ($n=6$) conceded that they have some influence through their local association: 'There is an organic association as well. Through these there is of course some way of getting your voice heard. But, yeah, as a single farmer, no, not really' (Interviewee A-62).

In addition to a perceived lack of political power, there is a notable concentration in the grocery sector and a further complication for farmers, especially small-scale farmers who do not produce enough for the large grocery chains to be willing to carry their products.

> I wanted to become involved because me and my colleague had discussed the need for a direct channel to the consumer, specifically a channel that made it possible to sell smaller quantities of various products (meat, potatoes etc.). If you try to approach the larger food store chains you have to be able to deliver a lot of goods right from the start and this is sometimes a problem for a smaller farm, especially if you are in a startup phase. (Interviewee E-9)

While the interplay between the institutionalization of the agricultural sector and the grocery sector has removed some of the autonomy from the farmer, alternative food movements in Finland are perceived as a way for farmers to regain their autonomy and exercise some power over the way they distribute their products. The REKO Circles are a strategy for farmers to be able to grow a smaller quantity and continue to be able to access a receptive consumer market.

15.9 Narrating the Effects of Urban Agriculture

The perceptions of interview participants in both of these grassroots food distribution networks explained above can be organized around three substantive themes: cultural values; agency within the food system; and upholding environmental values. Study participants often reflected on their concerns about the quality of food available in the dominant food chain, especially in reference to the increasing amount of imported food. The increase in foreign origin food is not unique to Finland: it is a wide-reaching result of the globalization of the food system (Trobe, 2001). The accounts of the interviewees suggest that constructions of well-being and contentment with food choice are undergoing a redefinition in Finland. Here, the notion of benefits is drawn from participants' perceptions that highlighted the connections between food-growing and cultural values. The key benefits and reasons that members chose to participate in an alternative food chain were articulated as follows.

1. A concrete way to express cultural values connected to food, specifically the desire to access high-quality local or known food.
2. A strategy to increase individual influence in the food chain and avoid support of the dominant food chain.
3. An instrument to lessen the environmental impact of food production in general by supporting food produced in close proximity to population centres.

15.9.1 Expressions of cultural values

The first perceived benefit was most clearly illustrated through a discussion of the values considered when purchasing a food item. Interviewees described the set of values they attach to food and then connected these values to their choice to participate in a grassroots food distribution network. Among the 21 interviewees, ten different values were expressed:

- seasonality
- ethical production
- personal health
- sustainable production
- organic
- localness of food
- economic limitations
- freshness
- 'Finnishness'
- taste.

The locality of food products ($n=9$), economic consideration ($n=8$), and organic ($n=7$) were the three considerations which were most often mentioned when making a decision to purchase a food item. The following quotations provide insight into these values.

> I usually try to buy local and stuff that is in season. I try to avoid products that come from [a] far distance. Yeah, so it is mostly that it is local and I try to buy as much organic as I can. (Interviewee F-6)

The relationship between local food and organic food was further explored through a line of questioning which assessed whether the member would prefer to buy a bag of Finnish conventional carrots or Swedish organic carrots. This question revealed an interesting correlation between the perception of quality and proximity of production. Over half of the interviewees indicated that they would prefer the Finnish conventional carrots over Swedish organic carrots, due to a trust for the Finnish produce, even if it is conventional.

> Conventional Finnish ones [carrots] because they are Finnish and I consider Finnish carrots and most other Finnish produce to be clean enough that I don't have to buy organic. I trust the Finnish produce to be healthy and clean even if they are not organic. I prefer organic of course, but I don't mind eating regular ones. (Interviewee F-9)

15.9.2 Increase of individual agency within the food system

> Well, I think my first thought was perhaps that it [the CSA] is a bit too hippy for us, but I don't know. But then again it is nice. It is like organic farming and nice and we have [a] really ... twisted market in wholesale products in Finland. Because we basically [have just] two major [grocery] chains and there is not too much going on outside of those two. And I think it is nice to support some ... alternative ways of producing. (Interviewee F-2)

In the Finnish grocery system, there are two main grocery groups – the S Group and the K Group. S Group has a market share of 45.7% and K Group has a market share of 33.1% for a combined share of 78.8% (Kesko, 2015). While these are two different companies with two different boards of directors, interviewees often gave the impression that they felt the two companies worked very closely together. One interviewee described the idea of choice in Finland as whether you would prefer your meat to be in S paper or K paper; that was the extent of consumer choice involved. Given this grocery climate, many of the interviewees described their participation in the CSA as a way to distance themselves from the S- and K-dominated grocery market. One interviewee elaborated in this way:

> Well, I have tons of reasons [to participate as a CSA member]. I have political reasons, I have actual reasons, and then health reasons, I guess, as well. Mostly it is because I want to support the cause. Small local farms, biodynamic farming, and also the political side of things is that I want to resist these big chains. (Interviewee F-6)

This thought was reiterated and expanded by another CSA member, who also felt that participation in the CSA is a way to avoid the dominant grocery system in Finland. This concern was conveyed as follows.

> [S]ome of the money we spend here, I think it is like, not just for the food, it is the whole concept. And I am happy to pay, because that is – sort of feels a bit like I am a rebel. We are rebels somehow, because we are not choosing the S Market or Lidl or some other big grocery store, but we are doing this ourselves. And that is value for money. (Interviewee F-1)

It is interesting to note the language used in both of these quotations: 'fight' and 'rebel'. These words have powerful connotations and lend some insight into the strong feelings that some CSA and REKO members have about their relationship with the big supermarkets.

Within the group of interviewees, there were also some who took a much more pragmatic approach to their food purchases and were not as concerned about upholding certain values with the considerations they placed on food purchasing decisions. However, even while specifically talking about food values, the quotation below also hints at a dissatisfaction with the mainstream grocery establishment.

> I have a food in my mind, and I just buy whatever I need. It's – the selection in Finnish market isn't that great. You need potatoes, you buy potatoes. There are not too many varieties available. (Interviewee F-14)

These perceptions indicate there is a certain level of powerlessness that the interviewees seem to feel in regard to the commercial grocery system which compels them to seek out an alternative. The dissatisfaction with the oligopolization of the grocery system by the S and K group was pervasive in many discussions of Finland's agricultural geography. One key informant from the organic sector-focused interviews relayed the thought that the grocery store chains went so far as to determine the size of the farms in Finland, based on their refusal to accept orders below a certain amount.

The CSA and REKO Circles represent an alternative for their members to make their food purchasing decisions outside the dominant food distribution chain. They can express their dissatisfaction with the grocery system by not using it for a portion of their food purchases. As there are not very many options given the market dominance of the main grocers, the CSA and REKO open the option to purchase directly from the farm fields.

15.9.3 Environmental considerations, nationalism and the benefit of local food

The perception that Finland is 'clean' is one that was articulated in several of the interviews. One interviewee expressed the perception quite directly, later in the interview she did acknowledge that she might be mistaken in her belief, as she does not have specific facts to back it up, but the lack of data did not stop her from holding this belief that 'We have clean nature here. We have large areas with no factories which could poison the environment' (Interviewee F-7). Another interviewee expanded on this 'ecological hunch' that food from Finland is preferable to food produced and shipped from foreign countries. This interviewee added the idea of spatial proximity, indicating that it is not simply that food from Finland is better in some way, but that one must take into account where the food was actually produced and how far it had to travel to get to the consumer. These complicated considerations concerning the spatial proximity and production methods used were explained as follows.

> I think I would go for the Swedish [carrots] because they are not that far away, but . . . if I should [decide] between conventional Finnish carrots and for example Spanish organic carrots. I think I would go for the Finnish carrots. I think it is like an ecological hunch. I don't know how much . . . carbon that produces to . . . ship those carrots from Sweden compared to . . . if you had to ship them from Spain, but I think . . . of course, I don't really have so much . . . data – but there is some kind of hunch [about] what kind of food items it is good to . . . ship from abroad [or] consider growing them here. (Interviewee F-2)

This idea was extended but qualified in that their desire to buy Finnish was not linked to special feelings for Finland or the quality of specifically Finnish produce; rather, it was tied up with thoughts solely about the ecological footprint. While the term 'food miles' was never specifically used, the concept of the spatial component of production and the linking of ecological footprint to the transportation in production was explored by the CSA members (Coley *et al.*, 2009).

> I try to buy organic or close produced [local] products when possible. So as to decrease the ecological footprint. I am not, I don't buy local food in order to support Finnish. I don't have nationalistic consumer behaviour, but rather just in terms of decreasing the ecological footprint. (Interviewee F-22)

There was also a component of nationalism in food choices expressed, with interviewees musing that organic is not even as important as the 'Finnishness' of the food. The following statement does not link the desire to buy Finnish food to any reasons such as environment; it is a reflection of a nationalistic desire for Finnish food.

> It doesn't have to be organic, just, you know, I prefer Finnish. Finnish like every citizen wants to buy... their own food and I like the Finnish food. (Interviewee F-17)

Whether the specific motivation for Finnish-produced food was nationalistic or ecological in origin, the interviewees stressed the theme that closely produced food was an important benefit of participation in these alternative markets.

15.10 Conclusion

The Finnish food–Finnish consumer connections promoted through the efforts of the CSA and REKO Circles are steps towards what could be described as 'transformative action' (Gray *et al.*, 2014). Finland stands out for its proactive nature of land-use planning to create spaces in the urban landscape for food and agriculture interactions. Our study, grounded in numerous field observations and qualitative data, suggests that the Finnish urban agricultural sector is enlarging and working to overcome economic and environmental limitations with creative utilization of new strategies and social networks. The discussion of the CSA in Helsinki and REKO Circles uncovers facets of emerging Finnish grassroots movements that aim to shorten the distance between farmers and consumers and that, in turn, reinforce important cultural and social values. Indeed, connecting the farm to the city in Finland supports the concept of eating local, which is perceived as being integral to the Finnish food heritage and to maintaining cultural and environmental sustainability through the (re)localization of the Finnish foodshed.

References

City of Helsinki (2009) City of Helsinki Culinary Culture Strategy. Available at: http://www.helsinkifoodism.com/wp-content/uploads/2012/12/ENG_Helsinki-Culinary-Culture-Strategy.pdf (accessed 17 November 2016).

City of Helsinki (2014) Allotment gardens. Available at: http://www.hel.fi/www/Helsinki/en/culture/leisure/cottages/allotment-gardens (accessed 29 December 2014).

City of Helsinki (n.d.) Edible city. Available at: http://www.helsinkifoodism.com/en/syotava-kaupunki (accessed 29 October 2015).

Colasanti, K.J., Hamm, M.W. and Litjens, C.M. (2012) The city as an 'agricultural powerhouse'? Perspectives on expanding urban agriculture from Detroit, Michigan. *Urban Geography* 33(3), 348–369.

Coley, D., Howard, M. and Winter, M. (2009) Local food, food miles and carbon emissions: a comparison of farm shop and mass distribution approaches. *Food Policy* 34, 150–155.

COST (2015) COST actions. Available at: http://www.cost.eu/COST_Actions (accessed 27 July 2015).

Curtet, P. and Girard, L. (2012) France's small-scale organic farmers celebrate 10 years of boxing clever. *The Guardian*, 10 April. Available at: http://www.theguardian.com/world/2012/apr/10/france-farming-organic-vegetable-boxes (accessed 10 October 2015).

Dijkstra, L. and Poelman, H. (2012) Cities in Europe: The New OECD-EC Definition. Available at: http://ec.europa.eu/regional_policy/sources/docgener/focus/2012_01_city.pdf (accessed 27 September 2015).

European Commission (2014) Glossary of terms related to the Common Agricultural Policy. Available at: http://ec.europa.eu/agriculture/glossary/index_en.htm (accessed 4 February 2015).

Gallaher, C.M. (2012) Livelihoods, food security and environmental risk: sack gardening in the Kibera slums of Nairobi, Kenya. PhD dissertation, Michigan State University, East Lansing, Michigan, USA.

Gallaher, C.M., Kerr, J., Njenga, M., Karanja, N. and WinklerPrins, A.M.G.A. (2013a) Urban agriculture, social capital, and food security in the Kibera slums of Nairobi, Kenya. *Agriculture and Human Values* 30, 389–404.

Gallaher, C.M., Mwaniki, D., Njenga, M., Karanja, N. and WinklerPrins, A.M.G.A. (2013b) Real or perceived: the environmental health risks of urban sack gardening in Kibera slums of Nairobi, Kenya. *EcoHealth* 10, 9–20.

Gray, L., Guzman, P., Glowa, K.M. and Drevno, A.G. (2014) Can home gardens scale up into movements for social change? The role of home gardens in providing food security and community change in San Jose, California. *Local Environment* 19, 187–203.

Hay, I. (2005) *Qualitative Research Methods in Human Geography*. Oxford University Press, South Melbourne, Victoria, Australia.

Kesko (2015) Grocery trade. Available at: http://www.kesko.fi/en/company/divisions/grocery-trade (accessed 5 October 2015).

Kloppenburg, J. and Lezberg, S. (1996) Getting it straight before we eat ourselves to death: from food system to foodshed in the 21st century. *Society and Natural Resources* 9, 93–96.

Kloppenburg, J., Hendrickson, J. and Stevenson, G. (1996) Coming in to the foodshed. *Agriculture and Human Values* 13, 33–42.

McClintock, N. (2014) Radical, reformist, and garden-variety neoliberal: coming to terms with urban agriculture's contradictions. *Local Environment* 19(2), 147–171.

McLain, R.J., Hurley, P.T., Emery, M.R. and Poe, M.R. (2014) Gathering 'wild' food in the city: rethinking the role of foraging in urban ecosystem planning and management. *Local Environment* 19(2), 220–240.

Moore, S. (2006) Forgotten roots of the green city: subsistence gardens in Columbus, Ohio, 1910–1935. *Urban Geography* 27(2), 174–192.

Murrieta, R.S.S. and WinklerPrins, A.M.G.A. (2003) Flowers of water: homegardens and gender roles in a riverine Caboclo community in the Lower Amazon, Brazil. *Culture and Agriculture* 25(1), 35–47.

OSF (2014) Use of information and communications technology by individuals. Official Statistics Finland, Helsinki. Available at: http://www.stat.fi/til/sutivi/index_en.html (accessed 15 November 2016).

Raento, P. (2010) Stomaching change: Finns, food, and boundaries in the European Union. *Geografiska Annaler: Series B, Human Geography* 92, 297–310.

Rinne, T. (2014) Kaupunkiviljelypaikat Helsingissä kesällä 2014 [in Finnish]. City of Helsinki. Available at: http://www.hel.fi/static/hkr/kaupunkiviljelypaikat_2014.pdf (accessed 29 December 2014).

Shillington, L.J. (2013) Right to food, right to the city: household urban agriculture, and socionatural metabolism in Managua, Nicaragua. *Geoforum* 44, 103–151.

Tornaghi, C. (2014) Critical geography of urban agriculture. *Progress in Human Geography* 38, 551–567.

Trobe, H.L. (2001) Farmers' markets: consuming local rural produce. *International Journal of Consumer Studies* 25, 181–192.

White, M.M. (2015) D-town farm: African American resistance to food insecurity and the transformation of Detroit. *Environmental Practice* 13(4), 406–417.

WinklerPrins, A.M.G.A. (2002) Linking the urban with the rural: house-lot gardens in Santarém, Pará, Brazil. *Urban Ecosystems* 6, 43–65.

WinklerPrins, A.M.G.A. and de Souza, P.S. (2005) Surviving the city: urban home gardens and the economy of affection in the Brazilian Amazon. *Journal of Latin American Geography* 4, 107–126.

WinklerPrins, A.M.G.A. and de Souza, P.S. (2009) House-lot gardens as living space in the Brazilian Amazon. *FOCUS on Geography* 52, 31–38.

World Bank (2016) Internet users (per 100 people). Available at: http://data.worldbank.org/indicator/IT.NET.USER.P2 (accessed 23 February 2016).

16 The Appropriation of Space through 'Communist Swarms': A Socio-spatial Examination of Urban Apiculture in Washington, DC

Lauren Dryburgh*
American University, Washington, DC, USA

> When I started researching... urban beekeeping, to be honest, I was a bit intimidated. Urban apiaries remain the domain of the hardcore – the tattooed hipsters in Bushwick or other outer-borough neighborhoods [in New York City] and their communist swarms.
> (Levin, 2010, p. 1)

16.1 Introduction

About a month after Eliza De La Portilla and her family moved into their urban South Florida home, their neighbours came over to thank them. De La Portilla and her family had brought something into the community that was then unusual – honeybees. The long-time residents in the house next door were overjoyed to find that there were again bees in a space where buzzing had not been heard for quite some time (De La Portilla, 2013). The family of urban apiculturists was not surprised to hear that their bees were filling an ecological gap – declines of honeybee populations are well observed and widespread across the US and throughout the world today (Neumann and Carreck, 2010). This chapter will examine the role of beekeepers like De La Portilla in municipal spaces, analysing how these practitioners navigate both the political and urban ecologies of beekeeping in cities.

Scholars have noted that the emergence of urban agriculture (UA) and alternative food networks is often a reaction to the failings of the global neoliberal food system (McClintock, 2014; see Chapter 8, this volume); urban apiculture can so be linked. Thus, in order to understand the role of apiculture in cities, it is helpful to understand its decline at a larger scale. Although losses in honeybee populations have been sustained in Europe, the Middle East and Japan (Neumann and Carreck, 2010), these losses have been particularly well documented in the US; the number of managed bee colonies (which average 50,000 bees at the height of the season) (IBRA, 2015) in the US declined by 59% between 1985 and 2005 (Potts et al., 2010). Around the onset of Colony Collapse Disorder, first described in 2006, more beekeepers began reporting colony losses with different symptoms than in the past (Ellis et al., 2010). In a national survey of managed honeybee colonies conducted over 2013 and 2014, beekeepers reported losing 34% of their colonies; the survey also found that 66% of practitioners had losses higher than 19%, which is what they deemed acceptable (Lee et al., 2015). As global colony numbers decrease,

---

*E-mail: dryburgh.lauren@gmail.com

the number of pollinator dependent crops is increasing (by over 300% since 1961), leading to fears of a pollinator deficit in North American and European agriculture (Potts et al., 2010). This has already been documented in Maoxian County of southwest China, where the virtual eradication of wild bees due to pesticide use and lack of habitat requires the hand pollination of apple orchards and has forced some farmers to switch to other crops (Partap and Ya, 2012).

The causes of the honeybee decline in particular are myriad, and are not, as yet, completely agreed upon by scientists or beekeepers. Vampire-like varroa mites sip bee blood and spread infections, devastating both managed and 'feral' (or unmanaged) colonies (Danka, 2012; Dennis and Kemp, 2016). Both the active and inert compounds in agrochemicals such as pesticides and fungicides cause learning impairment in adult bees (Mullin, 2015). One class of pesticides, neonicotinoids, are thought to be particularly deadly to honeybees; these compounds have been shown to impair their ability to navigate (Fischer et al., 2014). In the US, over 1.5 million colonies are stressed by forced annual migrations to pollinate the almond monocrop in California (Pettis and Delaplane, 2010). In Europe, socio-economic trends are pushing beekeepers out of the practice, and because of these environmental stressors, honeybees simply cannot survive without them (Potts et al., 2010).

Bees are indicator species. Due to their relationship with flowering plants, they will show signs of disturbances in their ecosystems (McFrederick and LeBuhn, 2006). A clear example of this was documented in the Canadian Province of New Brunswick in the late 1990s. At that time, conservationists used the pesticide fenitrothion widely in the forests of eastern Canada to control an invasion of spruce budworms and protect the timber industry there. The unintentional drift of these chemicals from forests to nearby blueberry fields caused reduced pollinator diversity, and therefore, lower blueberry yields. The effects of this reduction in biodiversity continue (Jarvis et al., 2007).

This is an example of an ecological rift (Foster and Clark, 2009; McClintock, 2010). These rifts occur when an attempt to ameliorate a metabolic rift, or disengagement of people from nature and sites of production that often results in ecological degradation, is made with a short-term, equally degrading technological fix. Such an ecological rift is important to keep in mind when discussing urban apiculture because it is rifts such as those in New Brunswick that have contributed to the general decline in bee health – and the rise of beekeeping in cities in response. And similar ecological rifts also occur in urban settings where bees are managed. This chapter will address these through a Marxian conception of space, particularly the ideas of Lefebvre, which are important for a critical understanding of these tensions. These concepts will be used in an attempt to understand the socio-spatial place of the urban beekeeper, situating the Lefebvrian view of the 'hardcore' practice of urban apiculture as a means through which to appropriate space within a wider urban political ecology (UPE) framework. This allows an examination of how overarching uneven processes affect the urban apiculturist and urban bees.

Urban cultivators navigate the uneven in myriad ways. As McClintock (2014) notes, in addition to filling the gaps created by market failure and the state's abandonment of social policies, urban cultivators are also doing radical work – often supporting a resilient environment and local food security. In undertaking a practice that radically alters urban ecology in a far-reaching way – urban bees will forage for flora on average up to 1.2 km away, and their maximum range extends up to 12 km (Garbuzov et al., 2015) – apiculturists have an integral role in these contradictions, and are integral for healthy UA.

16.2 Urban Political Ecology and Lefebvre

UPE is a framework that is useful for unravelling and analysing these interconnected economic, political, social and ecological processes that form unjust and uneven urban environments (Swyngedouw and Heynen, 2003). UPE resists the nature/society binary (Heynen et al., 2006), contending that the only accurate conception of space is an understanding that recognizes that 'human activity cannot be viewed as external to ecosystem function' (Swyngedouw and Heynen, 2010, p. 82). Through this examination of the socio-natural, UPE is able to tease apart these uneven processes and power structures and the discourses that reinforce them (Swyngedouw, 1999).

Swyngedouw (1999) notes that, for Lefebvre (1991), nature and society are produced and malleable. Therefore, the discourses surrounding nature are also produced, and many of them are expressed through symbols and romantic sentiments, or perpetuated by scientists and other so-called authorities. Examining these discourses is central to the analysis of urban beekeeping in this chapter.

Also central to this analysis are Lefebvre's concepts of the appropriation of space and dwelling. To explore these, it is first helpful to understand his theory of the production of space. Lefebvre (1991) asserts that the production of space has three elements: perceived space (space as it appears in objects or form), conceived space (knowledge about space and its production), and lived space (the emotional and subjective experiences and practices attached to space) (Purcell, 2002). The right to the city, a popular mode for envisioning more emancipatory urban spaces (see Chapter 8, this volume), is the ability to participate in the production of space and to appropriate space for new uses (Purcell, 2002; McClintock, 2014). Purcell (2002, p. 101) puts this in less abstract terms: the right to the city is 'a call for a radical restructuring of social, political, and economic relations, both in the city and beyond'.

This is why the right to the city is often used as a response to neoliberal urbanism. Yet, as Purcell (2002) argues, it benefits from some interrogation, particularly through a critical UPE lens. A central question for Purcell is that of scale. The participation in the production of urban space that is so central to harnessing the right to the city occurs at many scales, large and small. This can be problematic in determining exactly to *which* space(s) one has access. There is also the question of *who* has the right to the city. Lefebvre's Marxist right to the city sees a dismantling of the capitalist city, and he indicates that the working class must be the force that performs this dismantling. Purcell suggests this idea might be expanded to encompass other identities and projects, such as dismantling the racist, or patriarchal or otherwise bigoted city simultaneously.

Purcell's extrapolation of Lefebvre's right to the city is important, and is reflected in the many ways urban beekeepers relate to space. Some urban beekeepers have achieved their right to the city in very Lefebvrian terms, while others are appropriating space at a slower pace and scale, but still working towards creating structural change and dismantling the unjust city. These apiculturists are appropriating space through what Lefebvre would refer to as 'dwelling'. Unlike the larger project of staking out one's right to the city, the act of dwelling is mostly small-scale and intimate. Dwelling involves caring for a space and rendering it one's own, though it is not 'confined to an individual dwelling', but rather it relates to 'multiple scales of human processes' (Stanek, 2011, p. 86). It arose out of Lefebvre's observations of the needs-based, homogeneous and standardized housing stock (or *grands ensembles*) erected in France in the 1950s. Lefebvre found that, among the inhabitants of the *grands ensembles*, the ability to create a unique space or way of living distinct from one's neighbours was extremely important. Unable to live truly separately from others residing in these estates, the people of the *grands ensembles* adapted through dwelling.

For example, a strike in the *grands ensembles* in 1962 won residents the ability to dry their clothes outside, position TV antennas the way they desired, and use trashcans and parking lots in the area. These might seem like small victories, but Lefebvre noted that the desire to dwell was strong enough to fuel the organized, collective appropriation of the space (Stanek, 2011). Dwelling, for Lefebvre, is both poetic and political. It is users of a space rearranging what they have carefully and precisely in an everyday act of resistance (Mommersteeg, 2014). It is an act of appropriation because it involves making an activity 'an oeuvre . . . or one's own . . . in a conflict . . . between constraining powers and the forces of appropriation' (Stanek, 2011, p. 87). Such small-scale acts of dwelling and appropriation of space are integral to establishing a right to urban life that supports the multitudes of interests and identities of those who inhabit these spaces. As this book and specifically this chapter illustrate, this is not unlike the situation of many urban agriculturists and beekeepers today.

Thus, Lefebvre's conception of dwelling, in concert with a critical UPE lens, is useful for examining the practice of urban apiculture. And just as apiculture is integral to agricultural production at its most industrialized, largest scales, so too is it integral to urban agriculture. It is precisely the shifting socio-ecological assemblages and space appropriation of urban cultivators of all types, coupled with the ecological necessity of the honeybee, which sustains perennially flourishing UA. From the relatively small vertical green spaces created by sack gardens in the Kibera slum in Nairobi, Kenya (see Chapter 14, this volume) to 600 m² house lot gardens of Brazilian Amazonian cities (WinklerPrins and de Souza, 2009) to the 16,000 m² Newark Street Community Garden in Washington, DC (NSCGA, 2015), cultivated urban green spaces of all sizes create pollination opportunities for urban bees. These pollinators, in turn, support these spaces. Therefore there is room and a need for urban bees, and this chapter will highlight the usefulness of the Lefebvre's concept of dwelling and critical UPE analysis for examining the current state of urban apiculture globally.

The following employs the method of phenomenological analysis to provide more perspectives from which to understand the socio-spatial phenomenon that is urban beekeeping. This involves a brief discourse analysis on scientific studies, government documents, and the online correspondence and blogs of urban beekeepers themselves, to better understand the power structures that dictate how space should be conceived in the city. In addition, the analysis that follows is partly in the form of an analytic auto-ethnography. With faculty, other current undergraduates, graduate students and alumni, I helped keep approximately 240,000 bees in four hives on top of one of the buildings on American University's campus in Washington, DC, and I will include my experiences within this group as part of my data.

16.2.1 Findings: scientist–beekeeper relationships

As urban beekeeping becomes a more widely recognized practice globally, scientists and experienced apiculturists have found more opportunities for knowledge transfer and interactions with novices or beekeepers utilizing different methods. Suddenly, these practitioners are the subject of a great deal of analysis and focus from various groups and disciplines, as this chapter exemplifies.

When examining apiculture in the GS, a colonial canon begins to emerge; what beekeeping is mentioned in English is composed almost entirely of development literature (see USAID, 2013) and travel blogs (see Sanford, 2005; Blomstedt, 2009; Hihat Honey, 2010). Additionally, there is some concern that European honeybees (*Apis mellifera*) and beekeeping methods from parts of the GN are overtaking local species and practices in some parts of the GS. In one memo, scientists from Laos warn against keeping the European variety of honeybee, *Apis mellifera*, as there is a danger this could limit the indigenous populations of *Apis cerana*, *Apis florea* and *Apis dorsata*, as has been the case in Nepal, India and China (Sengngam and Vandame, 2003). The extensive introduction of *A. mellifera* in Laos could lead to lack of floral forage for other species, undesired interbreeding, and pose pathological risks by spreading mites or disease (Sengngam and Vandame, 2003). Another obstacle is that, in Laos, 'beekeeping with *Apis mellifera* implies the use of an advanced and expensive technology', and very close monitoring to ensure honey flow (Sengngam and Vandame, 2003, p. 14). Rather than promote this expensive technology, the authors describe the benefits of the hive design traditional to Laos: a hollow tree trunk that is sealed at each end (Sengngam and Vandame, 2003).

This raises questions about the production and sharing of beekeeping knowledge in the GS. Scientists in Laos promote beekeeping (in the capital city of Vientiane and beyond) that builds upon existing traditions. Elsewhere, particularly in Africa and Central Asia, there is mostly emphasis on 'beekeeping for development', and this is almost always a rural endeavour (Bees for Development, 2015). It seems urban apiculturists from the GS remain largely unobserved. Perhaps this is due to their ubiquity in some places; as Washington, DC beekeeper Toni Burnham discovered when she visited Mérida, Yucatán, Mexico, urban beekeeping is so common that 'everybody's aunt seems to have a few hives' (Burnham, 2015, p. 1). This ability to keep bees so casually and unchallenged, to dwell in this sense, might sound utopian to beekeepers in the GN.

Urban apiculture is still very much a burgeoning (or re-emerging) phenomenon in the GN, however, and this is evident in the ways scientists and other knowledge producers address and observe urban beekeepers. Some are vocally promoting urban beekeeping. Researcher Noah Wilson-Rich is a huge proponent of urban apiculture in Boston. He estimates that at 62.5%, bee survival is usually higher in urban areas than it is in rural areas (where the survival rate is about 40%) as bees have better quality, more diverse, and longer blooming forage in the city, due to the urban heat island effect, and less exposure to pesticides (De La Portilla, 2013; Johnston, 2013). He promotes 'the idea of beekeeping in your own backyard or rooftop or fire escape', hoping that people will see 'how simple it is and how possible it is' (Wilson-Rich, 2012, p. 1). In doing so, he is promoting the appropriation of these spaces. He has an incentive to do this; in order to subsidize his research, he started the company Best Bees, 'where we deliver, install and manage honeybee hives for anybody who wants them in the city' (Wilson-Rich, 2012, p. 1). This highlights two tensions that come with beekeeping in urban spaces: it promotes the idea that supervision from an expert is necessary in order to keep bees in the city, and it underlines how inaccessible beekeeping can be. It is interesting to consider the contradiction that urban beekeeping is simple and possible, yet requires the mediation of an expert. Intentional or not, this may create a financial barrier for people who want to keep bees and presumably could otherwise do so owning little to no land.

Other scientists are sceptical that encouraging urban apiculture is beneficial for urban ecology at all. Biologists Francis Ratnieks and Karin Alton at the Laboratory of Apiculture and Social Insects at the University of Sussex report that in 2013 there were ten beehives/km² in Greater London – ten times the average in the rest of England. This meant that, between 2008 and 2013, the number of beehives in Greater London doubled from 1677 to more than 3500 (Clune, 2013). Although many people have taken up beekeeping to 'help the bees', Ratnieks and Alton write that this altruism has resulted in there being too many honeybees in London. This has led to malnourished bees, and could lead to the spread of diseases among these weakened bees and the hives they inhabit. The increased risk of American Foulbrood is of particular concern to these scientists. American Foulbrood is rare in the UK, but it is devastating and very contagious – in order to prevent spreading, infected hives must be burned. To protect the honeybee population, the researchers instead promote the increased growth of bee-friendly flora by Londoners – cultivating plants, rather than beehives.

Bee researchers in the US, such as Marla Spivak, a bee specialist at the University of Minnesota, are encouraging plant cultivation as well. Spivak is not just encouraging people to put any flowering plant in the ground, however. She is specific that these plants need to be free of pesticides (Meador, 2015). The reasons why are particularly apparent in Utah, where government officials are concerned that novice beekeepers are unknowingly allowing bee disease to spread in urban areas. They are particularly worried because hobbyist beekeepers in Utah often rent their hives out to farmers, and the pesticides used on conventionally farmed crops weaken hives. This puts the hives of novice apiculturists at increased risk for diseases the new beekeepers do not know how to identify or treat (Swan, 2012). This is the flip side of Wilson-Rich's message. Beekeeping may be 'simple' and 'possible', but scientists want to monitor urban hives to prevent the spread of disease that can occur under the watch of an inexperienced beekeeper.

These contradictory messages fit well within the contradictions of UA. In London, urban apiculturists are able to truly dwell in and appropriate the space in which they live for beekeeping by fulfilling their aspirations to do so. For Lefebvre, the admonishments of scientists towards beekeepers for overpopulating London, as well as their instructions to aspiring apiculturists to simply 'grow bee-friendly flowers instead' (Clune, 2013, p. 1) might make scientists part of the overarching power structure that seems unable to halt the continued appropriation of space in response. Beekeeping in London has become an instrument through which to join a new movement, and people are picking it up in droves. In this sense, the act of individual beekeepers has become, in a sense, emancipatory, creating social, and therefore, ecological change throughout London and other cities in the GN. These particular movements are mostly populated by relatively wealthy people, which ties into Purcell's questions about who is emancipated by this right to appropriate space.

Lefebvre noted that appropriation of space is extremely useful for learning about the 'production of new spaces' (1991, p. 167). When Londoners began appropriating more and more space for beekeeping, it shed light on some of the current socio-ecological limits of London and, perhaps, the need for that space to change.

16.2.2 Findings: Government–beekeeper relationships

Despite the socio-ecological shift in London, beekeeping is still legal there, but this is not the case elsewhere. There are many places where municipal governments put stringent regulations on beekeepers. In her article, 'Where Beekeeping is Still Illegal', Kim Flottum (2010), chairman of hobbyist group the Eastern Apicultural Society of North America, lists the so-called 'no buzz zones' in the US. These are places where beekeeping is either entirely illegal, or has restrictions on the practice that are prohibitive to practitioners. These include cities like Omaha, Nebraska, Fort Worth, Texas, and Norfolk and Fredericksburg, Virginia.

Omaha, Nebraska, in particular, is an interesting example. Beekeeping there is not entirely illegal, but it is restricted. The Omaha Ordinance for Beekeeping does not allow apiculture within 100 feet of public recreation spaces, or within 25 feet of any dwelling, street or roadway, or other private property. Hives in legal locations require a permit, except in agricultural areas, according to the Omaha City Council's law (2009). This is an interesting loophole, as it is impossible to restrict foraging bees to agricultural zones. Therefore, it seems odd that there would be a difference in regulation for adjacent agricultural and residential areas, restricting the number of pollinators in the city's ecosystem.

Beekeeping was officially legalized in Los Angeles, California, in 2015, but it took a great deal of public outcry to overturn the 136-year-old ban on apiculture in the city (Wagner, 2015). Until 2015, beekeeping was not allowed in most residential areas in Los Angeles, except those that had passed separate ordinances to allow it, such as the neighbourhood of Del Rey. Passionate urban beekeepers continuously petitioned the City Planning Department to allow the activity in all areas of the city, arguing that the ecological importance of beekeeping necessitates the support of the practice within the city (McFarland, 2013). Del Rey council member Bill Rosendahl was an early proponent of legalizing beekeeping across Los Angeles. Rosendahl, who is very passionate about urban beekeeping, empowered local organizations such as Backyard Beekeepers and received support from local conservation non-profits such as a group called Honey Love (Kudler, 2011).

In some ways, these petitioners are similar to those who rioted for their right to hang their laundry to dry outside in the French *grands ensembles* in 1962. Both demand the right to use space in a manner that almost seems too trivial to demand so strongly from an outsider's perspective. Yet, as per Lefebvre's definition, they are actively caring for their surroundings. Perhaps it is this care for a space, creatures within that space, and the wider urban socio-ecology that is seen as odd. Bees nearby are certainly a frightening prospect for some neighbours, there are steps beekeepers can take to mitigate the concerns of neighbours if allowed to do so.

Until recently, beekeeping was restricted in New York City as well. In 2009, for example, Kathleen Boyer was fined $2000 for keeping bees in her front yard in Brooklyn. The Department of Health and Mental Hygiene, who issued the fine, justified it on the basis that the bees were a public threat. After Health department research found that reports of bee stings were minimal in the city, they passed a law in 2010 that allows New Yorkers to keep honeybees (Navarro, 2010). Recently, there has been some concern that there is not enough forage in the city and this is making bees defensive (though how defensive is disputed). It is also reported to be reducing honey yields (Nessen, 2012). Once beekeeping becomes sanctioned, it seems, urbanites have an overwhelming desire to take part – unknowingly altering the surrounding ecosystem with their subjective everyday dwelling.

An early supporter of legal beekeeping in New York City was councillor David Yassky, a representative of Brooklyn Heights. One of Yassky's justifications for legalization (as reported in the Brooklyn Paper in 2009) was that it would 'stimulate just the kind of niche manufacturing sectors that will be critical to an economic turnaround' (Muessig, 2009, p. 1). Such a justification is interesting, as it reinforces how UA can at once be an alternative, a reviving force post-market failure – in this case, after the financial crisis that began in 2007 – and also a way to promote self-reliance, instead of addressing the structural issues that require this type of self-reliance in the first place.

Some of these structural issues were made apparent by a small news piece about a beekeeping police officer in New York City published on a *New York Times* blog in 2008. The article describes how an officer, Anthony Planakis, successfully captured a few bee swarms from around the city and brought them to his Connecticut apiary. One commenter, 'm', recalls seeing a documentary film 'comparing the gentle European bee with the more dangerous and potent Africanized bee', a common racialized characterization of African bees (*Apis mellifera scutellata*) (Fahim, 2008, p. 1; Coyne and Knutzen, 2013). 'Racism extends all the way to insects', continues 'm' (Coyne and Knutzen, 2013). Another commenter, 'Jason F.', concurs: 'thank heavens one of the bees wasn't mistaken for a black man or else all six to eight thousand of those bees would've been shot!' (Coyne and Knutzen, 2013). This exchange, of course, speaks volumes about who has a right to exist in the city. The structural racism these commenters touch on extends beyond the city's police force. It extends into a deeply racialized agricultural and scientific history, and seeps right into the very socio-ecology of New York City. It is difficult to imagine that the people writing these comments could ever feel comfortable appropriating a space for beekeeping in the city.

The climate is similar in Washington, DC. Beekeepers in the District have had, for the most part, success getting full governmental support for urban apiculture. Hives can be found on top of government agencies, university buildings and private homes in the city, as well as at the White House. The Sustainable Urban Agriculture Apiculture Act of 2012 was passed in 2013, and further amended in 2014 (Department of Energy and Environment, 2015).

16.2.3 Findings: beekeepers and communities

Yet, beekeeping has issues of access beyond government regulations. A difficulty for many is that it is expensive; for example, the two 'packages' (screened cages containing adult bees, a queen, and a food source) of bees we bought in 2014 to start new hives at American University cost about US$75 each. Other equipment such as protective gear, hive tools, smokers and hives themselves can be costly as well. This does not include the transportation it requires to retrieve nucs and equipment from the people who sell them – usually beekeepers who run larger operations outside of the city.

Keeping bees through a university helped me avoid some of these obstacles. One of the few obstacles I encountered were the reactions I received when I told people that there are bees on campus, which included fear, disgust and lack of interest. Generally, the slight risk of getting stung was enough to make most people I talked to politely decline my offers to show them the apiary or attend club meetings. Much like the case in New York City, a fear of bee stings sometimes inhibited us as an organization from doing as much bee advocacy and education as we would have liked. In a beekeeping certification course I attended at the University of the District of Columbia, we were warned to make sure our hives were well hidden, or at least not accessible to passers-by. Our instructors told of instances in which hives in the District had been knocked over, destroyed with bricks, or sprayed with insecticide – all by bee-fearing neighbours.

Perhaps this illustrates why the trope of the solitary, hidden, rebel-beekeeper seems so pervasive. In an article he wrote about the economic returns associated with urban apiculture, economist Joshua Levin admitted that he was

Fig. 16.1. A suited beekeeper using a hive tool to observe one of the four beehives at American University. (Photo by Sarah Feder, taken 2013.)

afraid to research the practice, believing it was the realm of 'hardcore' renegades with unruly 'communist swarms' (2010, p. 1). In his research, Levin discovered why one might take up urban beekeeping: it can actually be quite lucrative. It can take at least 2 years of losses before one sees significant gains, however, further compounding its inaccessibility for many.

This 'rebel' narrative also serves to reinforce the Lefebvrian idea that urban beekeepers are appropriating space to conform to their desires and aspirations, and therefore, truly dwelling in and inhabiting these often small, interstitial or niche urban spaces. Thus, despite Levin's negative comments, the idea of beekeepers presiding over 'communist swarms' is certainly fitting here. These bees fly for kilometres with no regard for property lines. They alter the entire city, and the beekeeper facilitates this far-reaching ecological change, defining dwelling spaces within the city.

Rather than 'communist', however, honeybees are usually referred to as eusocial. They are social insects that limit individual reproductive ability in order to communally raise the offspring of others (Nowak et al., 2010). Similarly, beekeepers create nodes of collective networks that can be mapped across the city. These networks, like those in DC, seek to educate and build resilience. Such networks have been powerful in Tokyo as well. There, a small volunteer organization called Ginza Honeybee Project has been bringing diverse groups of people up to its rooftop hives to participate in honey harvests – from elementary school students to the elderly. The directors of the project say that urban honeybees serve to revitalize the urban landscape both ecologically and socially by building community ties and deepening environmental awareness. In fact, it has also been impactful upon the beekeepers themselves. One Japanese apiculturist recalled a day when he watched a swallow eating his bees with alarm. When he saw she was feeding them to her offspring, however, this beekeeper said he suddenly felt the urban ecosystem around him become larger (Kobayashi, 2013). Clearly bees can help illuminate and redefine the ways we understand the spatial and ecological systems of the city.

The community building and knowledge that is part of these beekeeper networks also spans across countries and throughout regions. The fifth annual Conference of Caribbean Beekeepers, which took place in the capital city of Georgetown, Guyana, in 2008, had attendees from Trinidad and Tobago, Suriname, Grenada, St Vincent and Haiti (Stabroek News, 2008). The conference had a Brazilian bee expert in attendance as well (Stabroek News, 2009). One attendee was Georgetown apiary owner Lyndon Stewart, who told the local newspaper he 'believed that closer collaboration among beekeepers across the country . . . could transform what is at present a sluggish local beekeeping industry' (Stabroek News, 2009, p. 1).

16.3 Conclusions

Urban apiculture is an integral and unique thread in the fabric of UA and urban ecology. In some places, it is so commonplace it goes largely ignored. In other places, beekeepers must fight to appropriate their own niche spaces, while others must fight to be recognized at all. Some take refuge in their anonymity, in their ability to dwell in the interstitial. For some, this is a strategy for resilience.

There are space appropriators dwelling with care and practising UA who are slowly building resilient networks and establishing a right to the city in their own ways. These actors are navigating and altering urban ecologies as nimbly as a honeybee collects pollen for her hive. These are not necessarily Lefebvre's workers dismantling the capitalist city. They are producing new spaces, new knowledge-sharing communities.

Yet, though these networks are important, and despite the comparisons these creatures allow us to draw to more collectivist social and political structures, it is clear that McClintock's (2014) observation that UA is both radical and neoliberal holds true here. While apiculture serves as a way to build a like-minded community and promote food security and support urban ecosystems, beekeepers are simultaneously the subject of neoliberalization. While attempts to position the beekeeper as a 'hardcore', self-sufficient or regulated provider of products and services legitimizes the appropriation of space, it also legitimizes cutbacks to the welfare state. Increasingly, urban beekeeping is an act of self-sufficiency where the state has failed (whether for the development of a livelihood or an additional honeypot). As with other types of urban cultivation in much of the world, the state did not simply fail to provide social services, it failed to regulate (and often, promoted) the usage of chemicals and practices that have spurred new urban beekeepers into action as an attempt to heal the ecological rift.

Additionally, it is clear that many pressures affect the urban apiculturist. As this chapter has shown, authorities are able to dictate or influence where hives can be placed, what kind of hives can be used, how many hives one can have on how much land, what kind of honeybee species should be nurtured, and even if they should be nurtured there at all. It will be interesting to see how these limitations play out in the future. How will beekeepers continue to react or adapt to these regulations? Is there a sort of socio-spatial-ecological carrying capacity when it comes to urban bees and their cultivators? This leaves a great deal of room for further research.

As I consider why it is I help to keep bees, I realize it is because it makes me feel as though I am part of a larger socio-spatial-ecological network. No other activity in the urban space is quite like beekeeping. As bees forage, they utilize spaces all across the ecosystems of the city. They pollinate flowers regardless of postal code or whether they are weeds, cultivated plants or crops. It is imperative to acknowledge the far-reaching ecological network that bees traverse when examining urban beekeeping, as social insects that are part of a socio-spatial network. Bees and their patterns are the ultimate metaphor for appropriating and dismantling the greater structures around us, and help us imagine what the right to the city might look like.

References

Bees for Development (2015) What we do. Available at: http://www.beesfordevelopment.org/our-work (accessed 25 November 2015).

Blomstedt, W. (2009) Beekeeping in Jordan. *American Bee Journal*, August. Available at: http://www.jordanbru.info/beekeeping_in_jordan.htm (accessed 31 October 2015).

Burnham, T. (2015) Downtown: a world of urban beekeepers is looking for you. *Bee Culture: The Magazine of American Beekeeping*, 18 September. Available at: http://www.beeculture.com/downtown-a-world-of-urban-beekeepers-is-looking-for-you (accessed 30 October 2015).

Clune, M. (2013) Rise in urban beekeeping may have gone too far, scientists warn. Available at: http://phys.org/news/2013-08-urban-beekeeping-scientists.html (accessed 27 February 2015).

Coyne, K. and Knutzen, E. (2013) The Africanized bee myth. Root Simple website. Available at: http://www.rootsimple.com/2013/10/africanized-bees (accessed 29 March 2016).

Danka, R.G. (2012) Functionality of varroa-resistant honey bees (Hymenoptera: Apidae) when used in migratory beekeeping for crop pollination. *Journal of Economic Entomology* 105, 313–321.

De La Portilla, E. (2013) Are urban bees immune to colony collapse disorder? *Huffington Post*, 7 October. Available at: http://www.huffingtonpost.com/eliza-de-la-portilla/are-urban-bees-immune-to-_b_4047576.html (accessed 1 March 2015).

Dennis, B. and Kemp, W.P. (2016) How hives collapse: Allee effects, ecological resilience, and the honey bee. *PLoS ONE* 11(2). DOI: 10.1371/journal.pone.0150055.

Department of Energy and Environment (2015) Urban agriculture: apiculture regulations. Available at: http://www.dcregs.dc.gov/Gateway/NoticeHome.aspx?NoticeID=5616511 (accessed 29 March 2016).

Ellis, J.D., Evans, J.D. and Pettis, J. (2010) Colony losses, managed colony population decline, and colony collapse disorder in the United States. *Journal of Apicultural Research* 49, 134–136.

Fahim, K. (2008) An officer and a beekeeper. *New York Times*, 27 May. Available at: http://cityroom.blogs.nytimes.com/2008/05/27/an-officer-and-a-beekeeper/?_r=0 (accessed 29 March 2016).

Fischer, J., Müller, T., Spatz, A., Greggers, U., Grünewald, B. and Menzel, R. (2014) Neonicotinoids interfere with specific components of navigation in honeybees. *PLoS ONE* 9(3). DOI: 10.1371/journal.pone.0091364.

Flottum, K. (2010) No buzz zones: 90+ U.S. cities and towns where beekeeping is still illegal (update). thedailygreen.com, 9 August. Available at: http://preview.www.thedailygreen.com/environmental-news/blogs/bees/illegal-urban-beekeeping-0602 (accessed 30 October 2015).

Foster, J.B. and Clark, B. (2009) Ecological imperialism: the curse of capitalism. *Socialist Register* 40, 186–201.

Garbuzov, M., Schürch, R. and Ratnieks, F.L.W. (2015) Eating locally: dance decoding demonstrates that urban honey bees in Brighton, UK, forage mainly in the surrounding urban area. *Urban Ecosystems* 18, 411–418.

Heynen, N.C., Kaika, M. and Swyngedouw, E. (2006) Urban political ecology: politicizing the production of urban natures. In: Heynen, N.C., Kaika, M. and Swyngedouw, E. (eds) *In the Nature of Cities: Urban Political Ecology and the Politics of Urban Metabolism*. Routledge, London, UK.

Hihat Honey (2010) Krygyzstan and transhumant beekeeping. 20 June. Available at: https://hihathoney.wordpress.com/2010/06/20/krygyzstan-and-transhumant-beekeeping (accessed 31 October 2015).

IBRA (2015) Frequently asked questions. International Bee Research Association. Available at: http://www.ibrabee.org.uk/index.php/information-services/faq (accessed 1 March 2015).

Jarvis, D.I., Padoch, C. and Cooper, H.D. (2007) *Managing Biodiversity in Agricultural Ecosystems*. Columbia University Press, New York, USA.

Johnston, S. (2013) Save a life, become an urban beekeeper. Ziptopia. Available at: http://www.zipcar.com/ziptopia/future-metropolis/noah-wilson-rich/urban-beekeeping (accessed 1 March 2015).

Kobayashi, E. (2013) Tokyo honey: a role for urban bees. Our World, 11 December. Available at: http://ourworld.unu.edu/en/tokyo-honey-a-role-for-urban-bees (accessed 1 March 2015).

Kudler, A.G. (2011) Mar Vista leading the charge to legalize backyard beekeeping in LA. Curbed, 25 August. Available at: http://la.curbed.com/archives/2011/08/mar_vista_leading_the_charge_to_legalize_backyard_beekeeping_in_la.php (accessed 1 March 2015).

Lee, K.V., Steinhauer, N., Rennich, K., Wilson, M.E., Tarpy, D.R., Caron, D.M., Rose, R., Delaplane, K.S., Baylis, K. and Lengerich, E.J. (2015) A national survey of managed honey bee 2013–2014 annual colony losses in the USA: results from the bee informed partnership. *Apidologie* 46, 292–305.

Lefebvre, H. (1991) *The Production of Space*. Blackwell Publishing, Oxford, UK.

Levin, J. (2010) Urban beekeeper economics: can you beat the stock market with a rooftop swarm? Open Forum, 19 November. Available at: https://www.americanexpress.com/us/small-business/openforum/articles/urban-beekeeper-economics-can-you-beat-the-stock-market-with-a-rooftop-swarm-joshua-levine (accessed 1 March 2015).

Meador, R. (2015) Marla Spivak: to grasp our bees' plight and prospects, stay focused on food. Minipost, 11 March. Available at: https://www.minnpost.com/earth-journal/2015/03/marla-spivak-grasp-our-bees-plight-and-prospects-stay-focused-food (accessed 29 March 2016).

McClintock, N. (2010) Why farm the city? Theorizing urban agriculture through a lens of metabolic rift. *Cambridge Journal of Regions, Economy and Society* 3, 191–207.

McClintock, N. (2014) Radical, reformist, and garden-variety neoliberal: coming to terms with urban agriculture's contradictions. *Local Environment* 19(2), 147–171.

McFarland, C. (2013) Legalize urban beekeeping in Los Angeles!! Honey Love, 4 March. Available at: http://honeylove.org/legalize-urban-beekeeping-in-los-angeles (accessed 1 March 2015).

McFrederick, Q.S. and LeBuhn, G. (2006) Are urban parks refuges for bumble bees *Bombus* spp. (Hymenoptera: Apidae)? *Biological Conservation* 129(3), 372–382. DOI: 10.1016/j.biocon.2005.11.004.

Mommersteeg, B. (2014) Space, territory, occupy: towards a non-phenomenological dwelling. MA thesis, The University of Western Ontario, London, Ontario, Canada.

Muessig, B. (2009) How sweet it is! Brooklyn Paper, 6 February. Available at: http://www.brooklynpaper.com/stories/32/6/32_6_bm_beekeepers.html (accessed 29 March 2016).

Mullin, C. (2015) Effects of 'inactive' ingredients on bees. *Current Opinion in Insect Science* 10, 194–200.

Navarro, M. (2010) Bees in the city? New York may let the hives come out of hiding. *New York Times*, 14 March. Available at: http://www.nytimes.com/2010/03/15/science/earth/15bees.html?_r=0 (accessed 1 March 2015).

Nessen, S. (2012) Two years after legalized beekeeping, city may be running short on forage. WNYC News, 25 June. Available at: http://www.wnyc.org/story/218358-urban-bees-may-be-running-out-foraging-ground (accessed 1 March 2015).

Neumann, P. and Carreck, N.L. (2010) Honey bee colony losses. *Journal of Apicultural Research* 49, 1–6.

Nowak, M.A., Tarnita, C.E. and Wilson, E.O. (2010) The evolution of eusociality. *Nature* 466, 1057–1062.

NSCGA (2015) History. Newark Street Community Garden Association. Available at: http://newarkstcommunitygarden.org/who-we-are/history/ (accessed 31 October 2015).

Omaha City Council (2009) Article II. Beekeeping. Available at: https://www.municode.com/library/ne/omaha/codes/code_of_ordinances?nodeId=PTIIMUCO_CH18NU_ARTIIBE (accessed 18 November 2016).

Partap, U. and Ya, T. (2012) The human pollinators of fruit crops in Maoxian County, Sichuan, China. *Mountain Research and Development* 32(2), 176–186. DOI: 10.1659/MRD-JOURNAL-D-11-00108.1.

Pettis, J.S. and Delaplane, K.S. (2010) Coordinated responses to honey bee decline in the USA. *Apidologie* 41(3), 256–263. DOI: 10.1051/apido/2010013.

Potts, S.G., Biesmeijer, J.C., Kremen, C., Neumann, P., Schweiger, O. and Kunin, W.E. (2010) Global pollinator declines: trends, impacts and drivers. *Trends in Ecology and Evolution* 25, 345–353.

Purcell, M. (2002) Excavating Lefebvre: the right to the city and its urban politics of the inhabitant. *GeoJournal* 58, 99–108.

Sanford, M.T. (2005) Beekeeping in Brazil: a slumbering giant awakens, Parts I – IV. *American Bee Journal*, vols 144–145.

Sengngam, B. and Vandame, J. (2003) Development of Beekeeping in LAOS: Various Strategic Choices. Available at: http://www.apiflordev.org/documents/development_of_beekeeping_in_Laos.pdf (accessed 31 October 2015).

Stabroek News (2008) Local beekeepers want Brazilian support to boost apiculture industry. Stabroek News, 28 November. Available at: http://www.stabroeknews.com/2008/business/11/28/local-beekeepers-want-brazilian-support-to-boost-apiculture-industry (accessed 30 October 2015).

Stabroek News (2009) Honey can earn sweet returns for Guyana – beekeeper. Stabroek News, 23 October. Available at: http://www.stabroeknews.com/2009/business/10/23/honey-can-earn-sweet-returns-for-guyana-beekeeper (accessed 30 October 2015).

Stanek, L. (2011) *Henri Lefebvre on Space*. University of Minnesota Press, Minneapolis, Minnesota, USA.

Swan, N. (2012) Explosion in urban beekeeping raises concerns for honeybee population. Explore Utah Science, 24 October. Available at: http://www.exploreutahscience.org/science-topics/life/item/37-explosion-in-urban-beekeeping-raises-concerns-for-honeybee-population (accessed 1 March 2015).

Swyngedouw, E. (1999) Modernity and hybridity: nature, regeneracionismo, and the production of the Spanish waterscape, 1890–1930. *Annals of the Association of American Geographers* 89, 443–465.

Swyngedouw, E. and Heynen, N.C. (2003) Urban political ecology, justice and the politics of scale. *Antipode* 35(5), 898–918.

Swyngedouw, E. and Heynen, N.C. (2010) Urban political ecology, justice and the politics of scale. In: Bridge, G. and Watson, S. (eds) *The Blackwell City Reader*. John Wiley & Sons, Hoboken, New Jersey, USA.

USAID (2013) Beekeeping industry reinvigorated in Haiti. Available at: https://www.usaid.gov/results-data/success-stories/beekeeping-industry-reinvigorated-haiti (accessed 31 October 2015).

Wagner, L. (2015) Backyard beekeeping approved in Los Angeles. The Two-way, 14 October. Available at: http://www.npr.org/sections/thetwo-way/2015/10/14/448725988/backyard-beekeping-approved-in-los-angeles (accessed 30 October 2015).

Wilson-Rich, N. (2012) Every city needs healthy honey bees. TED, July. Available at: https://www.ted.com/talks/noah_wilson_rich_every_city_needs_healthy_honey_bees/transcript?language=en (accessed 29 March 2016).

WinklerPrins, A.M.G.A. and de Souza, P.S. (2009) House-lot gardens as living space in the Brazilian Amazon. *FOCUS on Geography* 52, 31–38.

17 Urban Agriculture and the Reassembly of the City: Lessons from Wuhan, China

Sarah S. Horowitz* and Juanjuan Liu
Guizhou Normal University, Guiyang, Guizhou Province, China

17.1 Introduction

In 1990, one-quarter of mainland China's population lived in urban areas. By the end of 2015, that figure jumped to 770 million people – accounting for more than half the country, and 10% of the world's population (National Bureau of Statistics of China, 2016). The unprecedented scale and speed of China's urbanization raises critical questions about the nation's food supply. On the one hand, China has remarkably high food self-sufficiency standards, and protective measures to prevent the over-conversion of agricultural land into development land. On the other hand, environmental degradation, the rise of the middle class, an increasingly land-intensive diet, and growing dependency on global trade threaten China's food security agenda. This poses a threat not only to domestic social, ecological and political stability, but also to global systems.

As countries such as China attempt to 'de-peasantize' the countryside in the name of building a modern agri-food system, and 'de-agriculturalize' the city to make way for a modern metropolitan way of life, opportunities arise for reconfiguring intra-urban agriculture as an asset to, rather than aberration of, urban development. Not only does urban agriculture provide sustenance for urban dwellers, but it can also mitigate some of the ecological, social, economic and health consequences of an increasingly urban and resource-stressed environment. For countries such as China, where intra-urban agriculture is not an object of planning policy – let alone academic inquiry – empirical research on the modes and motivations for growing food inside the city is a critical first step.

The objectives of this chapter are to:

1. map the dynamic relations between actors, institutions, processes and spaces involved in China's urban agriculture discourse;
2. identify theoretical and practical challenges to growing food in the interior of Chinese cities;
3. suggest ways in which various state and non-state actors might work together to reassemble intra-urban agriculture in order to maximize its positive social, ecological and economic contributions to the city.

More broadly, this chapter seeks to demonstrate how empirical and methodological orientations of urban assemblage, an evolving theoretical framework within critical urban studies, might be used to advance urban agriculture theory and praxis.

---

*Corresponding author; e-mail: sarah.s.horowitz@gmail.com

17.2 Theoretical Framing: The City as 'Assemblage'

The concept of 'assemblage' in critical urban theory views the city as a sum of many material and non-material components that are continually assembled, disassembled and reassembled along socio-spatial and political-economic lines. Urban assemblage has been described variously as a 'guiding sensibility', 'research object', and 'methodology' (Brenner et al., 2011, pp. 229–231), 'concept, process, orientation and imaginary' (McFarlane, 2011, p. 651), and 'a philosophy of method' (Rick, 2014, p. 25). Urban assemblage has been used in urban studies as a '*research object* to be studied in a broadly political-economic framework'; '*methodology* of urban political economy'; and 'a new *ontological* starting point that displaces or supersedes the intellectual project of urban political economy' (Brenner et al., 2011, p. 230, original emphases).

The multi-disciplinary, multi-scale, and frankly indeterminate parameters of urban assemblage have produced interesting theoretical debates about the dynamics between social and material 'things' of the city, but have been far less successful in suggesting how stakeholders might actually help 'reassemble' these things to benefit the social and material fabric of the city. This analysis of urban agriculture in China applies urban assemblage as an empirical orientation (Section 17.3) and methodological framework (Section 17.4) to help imagine how various stakeholders might reassemble various legal, physical and conceptual components of the city to leverage intra-urban agriculture as a tool for sustainable urban development (Section 17.5).

As an empirical orientation, urban assemblage is concerned not only with 'urban "things"' but also how 'the urban itself' is discursively produced (Brenner et al., 2011, p. 228). Some of the causes and consequences of urbanization in China are not unique to China, but reflect processes of urbanization and capitalist market expansion across the world. In other ways, the specific historical, geographic and political-economic context of China has resulted in an expression of 'urban' that is quite distinct from other places in the world and, consequently, so has the expression of urban agriculture. By interrogating China's particular construction of 'urban', we seek to contribute to 'new geographies of theory' that dislocate the Euro American centre of theoretical production (Roy, 2009, p. 820) pervading much of the literature on urban agriculture today.

As a methodology, we apply assemblage thinking to map the dynamic relations between actors, institutions, spaces and processes involved in the process of making cities. We embrace the inherently eclectic and non-linear nature of assemblage thinking, expressed through the shunting back and forth between macro- and microanalysis, official policy rhetoric and on-the-ground reality, and theory and practice. The combination of discourse analysis and field research offers a multi-scalar and multi-disciplinary perspective on urban agriculture through which suggestions for an 'urban reassemblage' are based.

17.3 Urban Agriculture Discourse in China: A Peri-Urban Phenomenon

While urban agriculture has garnered increasing support from urban planners and policy makers over the past decade, particularly in North America (Thibert, 2012; Morgan, 2013; Chapter 3, this volume), urban agriculture remains at the literal and figurative periphery of the urban sphere in China. The term *dushi nongye* in Chinese refers to both peri-urban and intra-urban agriculture, but scholarship on *dushi nongye* in China has focused almost exclusively on what, by Western definitions, would be considered peri-urban agriculture (Miao, 2003; Zhou and Yu, 2003; Cai et al., 2004; Ning et al., 2006; Zhang et al., 2009). The scant Chinese scholarship on intra-urban agriculture mostly focuses on case studies from the Global North (Cai et al., 2004; Zhang and Sun, 2011; Gao, 2012; Yi, 2012). Those who have explored the issue of intra-urban agriculture in China come primarily from the fields of architecture and landscape architecture, and describe structural design strategies for incorporating agriculture into built spaces in the city (Pu and Miao, 2004; Li et al., 2012; Wu, 2012).

A critical part of the explanation for the peripheral nature of urban agriculture in China is a semantic one. 'City' (*shi*) in China is an administrative unit, and may not be entirely 'urban' in the common imagination of the word. Administrative cities, of which there are more than 600 in China, almost all include prefecture-level cities and relatively small county-level cities within their jurisdiction. Most 'cities' in China therefore contain a mix of urban land (owned by the state) and rural land (collectively owned by a village council). What may seem 'urban' in appearance might actually be titled as rural land, and vice versa. According to OECD (2013), approximately one-third of the land in major cities of China is collectively owned 'rural' land.

The legal distinction between urban and rural land has important political-economic implications for the development of urban agriculture. As the sole demander and sole provider of converted farmland, municipalities can set prices and extract significant revenues from converting rural land to urban development land (Ding, 2004; Feng, 2011; OECD, 2013; Hu et al., 2014). The conversion of rural to urban land is thus an important revenue-making mechanism of the state. As such, the presence of rural land inside of urban built-up spaces is seen as a sign of poor planning and economic inefficiency.

The urban–rural divide in China is further reified through the *hukou* system, a household registration system that classifies people born in rural areas as 'agricultural', while those born in cities are 'urban' – regardless of one's family's occupations. This system was initiated by Mao Zedong and began during the command economy era as a means of limiting rural-to urban migration and controlling the flow of resources between agriculture and industry. It still used in China today to allocate education, healthcare and retirement pension benefits. Although the government intends to remove *hukou* restrictions in towns and smaller cities in coming years, the majority of the some 230 million migrants working in Chinese cities today without permanent urban *hukou* have restricted access, or none at all, to the social services provided by the cities, including health care and free education (Chan, 2013). Land use rights, which are an important social safety net for those with agricultural *hukous*, are revoked upon conversion to urban registration status. This has resulted in a large urban population with agrarian skills and no land to farm.

Beginning in the early 2000s, the state began adopting a development strategy known as the 'Urban Rural Integration Policy' (*chengxiang yitihua*). Paradoxically, the policy aims 'to make the city more like a city and its rural outskirts more like the countryside' (Lei, 2015). This is achieved by integrating peri-urban villages into cities, a process through which villagers' rural household registrations are converted to urban registrations, and collectively owned village land is converted to state-owned urban development land. This policy reflects China's larger economic transition strategy, from a system of agriculture supporting industry, to one of urban migrant workers supporting industry through wage labour (Schneider, 2015).

In order to ensure food security vis-à-vis urbanization, the State has adopted a series of land conservation policies, agricultural subsidies and price controls to ensure that agricultural production can meet the demands of an increasingly urban population. Among these policies is the Vegetable Basket Program (*cailanzi gongcheng*). Appearing first in the late 1980s, and most recently in the 12th Five-Year Plan (2011–2016), the Vegetable Basket Program requires municipal planning departments and agricultural bureaus to outline regional policies and plans – including agricultural production, processing and marketing chains – to ensure adequate regional supplies of fresh vegetables, meat, dairy and aquaculture. This has resulted in relatively robust regional foodsheds across much of China. According to estimates, the percentage of vegetables produced and consumed within municipal boundaries of Beijing, Shanghai, Nanjing and Chengdu are around 40%, 50%, 40% and 70%, respectively (Cai et al., 2012; Lang and Miao, 2013).

In the early 2000s, the term *dushi nongye* became popularized by scholars at the Academy of Social Sciences in Beijing and Shanghai as a specifically peri-urban project of 'modernizing' agriculture and improving its 'multi-functionality' (ecological, economic, productive and social benefits). Now, dozens of cities across China have some kind of 'modern urban agriculture plan', based more or less on the same general formula: ornamental landscaping in the intra-urban area, followed by speciality crop production zones, agro-tourism, high-tech 'demonstration' parks in the peri-urban areas, and food-processing facilities in the rural hinterlands.

In 2012, the Ministry of Agriculture announced that the development of 'urban modern agriculture' (*dushi xiandai nongye*) is of 'extremely high significance' (Ministry of Agriculture, 2012). A summarized translation of the promulgation is found in Table 17.1.

Table 17.1. The Ministry of Agriculture's views on the rapid development of urban modern agriculture. (Adapted from Ministry of Agriculture, 2012.)

Importance	Optimize the layout of agricultural production
	Guarantee the stable supply of fresh agricultural products for medium-sized and large cities, and guarantee a market for domestically produced agricultural products
	Improve the natural environment for city residents
	Improve the incomes of people employed in the agricultural sector
	Support the coordinated development of industrialization, urbanization and agricultural modernization
Goals	Realize agricultural modernization
	Stabilize national grain production
	Help with the construction of the Vegetable Basket Program
	Strengthen the quality of food safety monitoring
	Actively develop recreational agriculture and ecological agriculture
	Spur development related to the food industry, including food processing and services
	Improve information technologies associated with the production and distribution of urban agricultural products
Measures to adopt	Strengthen the leadership roles of municipal officials in implementing the Vegetable Basket Program
	Encourage innovation and the exploration of 'model' modern urban agricultural forms
	Strengthen policy and financial supports to improve safeguards and subsidies for farmers and agricultural enterprises
	Improve the flow of information regarding urban agriculture production, consumption, and the different forms of modern urban agriculture emerging, and pay greater attention to theoretical research that can help spur innovations and guide policies regarding the development of modern urban agriculture

Many of the Ministry of Agriculture's objectives of 'urban modern agriculture' align with those of intra-urban food cultivation – stabilizing urban supplies of fresh fruits and vegetables, improving the natural and recreational environment for urban residents, increasing innovation and economic growth in the agricultural sector and related service and technology industries, and improving food safety controls – and yet, intra-urban food production is not mentioned in any policy text in China. As a result, intra-urban agriculture does not enjoy any of the subsidies or policy supports offered to peri-urban agriculture.

Having outlined the key policies and actors involved in the discursive production of urban agriculture in China, the following case study of Wuhan examines what has been so far excluded from Chinese academic discourse and policy – the agricultural activities in the interior of cities.

17.4 Intra-urban Agriculture: Wuhan Case Study

Between 2010 and 2014, we conducted a survey of informal urban agriculture activities in the urban core of Wuhan, China.[1] Wuhan is located in central China, along the intersection of the Yangtze and Han rivers. With 10.6 million people, it is the sixth most populated city in China. The urban core of Wuhan, which occupies about 10% of the total administrative city area, is relatively dense, with around 13,000 people/km^2. The urban core exhibits a wide variety of land-use types, including residential, commercial, industrial, government buildings, public parks, greenbelts and public transit.

To uncover the modes and motivations for growing food in the city, we conducted more than 60 semi-structured interviews with cultivators across the city. In addition, we selected a diverse sub-sample of the city to investigate the diversity of land-based cultivation areas more closely. We used a combination of qualitative interview data and satellite imagery to compare actual land use with the official land use plans (see Wuhan Land Use Master Plan, 2012).

17.4.1 Field observations

A significant number of Wuhan residents grew food crops on residential balconies and rooftops throughout the core of the city. Older residential apartment buildings (built before the residential property boom of the 1990s) were much more likely than new high-rises to have food growing on balconies and rooftops. This reflects the physical suitability of such buildings for growing food (open balconies and open access to rooftop space), as well as the demographics of building occupants (who are generally older and have lower incomes).

Cultivators demonstrated an intricate knowledge of plant properties and creative adaptation of recycled materials and built space for providing additional structural support, shade, heat retention and grey water collection. There was a preference for vertical- over horizontal-growing plants because of space limits, as well as a preference for local varieties of vegetables and fresh herbs hard to find in typical supermarkets. Most cultivators reported using soil dug from surrounding areas and a combination of rainwater and grey water from their homes to water their plants. Some used a mixture of food scraps and store-bought organic fertilizer for improving soil nutrients. Several rooftop cultivators also kept large ceramic pots for holding household-produced human waste, which they used as fertilizer.

In addition to potted crops and raised beds on balconies and rooftops, we observed vegetable plots of varying sizes along sidewalks, road meridians, under bridges, beneath power lines, and in unfinished construction sites. In our sub-sample, located between the Yangtze River and the first and third stops of the Line One Metro, there were eight larger-sized, multi-family cultivated plots, ranging from approximately 700 to 10,000 m^2 in size. Cultivators on these plots used a combination of rainwater, burned trash and municipal sewage to irrigate and fertilize their crops.

17.4.2 Cultivators' motivations

Cultivators were primarily aged 50–75, and the majority were female. Almost all had grown up or spent at least part of their life farming. Around one-third of interviewees were retired state-workers living off government pensions. The remaining interviewees included a mix of older people from the countryside who had come to the city to take care of grandchildren, local residents engaged in temporary or part-time work in the city, and a handful of university

Fig. 17.1. Vegetable plot on top of a residential building in Hankou, Wuhan. Eleven families grow food on this roof, including chillies, squash, green beans, leafy greens, herbs, aubergines, a pomegranate tree and a fig tree. (Photo by author, October 2013.)

professors with young families who rented small vegetable plots near their university. Their annual incomes were at or slightly below the city average of 24,000 RMB (around US$3,800 in 2011 terms) (Wuhan Statistical Yearbook, 2012).

Ten of the respondents were originally smallholder farmers on or near the land they currently lived on, and had been farming on it prior to its conversion to urban development land. Growing and selling crops was their primary source of income, and they earned an average of 1000–3000 RMB per month (around US$150–450 in 2011 terms). This was higher than the average income of farmers in the agricultural areas of Wuhan municipality at the time, which was around 800 RMB per month (Wuhan Statistical Yearbook, 2012). Respondents cited transportation cost savings, the absence of market vendor rental fees (most cultivators sold through informal street markets), and access to free water and fertilizer (most irrigated their fields with municipal sewage water) as advantages of growing food in the city.

The remaining group of interviewees grew food exclusively for household consumption. Across the socio-economic spectrum, non-economic reasons were cited as equally or more important than economic reasons for growing food. 'Health' – including the quality and safety of food, as well as physical benefits of manual labour – was the most frequent explanation for growing food. 'Enjoyment', followed by 'cost savings' and 'habit' were the next most popular responses. Those growing food on their rooftops cited the cooling effects of rooftop crop coverings as another important motivation.

Based on our interviews and review of local news stories (Wu 2013, 2014; Xia, 2015), values related to the quality of everyday life appear to outweigh economic necessity as the primary motive for growing food in Wuhan. Food production in Wuhan appears to provide new urbanites, most of whom are from rural backgrounds, with a way to cope with feelings of uselessness and disconnection associated with being relocated to a new place. This resonates with the observations of recent immigrants in North America, for whom growing food has been described as a tactic for overcoming feelings of dependence and culture shock (McClintock, 2013) and has been observed in many studies of urban agriculture (WinklerPrins and de Souza, 2005; Chapters 11 and 12, this volume). Besides the intrinsic joy of growing and consuming one's own food, control over the quality of food was a key concern of cultivators across the socio-economic spectrum. This reflects Chinese consumers' cultural preference for fresh food, and perhaps more importantly, in light of the food safety scandals over recent years, consumers' growing distrust of China's agro-food system (Shi et al., 2011; Si et al., 2015). In this way, growing food in the city is not merely an economic coping strategy of the urban poor. Importantly, it is a strategy for coping with a broken food system – a crisis, more broadly, embedded in global capitalism and the industrialization of the agro-food system.

17.4.3 Observed vs planned land uses

From the perspective of land use policy, all of the land-based growing activities observed in this study (as opposed to balcony or rooftop gardens) were either illegal or occupying a space intended for something else. According to the 2006–2020 Wuhan Master Land Use Plan, the majority of cultivated areas in the urban core of Wuhan are located on 'non-developable urban land', including protected green space along waterways, roads, bridges and underneath power lines. The largest continuous cultivated area, a 10,000 m^2 area along the Yangtze River, is designated as a protected riparian buffer zone and floodplain in the Master Plan. According to respondents, local authorities removed the cultivators in the mid-2000s because they obstructed a protected ecological area. The cultivators returned to the site and replanted their crops the following year, and had not been disrupted by authorities as of our last visit in 2013.

The plots rented by university faculty members present another example of the informal regulation of non-developable land. This area had been converted from village land and reassigned as a protected urban park years before. Because development plans had been stalled, however, the developer had decided to rent out small plots of land for locals to garden on.

The rest of the land-based cultivated areas we observed in the urban core were located on 'urban development land', meaning that they should have been, or would soon become, sites

Fig. 17.2. Urban village in Hanyang, Wuhan. Due to uneven urban development, pockets of collectively owned village land exist within the interior of cities. These are the only 'legal' land-based farming spaces inside the urban core. (Photo by author, October 2013.)

for residential and commercial development. The majority of small food plots in this land-use category were located on stalled construction sites. The larger, multi-family plots were located in 'urban villages' (*chengzhongcun*) – a phenomenon whereby rural land is engulfed by the city, but the administrative status of the land remains collectively owned village land. Until the land is retitled and the city sells the land-use rights to a developer, the village council has the legal right to partition land to villagers for agricultural purposes. According to city plans, all of the urban villages in the urban core of Wuhan will be demolished in the coming decade.

17.5 Towards a New Assemblage of Intra-urban Agriculture

The disconnect between city plans, land-use laws, and on-the-ground realities in this study indicate that intra-urban agriculture in Wuhan prospers not because there are formal mechanisms to enable it, but because there are a lack of formal rules and regulations against it. This resonates with Roy's observation that informality is not an 'exception to planning', but rather, it is produced by the suspension of formal processes of the state (2005, p. 155). Be it the expropriation of land from citizens under extra-legal terms, or citizens' reappropriation of interstitial spaces, informal land-use regulation gives both state actors and urban citizens a degree of flexibility to pursue their own interests.

The merit of the informal nature of intra-urban agriculture in China is that it grants dispossessed farmers-cum-urbanites welfare benefits and a 'right to the city' under a system that otherwise does not recognize private land ownership or citizen participation in the production of urban space. In the face of increasing urban densification and the ageing demographic of urban cultivators, however, we expect that informal food production will disappear from China's urban centres unless formal mechanisms are adapted.

17.5.1 Municipal planning institutions

The strict bifurcation of urban and rural land systems in China poses a significant structural barrier to incorporating agricultural activities with urban land. One possible way around this issue is to embed food production into an already existing category of urban land use – that of urban green space. According to national land-use planning law, green space and public plazas must account for at least 10–15% of urban developed land (Urban Land Use, 2010). Adding food production as a subcategory of urban green space could help increase the quantity and quality of food grown in the city. This could be coupled with infrastructure developments to facilitate the safe and sustainable closing of urban water and nutrient cycles, including on-site grey water treatment, soil remediation, community compost programmes, and rainwater irrigation systems.

Given the limited availability and high cost of urban development land, rooftops present the most pragmatic site for implementing this new category of green space. There are already numerous government-funded rooftop greening initiatives across China, including subsidies for property developers who include green roofs on their buildings, and the incorporation of rooftop greening as a part of municipal master plans (Xin, 2011; Xu, 2011). However, these policies and subsidies are currently only applicable to grass and ornamental coverings. Beyond its social and productive functions, rooftop food production offers important ecological benefits to the city. According to Nakonz (2013), for example, using 10% of the available flat roof area in Beijing for urban agriculture would correspond to a 7% reduction in Beijing's carbon emissions. Municipal planning institutions can contribute to the 'agriculturalization' of city green roofs by reducing the legal barriers to developing rooftop spaces. Currently, residential rooftops are considered collectively owned by building residents and, under normal circumstances, cannot be occupied by any individual.

17.5.2 Agricultural departments

Many of the objectives of the Ministry of Agriculture's 'urban modern agriculture' policies align with those of intra-urban agriculture. Stabilizing urban supplies of fresh fruits and vegetables, freeing scarce land resources for other uses, improving the natural and recreational environment for urban residents, and improving food safety controls are all potential outcomes of formally adopting food production as an object of 'urban modern agriculture'. Agricultural departments can support intra-urban food production by including it as a part of comprehensive regional food planning mechanisms. By including it as a formal category of 'urban modern agriculture', intra-urban agriculture could potentially receive the same technical, financial and policy supports currently enjoyed by peri-urban farm operations.

17.5.3 Business enterprises

So far, business enterprises have played the most proactive role in 'formalizing' intra-urban agriculture in China (Liu and Horowitz, 2013). A small but increasing number of landscape architect companies, social enterprises and property management companies in major cities across China have begun selling the concept of 'community gardens' (*shequ nongye*) in residential and business complexes. These community garden services offer rental plots, materials and technical support to help urbanites grow their own food. One of the earliest examples of this is The Wuhan New World Complex Urban Farm Project, launched by a Hong Kong-based developer in 2012 in a high-end apartment complex in Wuhan. The programme consists of 26 vegetable plots, each about 4 m^2 in size. Residents can apply to rent a plot for a fee of 300–400 RMB (around US$50–65) per month. Service staff are available on-site to provide residents with tools and technical assistance. Like most community gardening services in China, these companies target upper-middle-class families, and promote the health and recreation benefits of growing one's own food. Business entrepreneurs and landscape architects involved in these endeavours have been crucial in the promotion of a 'new face' of urban agriculture in China, one that associates growing food with a contemporary lifestyle, rather than a displaced rural livelihood (Evans et al., 2012).

In addition to residential buildings, an increasing number of commercial and industrial buildings have been opening their rooftops to food production. The Cheng Bang Agricultural Technical Development Company in Suzhou, for example, built a 7000-m^2 farm on top of a garment factory in Suzhou in 2012, and has aspirations to cover all of 118 of China's Red Star Macalline Furniture Mall roofs with farms in the future (Chen, 2013). The food-growing potential of industrial buildings in China is tremendous not only in terms of area, but also because of the relatively uncomplicated nature of use rights. Legally constructing on residential rooftops requires the consensus of building residents and property managers, while commercial and industrial rooftops can be altered at the discretion of the company. We predict that the promotion of rooftop farming by commercial enterprises as a 'green' or 'socially responsible' company practice will continue to play a leading role in the valorization and formalization of intra-urban agriculture.

17.5.4 Urban cultivators

'Upgrading' informal urban cultivation through the formalization of markets and state policies can have positive outcomes for urban citizens and environments (Pothukuchi and Kaufman, 1999; Viljoen et al., 2005; Thibert, 2012). But such upgrades will not result in an optimum reassembly of urban agriculture if they fail to include the participation of people who make up the majority of urban cultivators in China today – working-class elders and farmers-turned-urbanites. Formalization in the form of privatization has the potential to valorize some forms of urban agriculture over others, and intensify processes of accumulation and dispossession that marginalize many urban cultivators in the first place. Architects, landscape architects and business entrepreneurs can learn from the techniques and inherited agricultural knowledge of informal urban cultivators, and offer opportunities to engage them in the design and operation of such spaces in return. In this way, intra-urban agriculture can achieve its social function of re-embedding the moral economy of exchange into social relations, and invite stakeholders not traditionally involved in the project of city-building to find a voice in this process. For cultivators interested in making an earning from food production, collaboration with professionals in design and business fields can open new entrepreneurial opportunities.

17.5.5 E-commerce and online communities

Online platforms are playing an increasingly important role in the reassemblage of urban agriculture in China. Taobao, China's number one online shopping platform, has everything needed to start one's own garden – from organic soil and self-watering systems to balcony-sized containers and DIY aquaponic kits. Even live worms and aerobic bacteria for home composters can be purchased cheaply on Taobao. A search in Chinese for 'balcony vegetable garden' into Taobao's search engine comes up with more than 2000 products. According to a public relations officer at Taobao, more than 6000 people a day went to online shops at Taobao to buy seeds and garden tools in 2012, and searches for vegetable seeds grew nearly threefold between 2011 and 2012. The company attributes this primarily to the growing interest among urbanites of growing their own food (Zheng, 2012).

Social media spaces, such as blogs and the popular mobile app 'WeChat', are enlarging the space for exchange between urban cultivators. There are hundreds of Chinese online communities and chat forums dedicated to urban agriculture, from urban beekeeping unions to compost tea enthusiasts. In the absence of an articulated urban agriculture 'movement' in China, these online spaces are critical in helping to shape new civic discourses on urban agriculture.

17.5.6 Educators

Educators, particularly in the fields of urban studies, architecture, and agricultural sciences and environmental sciences are well-positioned 'reassemblers' of urban agriculture discourse at multiple scales. University classrooms (most of which in China, conveniently, are located in cities) can be great incubators for exploring new models of urban agriculture. Community action projects that engage students and faculty from different disciplines with community members, business enterprises and government departments can bring about new insights and design ideas. In our own experience leading the construction of an aquaponic rooftop farm at Huazhong University of Science and Technology in Wuhan, the rooftop itself becomes an important 'assemblage' of its own – a place for bringing together new materials, people and ideas. Students from landscape architecture, urban planning and architecture majors had never considered rooftop farming as a legitimate practice in their field before the start of the project. They expressed greater enthusiasm for incorporating food production into the urban built environment as a result of their experience. Local farmers, engineers, business entrepreneurs and teachers involved in the project also reported finding new meanings in and methods for growing food in urban places. Academic research and collaborative action projects across disciplines – and countries – can help reassemble new networks, knowledge and modes of engagement that are critical to innovating new forms of agro-friendly city building.

17.6 Conclusion

If China stays committed to its food self-sufficiency policies, it will have to radically change the way it imagines the role of cities: not only as a site of consumption for food, but also as a place of production.

To date, mainstream urban agriculture discourse in China has been a state-defined project concerned with growing food *for* the city, rather than *within* it. Based on empirical evidence from Wuhan, intra-urban agriculture in China remains a largely informal practice, dominated primarily by older, working-class individuals growing vegetables on balconies, rooftops, and in the non-built interstitial spaces in the city. While some urban cultivators sell to market, the majority grow food for self-provisioning. There is also increasing interest among middle- and upper-middle-class young families in growing food on and around residential apartment complexes as part of a healthy and contemporary urban lifestyle. Across the socio-economic spectrum, non-monetary values associated with growing food for home provisioning, including freshness, food safety and recreation, are cited as more important than monetary reasons for growing food. The increasing socio-economic diversity of urban food cultivators challenges inherited assumptions about urban food production as the plight of the poor.

While NGOs, land trusts, and other public interest groups have played a key role in the popularization and institutionalization of urban agriculture in other countries, China's centralized, top-down power structure, combined with the urban–rural dual nature of land policy, necessitates a different approach. Municipal planners have an opportunity to enable and strengthen these informal urban agricultural practices through including intra-urban agriculture in formal planning agendas, and adapting intra-urban food production as a legitimate category of urban green space. Agricultural departments can give much needed validity to current intra-urban practices by formally recognizing it as a mode of 'urban modern agriculture'. In the private sector, architects, landscape architects and business enterprises will be needed to design and retrofit urban spaces for the production of food, and change the 'face' of urban food production from backwards-looking to modern. Business enterprises have the potential to popularize intra-urban agriculture as a form of urban greening and 'corporate responsibility' by creating productive food spaces on company rooftops. The elderly 'grandmas and grandpas', with decades of farming experience, can offer important agricultural knowledge that can be integrated with urban growing techniques and technologies, and those interested in farming in the city for a living can explore new technologies and potential partnerships for improving their economic outcomes. Scholars and educators can support this work by facilitating community action projects and exchanging ideas with others from diverse professional fields and countries of expertise.

Urban assemblage can advance urban agriculture praxis and theory by helping map the nature of and relationship between human and non-human actors of the city. As the boundaries of where, who and what 'urban' consists of expand, so, too, do the opportunities for possible intervention. Urban assemblage is a particularly useful framework for regions where intra-urban agriculture is considered a largely informal, 'illegitimate' urban form. This is because urban assemblage understands informality as 'fundamental to understanding the productivity of cities' rather than 'an aberration or problem that can or should be erased' (Dovey, 2012, p. 354). Comparative case studies of urban agriculture across countries, strengthened by the use of a common framework and vocabulary, can help alter the course of 'de-agriculturalization' of cities in transitional economies. Simultaneously, cities built through more neoliberal planning mechanisms can learn from the regional foodshed planning strategies of more centrally planned economies. The reassembly of 'geographies of theory' at the global level has great potential to help reassemble intra-urban into a more productive, equitable, healthy and ecologically sustainable mode of urbanization at a local level.

Note

[1] Original interviews with urban cultivators were conducted by Juanjuan Liu between 2010 and 2011 for her PhD dissertation, which provides a comparative case study of urban agriculture practices in Wuhan, China and Seattle, Washington (Liu, 2011). Sarah Horowitz conducted additional fieldwork and secondary research between 2013 and 2014 under the auspices of the US Fulbright Student Research Program.

References

Brenner, N., Madden, D. and Wachsmuth, D. (2011) Assemblage urbanism and the challenges of critical urban theory. *City* 15, 225–240.
Cai, J.M. and Luo, B.Y. (2004) A look at the future of urban agriculture as a part of urban planning, based on international trends [in Chinese]. *Urban Planning* 9, 22–25.
Cai, J.M., Guo, H. and Muller, I. (2012) Urban food supply under constrained land resources in Beijing: potential and optimization. *Journal of Resources and Ecology* 3(3), 269–277.
Chan, K.W. (2013) A road map for reforming China's hukou system. China Dialogue, 22 Octobor. Available at: https://www.chinadialogue.net/article/show/single/en/6432-A-road-map-for-reforming-China-s-hukou-system (accessed 15 December 2014).
Chen, X. (2013) Rooftop farm in Suzhou: private enterprise creates 3000 mu of land [in Chinese]. Renmin, 25 October. Available at: http://js.people.com.cn/html/2013/10/25/263930.html (accessed 15 December 2014).
Ding, C. (2004) Farmland preservation in China. *Land Lines* 16(3). Available at: http://www.lincolninst.edu/pubs/913_Farmland-Preservation-in-China (accessed 18 November 2016).
Dovey, K. (2012) Informal urbanism and complex adaptive assemblage. *International Development Planning Review* 34(4), 349–367.
Evans, S., Valsecchi, F. and Pollastri, S. (2012) *Eco-urban Agriculture: Design for Distributed and Networked Urban Farming in Shanghai*. Cumulus 2012 Helsinki Conference. Available at: http://www.inuag.org/sites/default/files/Eco-Urban%20Agriculture-Design%20for%20distributed%20and%20networked%20urban%20farming%20in%20Shanghai.pdf (accessed 10 January 2015).
Feng, J. (2011) A balancing act? – an empirical examination of whether the dynamic balance policy has helped China reduce cultivated land loss amid rapid urban land expansion. PhD dissertation, University of Maryland, College Park, Maryland, USA.
Gao, N. (2012) Planning thought and spatial models for food urbanism [in Chinese]. PhD dissertation, Zhejiang University, Hangzhou, China.
Hu, R., Qiu, D.C., Xie, D.T., Wang, X.Y. and Zhang, L. (2014) Assessing the real value of farmland in China. *Journal of Mountain Science* 11(5), 1218–1230.
Lang, G. and Miao, B. (2013) Food security for China's cities. *International Planning Studies* 18(1), 5–20.
Lei, S. (2015) The development of modern urban agriculture in Shanghai. Press release, 1 June. Available at: http://en.shio.gov.cn/presscon/2015/06/01/1153268.html (accessed 10 January 2015).

Li, B.F., Liu, J.J. and Zhang, W.N. (2012) Research on strategies of agro-housing based on theory of urban agriculture: a case study in Wuhan [in Chinese]. *The Architect* 12, 48–53.

Liu, J.J. (2011) Research on the feasibility of and strategies for urban agriculture in China's built-up areas [in Chinese]. PhD dissertation, Huazhong University of Science and Technology, Wuhan, China.

Liu, J.J. and Horowitz, S. (2013) Opportunities and challenges for the development of China's community gardens. *Journal of Landscape Research* 5(11), 47–48.

McClintock, N. (2013) Radical, reformist, and garden-variety neoliberal: coming to terms with urban agriculture's contradictions. *Local Environment* 19(2), 147–171.

McFarlane, C. (2011) The city as assemblage: dwelling and urban space. *Environment and Planning D: Society and Space* 29, 649–671.

Miao, L. (2003) Urban agriculture: an inevitable choice for urban sustainability [in Chinese]. *Small City Construction* 10, 92–94.

Ministry of Agriculture (2012) The Ministry of Agriculture's views on the rapid development of urban modern agriculture [in Chinese]. 30 August. Available at: http://www.moa.gov.cn/govpublic/SCYJJXXS/201208/t20120830_2901993.htm (accessed 15 January 2015).

Morgan, K. (2013) The rise of urban food planning. *International Planning Studies* 18(1), 1–4.

Nakonz, J. (2013) Food footprint, urban agriculture and climate change in mitigation in China. *University of Lausanne and Low Carbon City China Alliance*.

National Bureau of Statistics of China (2016) China's economy realized a moderate but stable and sound growth in 2015. National Bureau of Statistics of China, 19 January. Available at: http://www.stats.gov.cn/english/PressRelease/201601/t20160119_1306072.html (accessed 20 February 2016).

Ning, C.Q., Wei, P.W. and Xing, J.M. (2006) Thoughts on city planning for urban agriculture [in Chinese]. *Urban Development Research* 2, 69–71.

OECD (2013) *OECD Economic Surveys: China 2013*. OECD Publishing. Available at: http://www.oecd-ilibrary.org/economics/oecd-economic-surveys-china-2013_eco_surveys-chn-2013-en (accessed 10 Dec 2015).

Pothukuchi, K. and Kaufman, J.L. (1999) Placing the food system on the urban agenda: the role of municipal institutions in food systems planning. *Agriculture and Human Values* 16(2), 213–224.

Pu, Q.D. and Miao, F.F. (2004) Preliminary exploration of urban greenspace agriculture [in Chinese]. *Small City Construction* 7, 77–78.

Rick, O.J.C. (2014) Cycling the city: locating cycling in the continued (re)structuring of North American cities. Ph.D dissertation, University of Maryland, College Park, Maryland, USA. Available at: http://drum.lib.umd.edu/handle/1903/15267 (accessed 25 November 2016).

Roy, A. (2005) Urban informality: toward an epistemology of planning. *Journal of the American Planning Association* 71, 147–158.

Roy, A. (2009) The 21st-century metropolis: new geographies of theory. *Regional Studies* 43, 819–830.

Schneider, M. (2015) What, then, is a Chinese peasant? *Nongmin* discourses and agroindustrialization in contemporary China. *Agriculture and Human Values* 322, 331–346.

Shi, Y. Cheng, C.W., Lei, P., Wen, T.J. and Merrifield, C. (2011) Safe food, green food, good food: Chinese community supported agriculture and the rising middle class. *International Journal of Agricultural Sustainability* 94, 551–558.

Si, Z., Schumilas, T. and Scott, S. (2015) Characterizing alternative food networks in China. *Agriculture and Human Values* 32(2), 299–313.

Thibert, J. (2012) Making local planning work for urban agriculture in the North American context: a view from the ground. *Journal of Planning Education and Research* 32(3), 349–357.

Urban Land Use (2010) Code for classification of urban land use and planning standards for development land. Ministry of Housing and Urban–Rural Development of the People's Republic of China [in Chinese]. 24 December. Beijing, China.

Viljoen, A., Bohn, L. and Howe, J. (2005) *Continuous Productive Urban Landscapes: Designing Urban Agriculture for Sustainable Cities*. Architectural Press, Oxford, UK.

WinklerPrins, A.M.G.A. and de Souza, P.S. (2005) Surviving the city: urban home gardens and the economy of affection in the Brazilian Amazon. *Journal of Latin American Geography* 4, 107–126.

Wu, W. (2012) Agriculture in urbanizing communities: thoughts on and applications of an international development trend in planning [in Chinese]. *Urban Development Research* 19(11), 35–41.

Wu, M. (2013) Urbanization of landless peasants [in Chinese]. Chutian City News. Available at: http://ctdsb.cnhubei.com/html/ctdsb/20130530/ctdsb2055274.html (accessed 15 January 2015).

Wu, M. (2014) Vegetables growing in community greenspace [in Chinese]. Chutian City News. Available at: http://ctdsb.cnhubei.com/html/ctdsb/20140507/ctdsb2335455.html (accessed 15 January 2015).

Wuhan Land Use Master Plan (2012) Wuhan Land Use Master Plan, 2006–2020 [in Chinese]. Wuhan Land Resources and Planning Bureau, Wuhan, China. Available at: http://www.wpl.gov.cn/pc-42243-485-0.html (accessed 25 November 2016).

Wuhan Statistical Yearbook (2012) [in Chinese]. Available at: http://www.whtj.gov.cn/downlist.aspx?id=2012111010461248&tn=2012111010454262 (accessed 10 December 2014).

Xia, Z.H. (2015) Urban greenspace occupied by vegetables [in Chinese]. Chutian City News, 22 January. Available at: http://ctdsb.cnhubei.com/html/ctjb/20150122/ctjb2536469.html (accessed 25 January 2015).

Xin, Z. (2011) More green terraces in Beijing soon. China Daily, 21 December. Available at: http://www.chinadaily.com.cn/cndy/2011-12/21/content_14296726.htm (accessed 10 December 2015).

Xu, J.Q. (2011) Cultivating a novel way to grow food. China Daily, 14 November. Available at: http://www.chinadaily.com.cn/bizchina/2011-11/14/content_14088437.htm (accessed 10 December 2015).

Yi, X.X. (2012) Western city planning's new topic: community food systems [in Chinese]. *Urban Planner* 6, 102–106.

Zhang, Y.S. and Sun, Y.B. (2011) Foreign urban agriculture and China's ecological urbanization strategy [in Chinese]. *Architecture Report* 4, 95–98.

Zhang, F., Cai, J. and Liu, G. (2009) How urban agriculture is reshaping peri-urban Beijing? *Open House International* 34(2), 15–24.

Zheng, J.R. (2012) Balcony farmers are taking root. China Daily, 23 June. Available at: http://www.chinadaily.com.cn/business/2012-06/23/content_15519074.htm (accessed 5 January 2015).

Zhou, N.X. and Yu, K.J. (2003) The natural combination of agricultural land and city [in Chinese]. *Planner* 3, 83–85.

18 The Contribution of Smallholder Irrigated Urban Agriculture Towards Household Food Security in Harare, Zimbabwe

Never Mujere
University of Zimbabwe, Harare, Zimbabwe

18.1 Introduction

Agriculture is no longer seen as just a rural activity but is now embraced as part of a livelihood and an important income-earning strategy in most cities. It has become an important economic and social activity, contributing to the aesthetic value of the city and socio-economic uplift of urban households. Most urban dwellers engage in urban agriculture (UA) as a survival strategy due to challenges of absolute and relative growth in urban poverty and food insecurity (Gallaher et al., 2013; Korir et al., 2015; Chapters 7 and 12, this volume). A significant proportion of low-income urban households face serious difficulties in accessing adequate basic foodstuffs, which are sold at prices beyond what consumers can afford. High food prices have drastically reduced people's purchasing power and raised the spectre of food and income disequilibrium at the household level. The urban food crisis is further worsened by a massive population shift from rural to urban areas. As a result, some urban authorities have reserved patches of land for urban agriculture in their municipalities (Olufemi and Alabi, 2012; Opitz et al., 2016). Urban farmers engage in the production, processing and selling of food products within and around cities or towns.

There are various viewpoints advanced for the emergence of urban agriculture. The Marxian view states it is a survival strategy of the urban poor, hence a means through which they are forced to bear the social cost of capitalistic development (Merrifield, 2014). Others like city developers and planners argue that urban farming is a rural cultural artefact of a recently urbanized population caused by rural–urban migration and the geographical expansion of city boundaries. However, the informal sector advocates argue that urban agriculture is an economic activity responding to incentives in the economy (Kiguli, 2005).

Urban agriculture is increasingly recognized as an important activity for ensuring food security, alleviating poverty and as an alternative strategy for greening the city. Some urban authorities have partnered with non-governmental organizations and private corporations to support the establishment of smallholder irrigated gardens (Crush and Frayne, 2010). However, there is little knowledge on the scope, performance and viability of such gardens.

This study aims to examine the contribution of an urban irrigated agriculture project towards household food security and poverty alleviation in Harare City, Zimbabwe. It also considers the types

---

*E-mail: nemuj@yahoo.co.uk

of crops grown, crop productivity and challenges faced by the urban farmers. The research employed qualitative methodologies, which include observations and interviews with farmers and key stakeholders. Findings from the study demonstrate that although urban irrigated smallholder agriculture contributes to livelihoods, there are numerous socio-economic factors that impede people from participating more in the project. Additionally, some people have withdrawn from the project, due to security threats from perceived political affiliations.

18.2 Dimensions of Urban Agriculture

Urban agriculture is the growing of crops and the raising of animals within and around cities. It is an industry that produces, processes and markets food and fuel, largely in response to customer demand within a town, city or metropolis and peri-urban areas (Mbiba, 2000; Crush *et al.*, 2011). In most cities of the Global South an important part of urban agricultural production is for self-consumption, with surpluses being traded. Products are sold at the farm gate, by cart in the same or other neighbourhoods, in local shops, on local (farmers') markets or to intermediaries and supermarkets. Fresh products are mainly sold, but part of it is processed for own use, cooked and sold on the streets, or processed and packaged for sale to one of the outlets mentioned above.

Urban farmers apply intensive production methods using and reusing resources to yield a diversity of crops and livestock. Various levels of technology are applied at individual, family, group, cooperative and commercial farms. Scales of productions range from micro- and small farms to medium-sized and some large-scale enterprises (Crush and Frayne, 2010).

Urban agriculture activities may take place in locations inside cities (intra-urban) or at the fringes of cities (peri-urban). As illustrated in Chapter 1 (Table 1.1) and in Table 18.1 below, UA activities may take place within residential house lots (on-plot) or on land away from the place of residence (off-plot), on private land (owned or leased) or on public land (parks, conservation areas, along roads, streams and railways), or semi-public land such as school yards, grounds of schools and hospitals. Backyard farming is possible in low-density or high-income residential areas or in less densely populated cities, where the ground around the houses is devoted to the growing of crops and rearing animals. Most low-income urban households, on the other hand, practise off-plot farming because they do not own houses with extensive land around them (Mbiba, 2000; Crush *et al.*, 2011).

Table 18.1. Types and characteristics of urban agriculture. (Derived from Mbiba, 2000 and Crush *et al.*, 2011.)

Feature	On-plot	Off-plot	Peri-urban
Location	On property in both high- and low-density areas	Public open spaces, utility service area all over the city, and on allotments	Outside city boundary in rural areas
Consumption mode	Mainly subsistence, more commercial in low-density areas	Mainly subsistence, slightly more marketed output than on-plot production	Subsistence in the smallholder sector but marketing on the increase
Crops produced	Up to 0.005 ha and can be as high as 0.4 ha in low density areas	Average 0.02 ha and up to 0.8 ha per household Cultivator	1.2 ha for smallholders and 5 ha or more for large-scale producers
Livestock	Negligible	Negligible	Poultry, pork, milk, beef, chicken, etc.
Households involved	80% of properties in summer and 60% in winter; 70% property owners, 30% lodgers	At most 25% of Harare households; property owners dominate	Those with land-access rights

18.3 Urban Food and Income Sources in Zimbabwe

In Zimbabwe, as in the rest of southern Africa, access to maize as a staple crop is associated with food security at both the national and household levels. Thus, the majority of urban households produce maize as part of their survival strategy. However, their farming activities are not large-scale. On average, the households cannot support themselves entirely with the maize they produce on urban agricultural plots. Vegetables on the other hand, are produced sufficiently for subsistence, with the surplus sold.

Urban areas in Zimbabwe have various sources of food. Purchases are the major source of a range of staple food stuffs consumed by households followed by own production (Table 18.2). With regards to vegetables, subsistence crop production (55%) is the major source of vegetables consumed by households followed by purchases (40%).

Urban households derive their livelihoods from a wide range of activities. Chief among them is self-employment and formal employment. Self-employment included activities such as cross-border trade, petty trade, retailing and artisanal work (Table 18.3).

Table 18.2. Major sources of food in urban areas. (Source: ZimVAC, 2009, p. 16.)

Source	Percentage
Purchases	70
Own production	17
Gift	10
Food assistance	1
Casual labour	1
Others	1
Total	**100**

Table 18.3. Major sources of urban income. (Source: ZimVAC, 2011, p.23.)

Activity	Percentage
Self-employment	43
Formal employment	30
Agricultural-related activities	5
Remittances	2
Other	20
Total	**100**

18.4 Urban Agriculture and Food Security

Food security is a situation existing when all members of a community have access to culturally acceptable, nutritionally adequate food through local, non-emergency sources at all times. Thus food security exists when all people, at all times, have access to sufficient, safe and nutritious food to meet their dietary needs and food preferences for an active and healthy life (Opitz et al., 2016). Mbiba (2000) asserts that urban agriculture improves both access and food intake of fresh foods, especially among populations suffering from food insecurity, through their own self-provisioning, which reduces market expenditure. Urban agriculture is therefore one of the survival strategies employed by urban residents to reduce poverty and improve their food security. Food security entails that the population has enough supply and access to staple foods. The tremendous influx of people from rural areas to urban areas is increasing the need for fresh and safe food (Kutiwa et al., 2010). Access to nutritious food is another perspective in the effort to produce food and livestock production in cities. Figure 18.1 shows how urban agriculture can contribute to food security.

Urban food security depends largely on whether a household has adequate effective purchasing power, given the prevailing prices and incomes (Kutiwa et al., 2010). The period of time during which the produce of urban agriculture is sufficient for a household is an indication of its food or consumption gap. The majority of the households depend on harvested crops from urban agricultural farms for a period of 1–3 months. However, food sufficiency from own production is significantly associated with household size.

18.4.1 Coping with food insecurity

Food-insecure households embark on a variety of consumption and non-consumption strategies to cope with food insecurity. Households employ at least one of the following consumption coping strategies as a result of facing short-term food-access challenges: limiting the size of

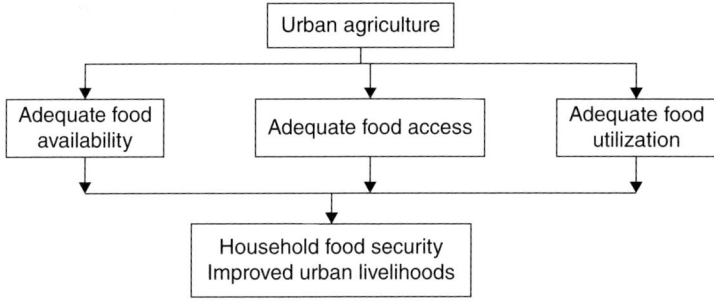

Fig. 18.1. Urban agriculture as an activity to ensure household food security. (Adapted from Kutiwa *et al.* (2010).)

portions, reducing the number of meals, buying food on credit, relying on less preferred foods, eating at another person's house, begging, having adults consuming less food than the young and having less varied diets. Non-consumption coping strategies include remittances from relatives or friends from outside the household (female-headed households receive more remittances than male-headed households), sale of assets to buy food, casual labour, urban agriculture, petty trading, cross-border trading and self-employment (ZimVAC, 2011).

18.4.2 Proportion of food insecure urban households and the food poverty line in Zimbabwe

Table 18.4 shows the levels of food-insecure households between 2006 and 2011 in Zimbabwe's ten provinces. On average, 13% of the urban households were food-insecure (that is, limited access to adequate food) by 2011 (ZimVAC, 2011). The highest level of urban food insecurity was in Mashonaland Central, followed by Bulawayo and Matebeleland North. The lowest levels were found in Harare (5%) followed by Matabeleland South (7%). The food poverty datum line (FPL) represents the minimum consumption expenditure necessary to ensure that each household member can (if all expenditures were devoted to food) consume a minimum food basket representing 2100 kilocalories. An individual whose total consumption expenditure does not exceed the food poverty line is deemed to be very food-poor. On average, 15% of households were found to be living below the food poverty line in 2011. Mashonaland Central (26%) had the highest percentage of household who were below the FPL while, Harare (6%) had the lowest percentage of the household that were living below the FPL.

The ZimVAC (2011) assessments found the following as common characteristics of food insecure households.

- Food-insecure households tended to have more household members (six people) than food-secure households (five people).
- Households with at least one orphan due to AIDS or otherwise were more likely to be food-insecure than those without orphans.
- Households with at least one member who was chronically ill were more likely to be food-insecure than those without.
- Households that reported receiving remittances from elsewhere in the country or outside the country were more likely to be food secure than those that were not doing so.
- Elderly-headed households were more likely to be food insecure than those with younger heads of households.

18.5 Policy and Legal Framework for Urban Agriculture in Zimbabwe

In Zimbabwe, urban development is often guided by development plans prepared by local authorities as provided for by the Urban Councils Act and the Regional, Town and Country Planning Act. The conventional land uses provided for in such plans are housing, industry, commerce, open spaces and infrastructure (roads, electricity,

Table 18.4. Percentage of food insecurity and households below the Food Poverty Line (FPL). (Source: ZimVAC, 2006, 2009, 2011.)

Province	2006	2009	2011	% of households below FPL, 2011
Harare	–	31	5	6
Bulawayo	35	28	17	17
Mashonaland Central	–	26	23	26
Mashonaland East	14	27	13	17
Mashonaland West	28	29	15	18
Manicaland	33	47	14	17
Matebeleland North	–	45	16	22
Matebeleland South	20	20	7	8
Masvingo	–	34	10	12
Midlands	17	36	13	13

telephone, etc.). As outlined above, urban agriculture is practised within residential plots (on-plot) or outside of them (off-plot). In the low-density areas, lot sizes are bigger and there is often enough space for limited agricultural activities. Some of the bigger low-density areas are classified as agro-residential, meaning that occupants are allowed to practise urban agriculture. The situation is quite different in the high-density areas, where plot sizes vary from about 0.015 to 0.03 ha. There is hardly any space left for urban agriculture, and yet the activity is needed for the purposes of sustaining the livelihoods of people in these areas.

The current situation in Zimbabwe is that agriculture is not classified as an urban activity. Hence, city planners do not plan for urban and peri-urban agriculture. Agriculture is legally permitted in the peri-urban areas, a zone which is normally dominated by titled properties (official tenure of the land), which are well beyond the reach of the poor. The major challenges in the development of urban and peri-urban agriculture are the issues relating to access to land and the management of the activity. Most households involved in the activity in Harare had accessed land through making a first claim on an open piece of land as squatters (Hungwe, 2006; Mwendera et al., 2013). This is particularly in relation to access to land by the majority of people in the high-density areas and other low-income people in the low-density residential areas. This process of self-allocation of plots intensified as settlers went beyond the 'open spaces in urban areas' to include farms in the peri-urban areas of major cities.

The policy framework for the development of urban agriculture is slowly starting to be supportive of the activity. For example, major cities like Harare, Bulawayo and Gweru have developed Master Plans that make provision for urban farming in designated zones. A good example is the Harare Combination Master Plan of 1992, which provided for intensive agricultural smallholdings within the city and the peri-urban areas surrounding it (City of Harare, 1992). Both central and local governments generally acknowledge the role of urban agriculture in poverty alleviation and enhancement of urban food security, employment creation and economic development. However, minimal technical capacity and meagre financial resources retard the implementation of policy proposals. Nevertheless, local and international NGOs are also playing a major role in providing services to support urban farming. They provide services to supplement government effort.

The development of smallholder irrigation projects in urban areas of Zimbabwe is a new phenomenon, which has not been well researched. Hence, there is still little known about the viability of such initiatives. It is important to have social evaluations of urban smallholder irrigated agriculture to improve the understanding of the factors that affect their viability. Therefore, the following section discusses the contribution of irrigated urban agriculture to household food security in the city of Harare, Zimbabwe.

18.6 Green Valley Garden

My empirical case study examines the operation of Green Valley (irrigation) Garden (GVG) in Harare. The garden is located in Glen View 7, a high-density suburb, a peri-urban setting, some

22 km southwest of Harare city centre. The Green Valley irrigation garden was opened in 2010 and covers an area of 6 ha, intended to be occupied by 300 plot-holders. Crops grown during the summer season are maize and groundnuts, while in winter the main crops grown are sugar beans, wheat, cabbage, rapeseed, onions and lettuce.

18.6.1 Planning and management of the garden

Green Valley Garden was established by the City of Harare with the assistance from Solidarités International, an international non-governmental organization. Solidarités International has been running food-security projects in Zimbabwe since February 2010, focusing on these community gardens in Harare. The purpose of the initiative was to improve the food security of the urban poor. The city of Harare provided 6 ha of land in Glen View 7 to elderly people. Solidarités erected a security fence, drilled two boreholes, installed an electric water pump, and eight 2000-litre water tanks and water pipes. The Zimbabwe Electricity Supply Authority (ZESA) provided electricity to the garden. The beneficiaries are responsible for maintaining the garden.

Green Valley Garden is managed by an irrigation management committee (IMC) and its sub-committees selected by farmers. The sub-committees are responsible for marketing produce, water management, discipline and crop production. The IMC and its subcommittees work under agreed by-laws aimed at effectively running the garden. Committee members are elected every year. The farmers collectively raise funds for operations and maintenance work (IMC staff, personal communication, 2015). The IMC comprises the chairperson, vice chairperson, secretary, vice secretary, treasurer, and subcommittee members for marketing, irrigation, discipline and production.

18.6.2 Beneficiary selection

The project beneficiaries include HIV/AIDS patients, widows, orphans and the elderly. These were selected to benefit from the project in a bid to improve their nutrition and food security status. Each beneficiary pays US$20 as a joining fee and US$6 monthly subscription for operation, maintenance and electricity bills. Initially, 300 beneficiaries were selected and allocated 0.2 ha of land each. By the year 2015, only 48 farmers (comprising 24 males and 24 females) remained in the garden. Thus, the remaining people each have more land.

The considerable drop-out was as a result of the menial and labour-intensive tasks involved in irrigation activities. Given that most beneficiaries are old and suffer from ill-health, this was a significant issue to have adequately anticipated. The by-laws do give members the freedom to leave the garden, which is what many have done after a trial period. Some farmers left the project or refused to join the garden because they thought that the initiative was political and did not want to be affiliated with it.

18.6.3 Agricultural performance

Farmers generally grew high-value horticultural crops to raise enough money for domestic needs, school fees, medical fees, farm operation and maintenance purposes. Lucrative crops are mostly vegetables, including green beans, squash, peas, cabbage, carrots, tomatoes, onions, potatoes, beans, tomatoes, courgettes (zucchini or baby marrow) and baby corn (Table 18.5). Maize is the most important crop because it is the staple crop grown mainly for food security. Farmers plant three times a year on the same field. The by-laws make it mandatory for farmers to plant the same crop type at the same time, but they are free to apply fertilizer to their crops at will. In the future, farmers are planning to grow lettuce because its market demand is high. Six farmers piloted growing the crop to test the

Table 18.5. Average crop yields at Green Valley Garden.

Crop	Yield (t/ha)
Tomatoes	17
Maize	2.5
Carrots	1.8
Peas	1.5
Potatoes	1.5
Green beans	1.5

market, and received high profits as compared to other crops.

18.6.4 Percentage income and costs of production

Individual farmers source their inputs. The procurement committee is responsible for identifying where they are available in the city. The highest costs are incurred in purchasing fertilizers. Fertilizer costs make up some 55% of total costs; with electricity costs comprising another 20%. However, costs for manure are the least (0.5%) of the total costs. This is because manure is locally available and is sold at a cheap price. Table 18.6 shows the percentage cash costs for crop production.

High income derives from the sale of potatoes, green beans, carrots and peas, with other vegetables (including cabbages, rapeseed and lettuce) and tomatoes comprising 8% of the total income (Table 18.7).

Table 18.6. Percentage costs of crop production.

Item	Percentage
Fertilizer	55.0
Electricity	20.0
Seeds	9.8
Labour	6.6
Chemicals	4.5
Packaging	1.4
Transport	1.2
Security guards	1.0
Manure	0.5
Total	**100**

Table 18.7. Percentage of income from sale of garden produce.

Crop	Percentage
Peas	44.7
Green beans	20.8
Carrots	12.1
Potatoes	8.8
Onions	4.5
Vegetables	3.6
Tomatoes	4.4
Maize	1.1
Total	**100**

18.6.5 Marketing

Besides using produce for subsistence purposes, Green Valley Garden produces high-value horticultural crops for sale. The garden has organized marketing strategies. The marketing or sales manager identifies potential markets before a crop is planted. Farmers hire transport collectively and travel to the markets in groups. The main markets for the farmers are local wholesalers, retail shops and residents. Nevertheless, farmers are free to sell and consume their produce. Some buyers come to the garden to buy produce of their choice.

18.6.6 Benefits derived from the garden

During the survey, farmers were asked to rank the three most important benefits they derive from the garden. Not unlike other urban agriculture projects, the majority of the respondents suggested that the garden provided them with household food security throughout the year. Social cooperation was ranked second, with the least important stated as being employment opportunities in the garden.

18.6.7 Challenges faced in Green Valley Garden

The research findings reveal that the irrigation effort is facing a number of challenges, which seriously affect farmers' ability to derive maximum productivity from their irrigation resources. The challenges include socio-economic challenges (inadequate cash to purchase inputs, pay for power or labour, marketing problems and theft), erratic water supply, political intimidation, and transport of inputs and produce. In order of greatest impact they are: transport for products (46%); social issues (30%); access to water (13%); and politics (11%).

The most significant issue is that of farmers lacking access to adequate transportation to bring their produce to the markets. In some cases, customers ask farmers to bring produce within a short time. Farmers often fail to meet customer needs at short notice because they lack transport facilities and money to hire trucks.

The political conflicts between the ruling and opposition parties in Harare have challenged smooth farming activities in the garden. Some members in the garden have been accused of belonging to either of the political parties, hence suffer intimidation. Some of the beneficiaries left the garden, accusing the initiative of being political and intended to buy votes.

Water supply is also problematic at the garden. During the field visit, it was observed that fields far away from the water tanks face more challenges to receive adequate water than those close to supply tanks. Frequent power cuts have resulted in zero water supply to the garden.

Additionally, Green Valley Garden faces problems of theft of produce. Thieves frequently cut the security fence and steal crops from the garden. Farmers with plots close to the fence suffer more losses as a result of theft than those with plots in the middle of the garden. The two security officers manning the garden during the night are unarmed and are limited in their ability to deter thieves.

18.7 Conclusion

The study considered the contribution of smallholder irrigation agriculture towards urban household food security by reviewing the Green Valley Garden in the City of Harare, Zimbabwe. The garden is farmer-managed and it was established in 2010 by the City Council and the donor community. Initially, 300 beneficiaries comprising orphans, elderly, widows, and those affected by HIV and AIDS were selected. Currently, there are only 48 farmers farming the garden due to political intimidation and lack of resources to continue farming. Thus, there are structural challenges that make it difficult for potential participants to participate in the garden. All of this undermines the good intentions of the NGO and illustrates the dire need for this sort of farming for Harare residents.

Green Valley Garden has the potential to significantly enhance household food availability, although farmers face a lot of challenges in order to benefit from it. The management system could be enhanced through training of the beneficiaries, so that adequate resources are available and they are able to manage the operation of the garden in a sustainable manner. The stakeholders could provide initiatives to equip the beneficiaries with the managerial skills and resources in order to avoid the problems recurring, but most importantly will be attention to the political complexity of the garden and the threat potential beneficiaries feel by participating in it.

The study generated information useful for directing the future planning and development of smallholder gardens. Research findings can build the capacity of stakeholders to work on mitigation measures and improve irrigation performance and the livelihoods of beneficiaries.

References

City of Harare (1992) *Harare Combination Master Plan*. City of Harare, Harare, Zimbabwe.

Crush, J. and Frayne, B. (2010) *The Invisible Crisis: Urban Food Security in Southern Africa*. African Food Security and Urban Network (AFSUN), Cape Town, South Africa.

Crush, J., Hovorka, A. and Tevera, D. (2011) Food security in Southern African cities: the place of urban agriculture. *Progress in Development Studies* 11(4), 285–305.

Gallaher, C.M., Kerr, J.M., Njenga, M., Karanja, N.K. and WinklerPrins, A.M.G.A. (2013) Urban agriculture, social capital, and food security in the Kibera slums of Nairobi, Kenya. *Agriculture and Human Values* 30(3), 389–404.

Hungwe, C. (2006) *Urban Agriculture as a Survival Strategy: An Analysis of the Activities of Bulawayo and Gweru Urban Farmers (Zimbabwe)*. City Farmer, Ottawa, Canada.

Kiguli, J. (2005) Contribution of urban agriculture to food security in Uganda. *Paper Presented at the Africa Local Government Action Forum (ALGAF) Phase V: Session VI. July 2005*. World Bank Institute, Harare, Zimbabwe.

Korir, S.R.C., Rotich, J.K. and Mining, P. (2015) Urban agriculture and food security in developing countries: a case study of Eldoret Municipality, Kenya. *European Journal of Basic and Applied Sciences* 2(2), 27–35.

Kutiwa, S., Boon, E. and Devuyst, D. (2010) Urban agriculture in low income households of Harare: an adaptive response to economic crisis. *Journal of Human Ecology* 32(2), 85–96.

Mbiba, B. (2000) The integration of urban and peri-urban agriculture into planning. *Urban Agriculture* 4, 1–2.

Merrifield, R. (2014) An urban political ecology of food insecurity in Washington, DC. A Bachelor of Arts (BA) honors thesis in Environmental Studies, Davidson College, Davidson, North Carolina.

Mwendera, E., Chilonda, P. and Chigura, P. (2013) Options for operation and maintenance partnerships: a case study of Rupike irrigation garden, Zimbabwe. *Sustainable Agriculture Research* 2(3), 136–137.

Olufemi, K.R. and Alabi, R.K. (2012) A review of urban agriculture as a tool for building food security in Nigeria: challenges and policy options. *Journal of Sustainable Development in Africa* 14(3), 1–12.

Opitz, I., Berges, R., Piorr, A. and Krisker, T. (2016) Contributing to food security in urban areas: differences between urban agriculture and peri-urban agriculture in the Global North. *Agriculture and Human Values* 33, 341–358.

ZimVAC (2006) *Zimbabwe Urban Areas Food Security and Vulnerability Assessment*. Zimbabwe Vulnerability Assessment Committee. SIRDC, Harare, Zimbabwe.

ZimVAC (2009) *Urban Livelihoods Assessment*. Zimbabwe Vulnerability Assessment Committee, Harare, Zimbabwe.

ZimVAC (2011) *Urban Livelihoods Assessment*. Zimbabwe Vulnerability Assessment Committee, Harare, Zimbabwe.

19 Community Gardens as Urban Social–Ecological Refuges in the Global North

Joana Chan,[1,2*] Bryce B. DuBois,[2,3] Kristine T. Nemec,[1,4] Charles A. Francis[1,5] and Kyle D. Hoagland[1]

[1]University of Nebraska, Lincoln, Nebraska, USA; [2]Cornell University, Ithaca, New York, USA; [3]City University of New York, New York, USA; [4]University of Northern Iowa, Cedar Falls, Iowa, USA; [5]Norwegian University of Life Sciences, Ås, Norway

19.1 Introduction

The goal of this chapter is to use the lens of social–ecological resilience to understand three systems of community gardens, and how they may represent the rapidly growing community garden movement in different urban contexts in the Global North. Community gardens can be defined as community-managed open spaces, where groups of people grow food and/or ornamental plants (Holland, 2004; Campbell, 2014). While many community gardens are considered forms of urban agriculture due to their food production, a significant number of community gardens have additional purposes. Furthermore, in many community gardens that do grow food, food production is often secondary to other objectives, such as neighbourhood stabilization, community empowerment and education (Campbell, 2014). These gardens are either collectively cared for as a whole, sets of plots that are individually cultivated, or a combination of these arrangements (Iaquinta and Drescher, 2010; Okvat and Zautra, 2011). Despite their communal nature, community gardens have also been found to be exclusionary (Kurtz, 2001; Chapter 5, this volume).

Interest in urban community gardening has grown in the Global North (GN) in response to the global economic downturn, shifting environmental and social values, and increasing food-related public health concerns (Hou et al., 2009). Communal gardens in the GN have historically been cast as 'supportive institutions', which exemplified 'local resiliency in times of crises' by cultivating opportunities to cope with various socio-economic challenges (Bassett, 1981; Lawson, 2005, p. 301; Barthel et al., 2010). For example, during the late 19th century, allotment gardens, or rented gardens from an allotment association, were sown across Europe to foster beautification, cultural integration, and improved public health for the burgeoning landless working class as industrialization propelled urban migration and public disease outbreaks threatening health in European cities (Rotenberg, 1995; Barthel et al., 2010).

During the First and Second World Wars, 'Liberty Gardens', 'War Gardens' and 'Victory Gardens' were cultivated extensively throughout the UK and US to supply fresh food, combat malnutrition, and bolster morale during a time of significant resource shortage (von Hassell, 2002; Lawson, 2005; Barthel and Isendahl, 2013). It is important, however, not to allow the 'crisis

*Corresponding author; e-mail: jchan@huskers.unl.edu

narrative' of community gardens as relief measures to obscure the integral role that these community open spaces and subsistence food-production sites play in the regular functioning of the urban environment (Moore, 2006). Allotment gardens are still prevalent in Europe; for example, in the UK there are an estimated 250,000 such allotments, with a waiting list of 100,000 people who desire to access their own small plot of land (Leapman, 2015).

One common feature of community gardens in these areas is the perception of their resilience, both biologically and socially. Reasons for participation vary, and we recognize that resilience is culturally and socially site-specific (Carpenter et al., 2001), thus case studies in different cities are used to explore how contemporary community gardens contribute to general and/or specific resilience on the local level. First, we present a case study of the burgeoning community gardens movement in a demographically changing Vienna, Austria, where gardening has a strong history and prominence in the urban landscape. While much of the community garden literature focuses on large cities (Guitart et al., 2012), we offer a case study that explores community gardening in the context of a small city in the agriculturally dominated Great Plains of the US: Lincoln, Nebraska. Finally, we turn to explore the specified resilience aspect of community gardens and present a case study of coastal community gardens in New York City, to understand the function and meaning of community gardens, post-Hurricane Sandy. We synthesize the findings and interpretations from these three case studies to address the local contributions of urban community gardens to social–ecological systems, and the theoretical contributions of community gardens research to the development of the social–ecological resilience framework.

19.2 Conceptual Framing

19.2.1 Resilience

The idea of resilience has become popular for conceptualizing urban systems and addressing the increasingly complex challenges, opportunities and uncertainties in cities (Berry, 2014). Social–ecological resilience is the capacity of a complex adaptive system, such as a farm, a neighbourhood, or a city to respond and adapt to disturbances and still maintain its essential structure and functions (Holling, 1973; Folke, 2006).

Resilience can be differentiated between specified resilience and general resilience (Folke et al., 2010). Specified resilience addresses the question of 'resilience of what to what?' (Carpenter et al., 2001) and concerns the resilience of a particular part of a system to certain identified disturbances (Folke et al., 2010). General resilience, on the other hand, deals with the system as a whole and involves coping with all types of uncertainty, both known and unknown (Folke et al., 2010). Properties that convey general resilience include diversity, modularity, tight feedbacks, ecosystem services, social capital, and overlapping governance (Carpenter et al., 2012). While there is a growing body of research on specified resilience related to risk management and vulnerability, there has been considerably less research conducted on general resilience (Adger et al., 2005; Carpenter et al., 2012). Addressing both aspects of resilience is important, as putting too many resources towards managing specific shocks and thresholds may inadvertently reduce a system's overall resilience to 'novel' surprises (Folke et al., 2010). As resilience is increasingly recognized as one key property of sustainable systems, understanding the social–ecological role of this broad phenomenon can inform governance and research related to community gardens and other urban civic ecology practices.

Although there is literature that discusses the capacity for community gardens to cultivate general community resilience (Okvat and Zautra, 2011), as well as specified resilience to historical urban food shocks (Barthel and Isendahl, 2013; Barthel et al., 2013c) and general social and natural disasters (Okvat and Zautra, 2014), there is limited empirical research conducted on community gardens and their contributions to general and specified resilience in a contemporary urban context.

19.2.2 Biocultural refugia

Resilience scholars Barthel and colleagues (Barthel et al., 2013a) borrow the term 'refugia' from population biology to describe a function of communal gardens as 'biocultural refugia' or valuable 'memory carriers', which can provide the means and resources necessary for responding and rebuilding after a social–ecological disturbance. In ecology, 'refugia' are locations of relict populations of once-widespread plant and/or animal species that were sheltered from disturbance or stress, such as natural disasters or disease outbreak (Stewart et al., 2010). The concept of biocultural refugia merges discourses within social memory, food security and biodiversity to refer to 'areas that harbor place specific social memories related to food security and stewardship of biodiversity' (Barthel et al., 2013a, b).

Prime examples of biocultural refugia are traditional smallholder farms that cultivate and develop a diversity of crop varieties and landscape features that are adapted to local biophysical conditions (Barthel et al., 2013a). During a time of rapidly degrading agrobiological diversity, traditional ecological knowledge, and increasing concerns over food security, the social–ecological or stewardship memories of how to grow food and manage green spaces can be critical to local systems renewal and reorganization in response to change and disturbance (WinklerPrins, 2006; Barthel et al., 2013a).

19.3 Study Areas

19.3.1 Vienna, Austria

Vienna, capital city of Austria, is a growing city, projected to increase from a population of 1.7 million in 2012 to 2 million by 2030 (WUA, 2012). In addition to an increasing population, the demographic composition of Vienna is changing with increasing immigration, aided by the fall of the Iron Curtain in 1988 and Austria's entrance into the European Union in 1995 (Fassmann and Reeger, 2008). In 2011, 21.5% of Vienna residents were immigrants (Statistics Austria, 2012). Vienna has a higher than average amount of open space compared to other major cities in the world, due to a history of its citizens valuing the protection of green space such as the Wienerwald, or Vienna Woods, which was protected from development in the late 1800s (Rotenberg, 1995, p. 27).

Gardens play an important role in the environment and society of Vienna. From the highly manicured historical baroque gardens reflecting the value placed on order during monarchical control, to the expansion of allotment gardens for self-sufficiency and health during times of war and economic hardship since the late 19th century, gardens are long-serving models of social and ecological relationships in Viennese society (Rotenberg, 1995). Community gardens have grown in the past decade, from the first one established in 2001 to more than 30 established throughout the city as of 2014 (Gartenpolylog, 2014).

19.3.2 Lincoln, Nebraska, USA

Located in the central midwestern region of the US, Lincoln is the capital city of the largely agrarian state of Nebraska. Lincoln is a relatively small city, with a population of nearly 270,000 that is growing in size and diversity (US Census, 2013a). The first official community garden in Lincoln was established in 2003 as part of Community Crops, a Nebraska Sustainable Agriculture Society AmeriCorps project that transformed a vacant, trash-filled lot to spaces for 35 gardeners representing seven different countries (I. Kirst, personal communication; CROPS, 2014). Despite its relatively small size, Lincoln was host to around 20 community gardens in 2014. Many of these community gardens are established in partnership with and on land from local churches and other community organizations. The garden spaces primarily consist of a set of personal/family garden plots with communal areas for seating, composting, a bulletin board, and sometimes art displays and ornamental plants.

Community Crops estimates over 500 participants in their community gardens, with refugees and immigrants from 25 foreign countries composing 30% of their membership (I. Kirst, personal communication). The countries with the largest numbers of refugees or immigrants represented include Myanmar, Iran, Iraq, Mexico, Nigeria and Sudan (I. Kirst, personal communication). This high level of ethnic diversity is attributable to Lincoln being an official refugee resettlement location (Pipher, 2003) and to the fact that Lincoln hosts a large public university, with more than 2500 international students and scholars from more than 150 countries

(University of Nebraska-Lincoln, 2015). As community gardens reflect their local and regional socio-political conditions and most of the research conducted on community gardens has focused on major metropolitan areas (Guitart et al., 2012), this case study on community gardens in Lincoln offers insight into the possible differences among the roles of community gardens in smaller urban contexts.

19.3.3 New York, New York, USA

With a population of over 8 million, New York City (NYC) is the largest city in the US (US Census, 2013b). NYC's approximately 700 community gardens (as of 2014) make up one of the most active programmes in the US (von Hassell, 2002). From the 1970s to 1980s, communal gardens were cultivated throughout the city, especially in poorer neighbourhoods such as Harlem and the Lower East Side where buildings were subjected to abandonment, arson and demolition (von Hassell, 2002). The communal garden projects that grew out of NYC's urban decline are often cited as models of contemporary grassroots community gardens, as they are now known (von Hassell, 2002; Saldivar-Tanaka and Krasny, 2004). The early gardens leased the land for US$1 per year and functioned as a free revitalization programme for the city of New York (Schmelzkopf, 1995). The City further supported these efforts by establishing the city agency Operation Green Thumb in 1978 to oversee and encourage community groups, an agency that was later absorbed into the NYC Department of Parks and Recreation in 1995 (Eizenberg, 2012).

People involved in community gardens in NYC by and large represent the racial/ethnic makeup of the neighbourhoods they are embedded in, while the age and gender breakdowns vary. New York City community garden researcher Effrat Eizenberg (2013) noted that garden activists were concerned that the majority of gardeners are over 50 years old, a finding echoed in interviews with Green Thumb employees in 2005, who expressed concerns about the lack of participation by young people. Other studies have found similar results, although they note that young people are frequent visitors to gardens (Saldivar-Tanaka and Krasny, 2004).

19.4 Methods

In developing these case studies, we used a qualitative phenomenological and field-based research approach. As social–ecological resilience is an ecologically extrapolated concept, resilience research has been critiqued for overlooking critical issues of human agency and power which inform social action and response to change (Davidson, 2010). Utilizing humanistic and reflexive methods that empower individuals to share their experiences and highlight a diversity of voices, qualitative research may address these social deficiencies in resilience research and contribute to the development of a more complex and holistic understanding of social–ecological systems (Creswell, 2007).

We participated in observations of 18 community garden sites, and performed formal interviews with 37 community garden participants and managers. Fieldwork was conducted by the first and third authors in Vienna, by the first author in Lincoln, and the first and second authors in New York City. We recruited community gardener participants via purposive snowball sampling to reach a diverse representative sample of community gardeners based on gender, age, cultural background and gardening experience. We audio-recorded and transcribed all interviews, and took down descriptive and analytic field notes after each community garden visit. We also conducted archival reviews of news media, social media postings, and documents related to the community gardens. Interview transcripts were coded independently by the primary field researchers and then compared and discussed until consensus was reached regarding the final themes related to the social–ecological functions of the community gardens. To triangulate our data, we compared the themes developed from the interviews with those from participant observation field notes and our archival review to check for and resolve any thematic discrepancies (Denzin and Lincoln, 2011).

19.5 Results and Discussion

In this section, we explore and discuss research results on the social–ecological roles of community gardens and community gardening from each of the case studies within the context of:

1. the physical space of community gardens;
2. the experiential practice of community gardening; and
3. the social actors and networks that put the 'community' in community gardens.

19.5.1 Community gardens: diverse social–ecological systems as community places

Community gardens in our study sites were heterogeneous spaces that often embodied the changing demographic diversity of the neighbourhood or city at large. Diversity is a key property of generally resilient systems (Folke, 2006). Community gardening in these cases promoted the development of diverse, vibrant, community places – both socially and ecologically. Places are not only physical locations (spaces), but also the settings for social relationships, and the personal subjective and emotional attachments that people have to places (sense of place). Most of the community gardens displayed a high diversity of participants in terms of socio-economic status, ethnic background, and geography of origin. This diversity was displayed in both the biophysical structure of the individual gardens themselves as well as the multiple functions that these gardens provided to their neighbourhoods.

In all three case studies, the social diversity of community gardeners also manifested in the varied ecological mosaic of landscapes within community gardens, many of which reflected a biocultural diversity of plants that may have traditional culinary, medicinal or other utilitarian purposes (e.g. Cameroonian garden huckleberry and Chinese bottle gourd) and/or traditional agricultural management techniques (e.g. trench irrigation by Middle Eastern gardeners and vertical trellising by Asian gardeners). There was also an observed spatial diversity of plantings, a valuable contributor to insect and pathogen protection in the gardens, compared to large monoculture plantings of the same crops that often require heavy applications of pesticides to prevent economic losses.

In each of the case studies, there were community gardeners who identified the promotion of biological diversity as a primary goal within their community gardens. This was especially true of the community gardens studied in Vienna (Chan and Nemec, 2011, unpublished results), where community gardeners were actively involved in planting and saving seeds from heirloom and other open-pollinated cultivars to establish plant genetic diversity in their own gardens and encourage this with others.

Gärtnernwienochnie Augarten, for example, was awarded with the municipal 'Natural Green Oasis' (*Naturnahe Grünoase*) recognition for its extensive efforts to increase biodiversity and wildlife habitat. *permaBlühGemüse*, another community garden in Vienna, began as a permaculture discussion group and seed exchange that eventually established a space to garden and practice in 2001. The community garden was home to an extremely high diversity of plants; with records of more than 420 different varieties of plants, ranging from vegetables, fruits and herbs to wild plants and weeds.

In all three cities, community gardeners were also actively involved in promoting a more macro-level biodiversity, by cultivating specific plants and building physical structures to attract and provide habitat for wildlife such as birds, bats, frogs, small mammals and various insects. A *permaBlühGemüse* gardener shared how he actively created habitat to attract a diversity of wildlife to the garden:

> I try to attract all types of ... insects. I try to attract the hedgehog. This year I also had a lot of [sunchoke], [and] the plant [can] get quite high, so I cut it and I made a ... mini hedgehog home. ... As you saw in our place... the structure of our field is made of mulch [and] compost ... we see a lot of holes in the walls ... we are not sure what's living there ... but something is.

The multiple levels of diversity found in community gardens – human, floral and faunal – follow other studies that demonstrate community gardens as spaces of social and ecological diversity (e.g. Holland, 2004). This is not unlike the diversity found in home gardens in much of the Global South (Trinh et al., 2003; Coomes and Ban, 2004; WinklerPrins, 2006).

In addition to providing supporting ecosystem services such as plant genetic biodiversity, community gardens in the Vienna study demonstrate the capacity to provide a range of other local ecosystem services, as illustrated in greater detail in Table 19.1.

Table 19.1. Possible ecosystem services provided by *Gemeinschaftsgarten Augarten*, *Nachbarschaftsgarten Macondo*, and *permaBlühGemüse Garten*, Vienna, Austria. (Adapted from Krasny et al., 2014.)

Category of ecosystem services	Possible ecosystem services provided
Provisioning	Fresh food
	Ornamental plants
Regulating	Pollination
	Natural control of plant pests
	Food sources and habitat for beneficial insects
Cultural	Support diverse gardening knowledge systems
	Space for art, film and cultural projects
	Environmental education
	Foster social capital and trust
	Enhance sense of place, appreciation of nature, and spiritual and aesthetic use
Supporting	Create compost for soil enhancement
	Support plant genetic biodiversity

Because composting, for example, was a common practice in all three case studies, community gardens provide the supporting ecosystem service of soil enhancement. The flora and fauna of community gardens also offer regulating ecosystem services, such as pollination, natural control of plant pests, and food sources and habitat for beneficial insects. Most clearly, in terms of provisioning ecosystem services, community gardens provide fresh food in urban areas, sometimes serving as a critical source of food security, as discussed in depth below and in other chapters in this volume (Chapter 3, this volume; Chan et al., 2015, 2016).

Community gardens can also offer a breadth of cultural ecosystem services by providing spaces for neighbourhood art, cultural projects and environmental education. These spaces support general resilience by functioning as outdoor community hubs where family, friends and neighbours can meet, gather and hold events, such as potlucks and other activities.

In addition to being places that support the biocultural diversity and the ecosystem services that can bolster general resilience, community garden spaces are also key to supporting specific resilience; in this case, the resilience of neighbourhoods with community gardens to Hurricane Sandy. On the evening of 29 October 2012, Hurricane Sandy made landfall in the New York metropolitan area. The immense diameter of the storm (roughly 1000 miles) and landfall during a high tide and full moon led to an average tidal surge of over 35 cm in NYC. That surge led to 72 deaths in the Northeast, directly attributed to the storm, and was the second-costliest in US history (Blake et al., 2013). In NYC post-Sandy, community gardens served as spaces for art, or communal creative expression and meaning-making. In Campos Community Garden, gardeners displayed the 'eco-art' made by neighbourhood children as a whimsical act of beautification and symbolic resilience immediately after the storm.

Community garden art also functioned as a form of memorialization, as exemplified by the Sea Song Memorial Sculpture. Sea Song used pieces of Sandy storm driftwood and debris to create a structure that resembled a tree. The sculpture held small items of meaning contributed by community members and was flanked with colourful 'resilience flags' designed by residents of areas heavily impacted by Sandy, ranging from Rockaways to Coney Island to Haiti. The flags were decorated with water motifs, images of seahorses, crabs and fish native to the local watershed, and words that evoked solidarity, good will and strength, and these were hung throughout the garden space; all of the flags were memorialized in a tree-like form that supported reflection among neighbourhood residents.

The open, shared, safe spaces of the community gardens were especially important for some community gardeners in New York City post-Hurricane Sandy (Chan et al., 2015). The gardeners appropriated community garden spaces immediately after the storm for use as makeshift relief distribution sites, meeting places for sharing information and, later on, as gathering spaces for community healing. Recognized as spaces of community support, community gardens were places where gardeners and their neighbours could go to receive community assistance, as well as places people gravitated towards to provide or access community services. Another form of vital support was to check on residents who had not appeared outside their houses, and garden community members were especially attentive to the needs of the elderly, who may be have been trapped or reluctant to come out after the storm.

Immediately after Sandy, some coastal NYC community gardens were sites for convening, news sharing and communal cooking. Community gardens were appropriated for emergent community needs, such as staging grounds and distribution sites for food, clothing and solar-generated electricity immediately after Sandy. In the months after the storm, the Beach 91st Street garden in NYC also hosted a regular community-healing circle to help neighbourhood residents to collectively process the stress and trauma of the storm.

In the development of inclusion and refuge, however, many community gardens also embodied aspects of exclusion and enclosure (Kurtz, 2001; see also Chapter 5, this volume). In Lincoln, issues such as property vandalism and produce theft were sometimes mentioned as symptoms of this lack of integration with the neighbourhood. That community gardens are considered safe havens and spaces for socializing can also be a source of problems both for members and neighbours of the garden. One community garden coordinator spoke of enforcing more exclusiveness after a member began 'living' at their site:

> We have had some issues of homeless[ness]... connected with this gardener who have been sort of ... living here ... so you can see the 'No Trespassing' sign put up and the neighbor has been calling police this summer ... And so there have been some issues with that, but as far as I know, everybody has been safe, there has been no violence. It is just frustrating to come here and find a 12-pack of beer just scattered all over the front here. (First and L. St Gardener, Lincoln)

While many identified the garden as a place for social interaction, most gardeners also reported social challenges both within the community garden organization as a whole and between the garden and the surrounding neighbourhood. Community Crops staff and garden members stated that language barriers sometimes impeded effective communication of garden policies and news to and between participants. Thus, while there are many community development-related benefits that are nurtured by community gardens, there are also many challenges.

In Lincoln, community gardens were some of the few places where people could regularly interact with many people of different cultural backgrounds, as ethnically diverse spaces in an otherwise relatively racially homogeneous (largely white) city. Community gardening also functioned as a community-building mechanism for bridging social networks – connecting people to those of different social groups in a shared space over a common practice.

> I am Christian, he is Muslim, but we had wonderful discussions. [Laughs] ... we have different people with different philosophies and different viewpoints, but inside the garden, inside the fence of the garden, we are all one family, and so whether we disagree or we don't like somebody's philosophy or idea, we are still one family working together. And there are times that gets stressed a little bit [sic], but ... the goal of the community garden is for everybody, for adults, for children, for everybody to work together and to be nurtured. (Antelope Community Gardener, Lincoln)

19.5.2 Community gardening as a meaningful practice of reconnection

Besides functioning as significant physical places that support resilience, the community gardens in these studies also supported meaningful community garden practices – embodied actions ranging from gardening and creating art to encouraging participation in community events, which can help gardeners develop the capacity to cope with general challenges such as displacement and stress, as well as shocks such as natural disasters.

As most of the community gardeners in Lincoln were migrants to the city, either transnational immigrants or domestic migrants from rural areas, the act of growing food often connected people to meaningful places and traditional practices that were reminiscent of their homes and personal histories. While this agricultural knowledge and perpetuation of traditional agricultural practices is beneficial in terms of identity development and practical cultivation of the land, some traditional practices rooted in other agroecosystems may be maladapted to temperate semi-arid ecosystems in the US.

> We see, especially from the Middle East, trench gardening, where trench gardening isn't really needed here... a lot of the trench gardeners will fill them up daily when we really only need 2 inches... of water a week, so... really they could be trench gardening but only filling up them up once a week instead of every day, so they are way over-watered... but it's been a challenge to convey that to them and as a white woman from America telling us a farmer who has done it for 30 years that his way of farming is not good...(Community CROPS staff, Lincoln)

In all of the case studies, community gardening was also a means for the many gardeners who grew up farming or gardening to reconnect with a practice, landscape and range of agricultural products that were an integral part of their personal and/or cultural identity. This theme echoes what Foeken (2006) found in eastern Africa, where urban farmers said that their rural agricultural upbringing created a desire to engage in the familiar custom of growing one's own food and preserving one's agricultural identity after moving to the city. Echoes of this are also found in numerous chapters in this volume (e.g. Chapters 6, 8, 11, 12 and 17).

For some other Lincoln community gardeners who grew up gardening in the Midwest, growing their own vegetables not only granted them choice and access to the best possible food, but also perpetuated traditional rural practices of food preservation. The experience of community gardening thus functioned to preserve and perpetuate collective social–ecological memories associated with growing food and working with the local landscape (Barthel *et al.*, 2010).

In all three case studies, participants expressed that the act of gardening was therapeutic, compelling them to be more active and spend time outdoors, and restoring a connection to their physical, emotional and/or spiritual sense of self and well-being. One Viennese community gardener expressed the emotional health benefits of gardening:

> I was working a lot of time in the office and now when I'm back outside... it's changing how I feel... it's something to do with joy, and, and good living...If you have the possibility to be outside in daylight and not artificial light, that affects your body and...maybe your soul also. (permaBlühGemüse Gardener, Vienna)

The gardeners' reactions support previous research that has found positive feelings expressed by gardeners in private domestic gardens when describing gardening activities or simply being in the presence of the garden (Gross and Lane, 2007; see also Chapter 12, this volume).

The therapeutic nature of gardening and working with the land also supported the resilience of community gardeners themselves to the stresses experienced following Hurricane Sandy (Chan *et al.*, 2015). Community gardeners discussed the healing nature of returning to and restoring their gardens in a manner reminiscent of 'urgent biophilia' and 'restorative topophilia', in which people may feel compelled to engage in greening practices and restore a beloved natural place in post-disaster situations (Tidball, 2012). The practice of working in the garden was repeatedly highlighted as a source of catharsis and a reminder of normality during the stressful recovery after the storm in NYC. Our findings lend credence to Okvat and Zautra's (2014) assertions that community gardening has the capacity to nurture social resilience post-disaster by providing positive spaces and activities to help alleviate the stress of post-disaster situations.

Beyond individual and social reconnections, community gardeners also framed their practice of gardening as strengthening linkages with their food system and local environment. For some participants in Lincoln and NYC in particular, community gardens provided a significant source of food security and a more empowering supplemental nutrition alternative to food pantries and soup kitchens (Chan *et al.*, 2016). A community garden coordinator in Lincoln discussed how community gardening was a particularly empowering experience for some participants, allowing them to exercise some control over their livelihoods and the type and quality of food that they and their families consumed.

Besides motivations to respond to local degradation and lack of respect for social community and environment, the garden coordinators for all of the gardens discussed a global deterioration of social and ecological capital, compelling them to initiate actions that beautified the environment, preserved local biodiversity, and reconnected people to nature and each other. Thus, community gardens had not only an adaptive function to some disturbances, but also a transformative function in dealing with chronic social–ecological issues (Folke et al., 2010).

19.5.3 Community gardens cultivate supportive communities

Beyond the physical space of the community garden, and the embodied practices of community gardening, community gardens also contribute to local resilience by building community, fostering social capital, and functioning as supportive communities of practice that can help gardeners to learn from and manage both daily challenges and more acute disturbances. Communities of practice are groups of people (e.g. community gardeners) who share concern for a certain issue or activity (growing food, civic action, neighbourhood greening, etc.) and learn to improve their engagement with this concern through continual interactions (community gardening) (Wenger, 2012; see also Chapters 6 and 8, this volume).

Social learning is central to communities of practice and can be a means of building resilience by fostering adaptive changes and innovations (Pahl-Wostl, 2006; Wenger, 2012). Social learning describes the process of continued feedback and changes occurring between learners and their environment (Pahl-Wostl, 2006). Participants spoke of how social interactions in the gardens helped facilitate the development of knowledge related to gardening and the use and cultivation of plants. As communities of learning, community gardener groups discussed sharing and exchanging information, resources and manual assistance, which helped them to adapt and improve their gardening practices. Community gardens also offered opportunities for gardeners to exchange knowledge and to learn from each other about gardening methods and about different types of plants and foods that they had not been exposed to previously.

Participants in Vienna, Lincoln and NYC also spoke of how community gardening was a way for them to connect with and deepen their relationships with other people, such as family members and neighbours. By creating opportunities to have a shared time, space and activities together with their family and community members, community gardening helped to strengthen both bonding and bridging types of social capital among our case study participants. This conclusion supports similar findings on the contributions of community gardens and urban agriculture to social capital in the US (Glover et al., 2005), and in South Africa (Dunn, 2010).

Many immigrant community gardeners expressed concerns with social isolation either as new arrivals to the city, or because of language and/or socio-economic barriers. Community gardening functioned as a community-building mechanism for social bonding, strengthening the connections within social groups like families and ethnic groups.

> We are getting some of the [Cameroonian huckleberry leaf] to friends to share . . . It has created some relationship with the people that we never would have reached with[out] them . . . they would either come to my house, or if we give it out to a friend and [another] friend would see them cooking and say, 'Where do you get this?' . . . I say, 'Hey, if you need some, come on this day, go to the garden.' . . . It gives us the opportunity, especially immigrants, to interact and form some sort of social group within our self [sic]. (46th St Community Gardener, Lincoln)

The community relationships and social support networks that developed from strong communities of practice were not only sources of support in coping with general stresses, but also keys to recovery from major disasters, as in the case of Hurricane Sandy in NYC (Chan et al., 2015). The sense of community developed from these strong communities of practice was cited as one of the primary factors that initially inspired involvement in community gardening for many members. After Sandy, these community relationships and social support networks were mobilized and cited as central to recovery. This was illustrated most vividly at Beach 91st St garden, where the trust and camaraderie nurtured from neighbours working together in the community garden was identified as a source of mutual support and safety:

> [L]ook at the 50 people eating homemade chili over an open fire two days after one of the most devastating hurricanes in the East Coast. Standing around joking, having hot chocolate . . . when the National Guard can't even get

through yet . . . That is the best defense we have against fear. The best defense we have against looting, rioting, or any other kind of insecurity . . . And that is a direct result of the community garden. You know being a hub for safety, security. A blanket of support between neighbours. (Beach 91st Gardener, NYC)

In Campos Community Garden, gardeners mobilized their support networks within and beyond the garden to help ensure the well-being of local residents, facilitate the transfer of information in the community around the garden, and aid those stranded without electricity, food or water. The trust and cohesion that developed from neighbours working together in the community garden was identified as a source of collective efficacy, mobilizing support networks both before and after the storm to ensure the well-being of local residents and facilitate the transfer of aid in the community that was made possible because of the community garden space itself. These findings support Teig *et al.*'s (2009) claim that community gardens can develop collective efficacy by fostering the social cohesion that is crucial for catalysing actions that benefit communities.

19.6 Conclusions

Urbanization creates both opportunities and challenges for our social–ecological systems. Our exploration of three case studies of community gardens in the GN as described in this chapter provides a potentially valuable empirical contribution to the understanding of function and meaning of urban community gardens to resilience. The physical space, embodied practices and communities of practice associated with these community gardens reflect their local context to support vital social–ecological functions ranging from generating local ecosystem services to cultivating community empowerment and bolstering food sovereignty.

Based on the findings from this set of case studies of community gardens, we propose that community gardens can function as 'urban social–ecological refuges' to support resilience in cities. Building upon the idea of biocultural refugia, we suggest that community gardens can serve as potential *urban social–ecological refuges*, functioning not only as biocultural and cognitive refuges of social–ecological knowledge, practices and resources, but also as physical havens for safety and the development of social and spiritual support. As urban social–ecological refuges, community gardens have the potential to not only carry and transmit cognitive and biological means for facilitating resilience and response capacity by protecting ecological and biocultural diversity and social–ecological memories (as highlighted by community gardeners in Vienna and Lincoln), but also be restorative physical refuges for communities in disaster contexts, as demonstrated by community gardens in New York City following Hurricane Sandy.

As a nested and integral part of the social–ecological fabric in their communities, the vital functions of these urban social–ecological refuges extend beyond the physical boundaries of community garden spaces. The communities of practice that contribute to these social–ecological refuges, for example, are embedded in social networks and other communities of practice that allow for the exchange of information and resources that enable these places and their contained practices to also develop and adapt, based on the changing social and environmental context.

To build on this broad exploratory study and further develop the hypothesis of community gardens as urban social–ecological refuges, additional research is suggested to measure the biological diversity of urban community gardens and assess what social factors (such as governance and participation) and spatial factors (such as distance between gardens and gardeners, and gardens and other public and/or green spaces) best facilitate biocultural diversity and preserve social–ecological memory. As most research on community gardens has been conducted in the GN (Guitart *et al.*, 2012), further research in the Global South would help us to understand if and how community gardens may manifest as urban social–ecological refuges in different socio-economic and geographic contexts.

Acknowledgements

Joana Chan and Kristine T. Nemec received support through the NSF IGERT on Resilience and Adaptive Governance of Stressed Watersheds at the University of Nebraska-Lincoln (NSF #0903469). Bryce B. DuBois received support through USDA Federal Formula Funds CUAES (USDA #147-7815).

References

Adger, W.N., Hughes, T.P., Folke, C., Carpenter, S.R. and Rockström, J. (2005) Social–ecological resilience to coastal disasters. *Science* 309(5737), 1036–1039.

Barthel, S. and Isendahl, C. (2013) Urban gardens, agriculture, and water management: sources of resilience for long-term food security in cities. *Ecological Economics* 86, 224–234.

Barthel, S., Folke, C. and Colding, J. (2010) Social–ecological memory in urban gardens – retaining the capacity for management of ecosystem services. *Global Environmental Change* 20, 255–265.

Barthel, S., Crumley, C. and Svedin, U. (2013a) Bio-cultural refugia – safeguarding diversity of practices for food security and biodiversity. *Global Environmental Change* 23(5), 1142–1152.

Barthel, S., Crumley, C. and Svedin, U. (2013b) Biocultural refugia: combating the erosion of diversity in landscapes of food production. *Ecology and Society* 18(4), 71. DOI: 10.5751/ES-06207-180471.

Barthel, S., Parker, J. and Ernstson, H. (2013c) Food and green space in cities: a resilience lens on gardens and urban environmental movements. *Urban Studies* 52(7), 1321–1338.

Bassett, T.J. (1981) Reaping on the margins: a century of community gardening in America. *Landscape* 25(2), 1–8.

Berry, M. (2014) Thinking like a city: grounding social–ecological resilience in an urban land ethic. *Idaho Law Review* 50(2), 117–151.

Blake, E.S., Kimberlain, T.B., Berg, R.J., Cangialosi, J.P. and Beven, J.L. II (2013) *Tropical Cyclone Report: Hurricane Sandy*. Report AL182012. National Hurricane Center, Miami, Florida, USA.

Campbell, L. (2014) The sky is the limit for urban agriculture. Or is it? The Nature of Cities. Available at: https://www.thenatureofcities.com/2014/04/07/the-sky-is-the-limit-for-urban-agriculture-or-is-it-what-can-cities-hope-to-get-from-community-gardens-and-urban-agriculture (accessed 20 March 2015).

Carpenter, S., Walker, B., Anderies, J.M. and Abel, N. (2001) From metaphor to measurement: resilience of what to what? *Ecosystems* 4(8), 765–781.

Carpenter, S.R., Arrow, K.J., Barrett, S., Biggs, R., Brock, W.A., Crépin, A.-S., Engström, G., Folke, C., Hughes, T.P., Kautsky, N., *et al.* (2012) General resilience to cope with extreme events. *Sustainability* 4(12), 3248–3259.

Chan, J., DuBois, B. and Tidball, K.G. (2015) Refuges of local resilience: community gardens in post-Sandy New York City. *Urban Forestry and Urban Greening* 14(3), 625–635.

Chan, J., Pennisi, L. and Francis, C.A. (2016) Social-ecological refuges: reconnecting in community gardens in Lincoln, Nebraska. *Journal of Ethnobiology* 36(4), 842–860. DOI: http://dx.doi.org/10.2993/0278-0771-36.4.842.

Coomes, O.T. and Ban, N. (2004) Cultivated plant species diversity in home gardens of an Amazonian peasant village in northeastern Peru. *Economic Botany* 58(3), 420–434.

Creswell, J.W. (2007) *Qualitative Inquiry and Research Design: Choosing Among Five Approaches*. Sage Publications, Thousand Oaks, California, USA.

CROPS (2014) About Community Crops. Available at: http://www.communitycrops.org/about (accessed 15 September 2014).

Davidson, D.J. (2010) The applicability of the concept of resilience to social systems: some sources of optimism and nagging doubts. *Society and Natural Resources* 23(12), 1135–1149.

Denzin, N.K. and Lincoln, Y.S. (eds) (2011) *The SAGE Handbook of Qualitative Research*, 4th edn. Sage Publications, Thousand Oaks, California, USA.

Dunn, S. (2010) Urban agriculture in Cape Town: an investigation into the history and impact of small-scale urban agriculture in Cape Flats townships with a special focus on the social benefits of urban farming. Unpublished MA thesis, Department of Historical Studies, University of Cape Town, South Africa.

Eizenberg, E. (2012) Actually existing commons: three moments of space of community gardens in New York City. *Antipode* 44(3), 764–782.

Fassmann, H. and Reeger, U. (2008) Austria: from guest worker migration to a country of immigration. *IDEA Working Papers* 1, 1–39.

Foeken, D. (2006) *To Subsidise My Income: Urban Farming in an East African Town*. Brill, Leiden, The Netherlands.

Folke, C. (2006) Resilience: the emergence of a perspective for social–ecological systems analyses. *Global Environmental Change* 16(3), 253–267.

Folke, C., Carpenter, S.R., Walker, B., Scheffer, M., Chapin, T. and Rockström, J. (2010) Resilience thinking: integrating resilience, adaptability and transformability. *Ecology and Society* 15(4), 20.

Gartenpolylog (2014) Gartenkarte. Available at: https://gartenpolylog.org/gardens [in German] (accessed 15 September 2014).

Glover, T.D., Parry, D.C. and Shinew, K.J. (2005) Building relationships, accessing resources: mobilizing social capital in community garden contexts. *Journal of Leisure Research* 37(4), 450–474.

Gross, H. and Lane, N. (2007) Landscapes of the lifespan: exploring accounts of own gardens and gardening. *Journal of Environmental Psychology* 27(3), 225–241.

Guitart, D., Pickering, C. and Byrne, J. (2012) Past results and future directions in urban community gardens research. *Urban Forestry and Urban Greening* 11(4), 364–373.

Holland, L. (2004) Diversity and connections in community gardens: a contribution to local sustainability. *Local Environment* 9(3), 285–305.

Holling, C.S. (1973) Resilience and sustainability of ecological systems. *Annual Review of Ecology and Systematics* 4(1), 1–23.

Hou, J., Johnson, J. and Lawson, L. (2009) *Urban Community Gardens: Greening the City and Growing Communities in Seattle*. University of Washington Press, Seattle, Washington, DC, USA.

Iaquinta, D.L. and Drescher, A.W. (2010) Urban agriculture: a comparative review of allotment and community gardens. In: Aitkenhead-Peterson, J. and Volder, A. (eds) *Urban Ecosystem Ecology*. Agronomy Monograph Series 55, Madison, Wisconsin, USA, pp. 199–226.

Krasny, M.E., Russ, A., Tidball, K.G. and Elmqvist, T. (2014) Civic ecology practices: participatory approaches to generating and measuring ecosystem services in cities. *Ecosystem Services* 7, 177–186.

Kurtz, H. (2001) Differentiating multiple meanings of garden and community. *Urban Geography* 22(7), 656–670.

Lawson, L.J. (2005) *City Bountiful: A Century of Community Gardening in America*. University of California Press, Berkeley, California, USA.

Leapman, M. (2015) Allotments: a very British passion. *The Telegraph* [London], 10 August. Available at: http://www.telegraph.co.uk/gardening/4699817/Allotments-a-very-British-passion.html (accessed 21 November 2016).

Moore, S. (2006) Forgotten roots of the green city: subsistence gardening in Columbus, Ohio, 1900–1940. *Urban Geography* 27(2), 174–192.

Okvat, H.A. and Zautra, A.J. (2011) Community gardening: a parsimonious path to individual, community, and environmental resilience. *American Journal of Community Psychology* 47(3–4), 374–387.

Okvat, H.A. and Zautra, A.J. (2014) Sowing seeds of resilience: community gardening in a post-disaster context. In: Tidball, K.G. and Krasny, M.E. (eds) *Greening in the Red Zone*. Springer, New York, USA, pp. 73–90.

Pahl-Wostl, C. (2006) The importance of social learning in restoring the multifunctionality of rivers and floodplains. *Ecology and Society* 11(1), 10.

Pipher, M. (2003) *The Middle of Everywhere: Helping Refugees Enter the American Community*. Mariner Books, New York, USA.

Rotenberg, R. (1995) *Landscape and Power in Vienna*. Johns Hopkins University Press, Baltimore, Maryland, USA.

Saldivar-Tanaka, L. and Krasny, M.E. (2004) Culturing community development, neighbourhood open space, and civil agriculture: the case of Latino community gardens in New York City. *Agriculture and Human Values* 21, 399–412.

Schmelzkopf, K. (1995) Urban community gardens as contested space. *Geographical Review* 85(3), 364–381.

Statistics Austria (2012) Wanderungsstatistik 2011. Available at: http://www.statistik.at/web_en/publications_services/Publicationsdetails/index.html?includePage=detailedView&sectionName=Population&publd=542 (accessed 24 November 2016).

Stewart, J.R., Lister, A.M., Barnes, I. and Dalén, L. (2010) Refugia revisited: individualistic responses of species in space and time. *Proceedings of the Royal Society of London B: Biological Sciences* 277(1682), 661–671.

Teig, E., Amulya, J., Bardwell, L., Buchenau, M., Marshall, J.A. and Litt, J.S. (2009) Collective efficacy in Denver, Colorado: strengthening neighborhoods and health through community gardens. *Health and Place* 15(4), 1115–1122.

Tidball, K.G. (2012) Urgent biophilia: human–nature interactions and biological attractions in disaster resilience. *Ecology and Society* 17(2), 5.

Trinh, L.N., Watson, J.W., Hue, N.N., De, N.N., Minh, N.V., Chu, P., Sthapit, B.R. and Eyzaguirre, P.B. (2003) Agrobiodiversity conservation and development in Vietnamese home gardens. *Agriculture, Ecosystems & Environment* 97(1), 317–344.

University of Nebraska-Lincoln (2015) UNL enrollment climbs to all-time high of 25,260. Available at: http://news.unl.edu/newsrooms/today/article/unl-enrollment-climbs-to-all-time-high-of-25260 (accessed 22 November 2016).

US Census (2013a) Lincoln (city) QuickFacts from the US Census Bureau. Available at: http://www.census.gov/quickfacts/table/PST045215/3128000,31111,00 (accessed 15 September 2014).

US Census (2013b) New York (city) QuickFacts from the US Census Bureau. Available at: http://www.census.gov/quickfacts/table/PST045215/3651000,3128000,31111,00 (accessed 15 September 2014).

von Hassell, M. (2002) *The Struggle for Eden: Community Gardens in New York City*. Bergin and Harvey, London, UK.

Wenger, E. (2012) Communities of practice and social learning systems: the career of a concept. Available at: http://wenger-trayner.com/wp-content/uploads/2012/01/09-10-27-CoPs-and-systems-v2.0.pdf (accessed 15 September 2014).

WinklerPrins, A.M.G.A. (2006) Urban house-lot gardens and agrodiversity in Santarém, Pará, Brazil: spaces of conservation that link urban with rural. In Zimmerer, K.S. (ed.) *Globalization and New Geographies of Conservation*. University of Chicago Press, Chicago, Illinois, USA, pp. 121–140.

WUA (Wiener Umweltanwaltschaft) (2012) Urban planning, traffic and transport. Vienna City Administration. Available at: https://www.wien.gv.at/english/environment/ombuds-office/urban-planning.html (accessed 25 January 2013).

20 Global Urban Agriculture into the Future: Urban Cultivation as Accepted Practice

*Antoinette M.G.A. WinklerPrins**
Johns Hopkins University, Washington, District of Columbia, USA

20.1 Introduction

In May of 2016, the journal *Science* produced a special section entitled 'Urban Planet', acknowledging the rise of this way of life around the world as the dominant one on the globe (Science, 2016).[1] The timing could not have been more prescient as I was finalizing this edited volume focusing on urban agriculture. The objective of this volume has been to bring together the often quite disparate literatures on urban agriculture in the Global South and the Global North. What it has illustrated are the many similar ways in which UA is expressed in theory and practice in *both* the Global North and the Global South, and that cultivation in the city is increasingly seen as a normal activity of urban life. In Chapter 1 of this volume, I invoked Bassett's suggestion that cultivating the city makes urban life a 'palatable experience' (1981) for all, improving the quality of life through its greening of the city, provisioning of ecosystem services, providing meaningful work and activities for residents, and fostering community building. Here, in the concluding chapter, I wish to highlight the numerous takeaways implicitly or explicitly discussed by the authors in this volume by distilling them into three main points, while also offering some future directions. But first a note about terminology.

In Chapter 1 of this volume, I suggested that there be a semantic shift to using 'urban cultivation' instead of urban agriculture. Some authors made this shift in their chapters (e.g. Chapters 10 and 17, this volume). In working with the material in the book and having an awareness of the ever-increasing literature on the topic, I now feel convinced that this semantic shift needs to take place. As can be discerned from this volume, 'urban agriculture' is much more than simply growing crops on a plot of land, and more than can be encompassed by the term 'agriculture'. Conceptualizing urban agriculture (UA) as urban cultivation (UC) also removes the many assumptions and expectations that the term 'agriculture' implies, and better encompasses what is actually happening in cities. It also gets beyond the agriculture vs gardening dilemma that many practitioners and observers struggle with.

The overwhelming and obvious takeaway of the material in this volume is to re-emphasize that **urban cultivation is much more than producing food**. This is clear around the world, in the Global North *and* Global South. The second takeaway is that **urban cultivation should be seen as a normal and accepted livelihood in all urban areas** of today and the future, and that land rights and the regulatory environments need to become aligned with this reality. The third takeaway is to embrace **urban cultivation's critical**

---

*E-mail: winklerprinsconsulting@gmail.com

role in making cities sustainable, green, and thereby more resilient in the future. Urban cultivation provides green space, usable non-permeable surfaces, plants and habitats, in short, it provides ecosystem services that, in turn, enable urban environmental resiliency.

20.2 Urban Cultivation as Much More than Producing Food

The practice of urban cultivation, as illustrated through the material presented in this book and elsewhere, is clearly about much more than producing food. UC certainly does produce food, and often provides a level of food security that otherwise would be unavailable to city dwellers, especially to the marginalized, but this instrumental value is not its main contribution. Although economists and others will probably continue to focus on this narrow aspect of UC, it is necessary to reframe urban cultivation as a social process, an entry point into the broader food system. As pointed out in Chapter 2, the volume, 'following' urban agriculture, can lead to other areas of urban food networks and shed light on the relationships and practices that people use to access food. UC, through its ability to enable the building of community and trust, opens up even more access to food, and thereby provides food security indirectly, as well as other necessities of living in urban places. This is critical for residents of cities, especially those more vulnerable and marginalized from the mainstay of economic activity. Those who fixate on just the instrumental value of UC miss the point of its much broader capacity. Will cities be able to feed themselves *in toto* via UC as practised today? Probably not without continued linkages to rural production or via high-tech indoor and vertical solutions (see Section 20.4 below), but that should not be the objective of UC. This broader way of thinking of UC needs to be more explicitly embodied by planners and urban policy makers, who typically still focus on the much more narrow instrumental value of UC (e.g. Chapter 13, this volume; Redwood, 2009).

More important though, is the community-building that occurs via UC, with food production lower on the list of priorities of the outcomes of UC. Chapter 3 of this volume demonstrates the range of dimensions of UC, food access, economic savings and health, education, civic engagement, and the significant community development that can occur. Chapter 8 clearly demonstrates the power that comes from organizing via UC for engagement in fighting for social justice causes, such as improving land use for environmental health purposes and other actions that benefit the local community, that might otherwise have been treated as powerless had it not been for the power of organizing that stemmed from UC. The authors of Chapter 6 of this volume note that the value in both the food produced and especially the community built via UC are high, despite the honest drudgery of the work involved. A cautionary note also emerges from work presented in this book, specifically about the idea of 'community' which is not obvious, holistic or naturally occurring, and can be quite problematic (Chapter 5). UC has the potential to foster community, but it should not be assumed to be necessarily harmonious or perceived in similar ways by all involved. More threatening perhaps to the 'UC as social empowerment' movement is that the social and food justice narrative is sometimes co-opted by the hipster and locavore middle class and in their attempts to be socially just actually undermine processes that are fundamentally empowering to the marginalized (examples in Chapters 5 and 8, this volume).

Another important area to which UC contributes beyond food is that it provides a sense of purpose and a sense of place. New migrants, the elderly, or those left destitute due to structural economic change find meaning and purpose through urban cultivation. UC keeps people active and healthier, both physically and mentally, and is an activity that gives people something to do, to be engaged with and to feel valued for, which confers an important sense of purpose. The book provides examples from a wide range of places and people for whom UC is about engagement and activity (e.g. Chapters 4 and 19, this volume). Finally, UC creates social spaces that offer respite from the stresses of urban life, places to recharge and to socialize (Chapters 11, 12 and 19, this volume). The spaces of UC become substitute 'parks' in places where these may not otherwise exist or be safe to be in, and thereby provision another important component of the urban fabric.

20.3 Urban Cultivation as an Accepted and Normal Livelihood in Urban Areas

Cultivation in the city has long been regarded as a response to a crisis or misplaced activity, i.e. a rural activity that should not be practised in urban spaces. What is clear from the material presented in this book is that UC does occur in places that are not in crisis, and that cultivation is a desired practice by many who live in places classified as 'urban'. Despite this, land-use policy has long separated agricultural practice from cities and it is excluded from the notion of what a modern city should look like (Covert and Morales, 2014). This has rendered UC an informal and marginalized activity. This deeply entrenched division of what one does in the 'city' vs the 'countryside' is changing in perception and practice. In reality, people have always been cultivating in the city, but the time has come for it to be an explicitly accepted and normal part of urban living, happening without a precipitating crisis. There is a need to embrace UC as something integral to urban life and part of what a 21st-century city *should* contain, especially as the world is ever more an urban place (Science, 2016). Despite the reality of the urban as the dominant way of life, and with rates of acceleration that put predictions of the percentage of the world's population as urban at about 66% by 2050 (Science, 2016), the vast majority of the world's national governments remain 'urbanophobic' (Meyer, 2013). What this means is that governments do what they can to stem the flow of citizens to cities, and try to stabilize them in rural areas. Legislation is meant to keep people out of cities, yet this has not worked, as people find the potential opportunities in cities a very strong pull.

This urbanophobia needs to cease as it is a lost battle. Instead, city planners and other policy makers need to embrace the urban influx and legislate for it, not against it. This includes attention to UC, to fully understand and envision that it is an urban livelihood and not some sort of displaced or misplaced activity. UC needs to be accepted as a way of life in the city and not something that people 'move through' on their way to something thought to be more appropriate for city dwellers. As we see in some cities in the Global South, there is a tension between a vision of a modern world city, that many in the power structure of those cities want to create, and the reality of accommodating the needs of the many (new) city dwellers (e.g. Chapters 13 and 17, this volume).

The new vision of the 21st-century city should see UC as a component of it. Part of this acceptance is that the informality of UC in practice is everywhere (Chapter 12, this volume; Mukhija and Loukaitou-Sideris, 2014), not just in the GS, where it has assumed to have persisted because cities there are less 'modern' and not fully developed, and also that some of what is today an informal activity could become formalized, with its own regulatory environment (Covert and Morales, 2014). There will probably be economic consequences for formalizing and regulating UC, which may not be desired by all who practise UC, as some wish to stay on the fringe and be countercultural, but doing so would add a level of legitimacy that others would appreciate, as it would make their livelihood a formal one and contribute to urban food policy (Covert and Morales, 2014).

Part of no longer believing that 'cultivation' and 'urban' should be kept separate is acknowledging the multifaceted livelihoods of those who practise it, in both the GN and the GS. Currently, this is more accepted in the GS, where farming in the city is seen as something one does as part of a series of jobs to make ends meet (e.g. Chapters 11 and 12, this volume). But the same could be said of the residents of Westfield, Massachusetts (Chapter 8, this volume), or Helsinki, Finland (Chapter 15, this volume), or of the Brooklyn hipsters who have day jobs as programmers or designers and grow their own vegetables and raise chickens in their backyards, on rooftops, or in community gardens. UC as either a part- or a full-time occupation is a holistic livelihood that keeps people connected to the generative power of food cultivation.

20.3.1 Stabilizing access rights to spaces for urban cultivation

Normalizing cultivation in the city as an accepted livelihood necessitates that rights to spaces where it is practised are improved and stabilized. Most

UC still occurs on land, which is limited in cities and typically of high value. However, this is changing: see Section 20.4 below on vertical farms, which have a new set of access challenges. Therefore, its use for agriculture is rarely considered permanent, as it has a higher development potential, unless the land is considered undevelopable in some way. And those spaces are legally challenging also, as illustrated in Chapter 12 in this volume, which discusses the significant legal ambiguity about who governs the 'open' or interstitial spaces within Dar es Salaam, which are important for cultivation and as social spaces. Land rights to cultivate are also a challenge in strongly regulated places such as China (see Chapter 17, this volume), where there is a significant difference in land system planning between rural and urban areas.

Therefore, a significant challenge faced by urban cultivators in both the GN and the GS is that the space they are using for this purpose is rarely secure and fully owned, fostering a temporality to UC that in turn challenges its full acceptance as a legitimate and long-term urban activity (e.g. Hou, 2014). Even if owned (e.g. backyards or institutional spaces), the right to cultivate it (or keep poultry, livestock or bees on it) may not be granted or accepted by the norms of the surrounding area (e.g. Chapters 9 and 16, this volume; Covert and Morales, 2014). Notable are the differences in land rights between urban (agriculture) cultivation and peri-urban (agriculture) cultivation (Opitz *et al.*, 2016) as the edges of cities typically have different development pressures. This topic needs more attention in the future, as does the topic of urban land rights and its relationship to UC in general.

20.3.2 Attention to the regulatory environment for urban cultivation

Given the challenges of land and access rights for normalized UC, a related takeaway of this book is that urban planners and other urban land-use policy makers need to address the regulatory environment to overcome the real or perceived temporality of UC and related access rights. As documented in this book and other work on UC, it is usually an informal, organic bottom-up activity that is rarely regulated.

Exceptions are in Cuba (Chapter 9), where UC has been a state-sponsored activity, and in Australia (Chapter 10), where there has been significant proactive enabling legislation for UC. There is also remarkable silence on UC; for example, the authors of Chapter 15 in this volume note that agricultural policy in the European Union does not recognize urban agriculture at all, and excludes urban areas from its agricultural policies. In order to bring UC into accepted practice, along with stabilizing access rights, the regulatory (policy) environment needs to align with UC being an accepted part of today's urban places.

There is currently a range of recognition and formal acceptance of UC from the outright persecution and punishment of the illegality of UC (e.g. Chapter 12) to the legislated and subsidized forms of it (Chapter 10). I urge that policies be developed that, at minimum, no longer legislate *against* cultivation in the city, and that it be at least tolerated. Better yet, that cities and other regulatory bodies promote UA and regulate aspects of it that need guidance for the public good. For example, numerous cities in North America now permit the keeping of hens in backyards but, for public health reasons, limit the number and, for noise reasons, do not permit roosters. Cities are starting to grapple more specifically with legislation to regulate UC. Chapter 9 in this volume outlines the challenges that the City of Austin, Texas, has confronted as it has started to regulate UC more explicitly. Covert and Morales (2014) describe in detail the process by which Kansas City, Missouri, formalized what had been informal practice. There is also a need for reforming the way that city governments envision how cities are to be fed, and in fact embrace UC as a form of their own urban food policy that can have many other positive impacts (Covert and Morales, 2014).

20.4 Urban Cultivation's Contribution to the Sustainable City of the Future

Sustainability is a challenged yet deeply entrenched concept that is typically conceptualized as having three parts – environmental, social and economic. UC contributes towards all three aspects of urban sustainability.

From an **environmental perspective**, UC contributes to the greening of the city by converting more land into literal green spaces through cultivation of what would otherwise be empty lots or paved areas, including rights of way, abandoned lots, rooftops and simple ecosystems such as lawns. UC helps with waste management, as it can become a focus of organic waste recycling in the form of compost. Cultivation increases species diversity and adds texture to the landscape, which, in turn, contributes to increasing ecosystem services as these areas of cultivation help absorb storm water and become biological refugia on a micro- and macro-level, fostering biodiversity beyond the crops (e.g. Chapter 19, this volume; Pearsall et al., 2017). We learned in Chapter 14 of this volume that even green spaces as small as sack gardens can help green a place that is otherwise a completely built environment. Beyond its functional value, this increased greenness is also a form of respite and makes the city more pleasant to be in from an aesthetic and environmental health perspective (e.g. Chapters 8 and 12, this volume). UC provides not just any green, but a productive green that helps heal the metabolic rift (McClintock, 2010) and challenges deeply entrenched thinking about the nature/society divide, and thereby contributes to its dismantling. Cities and urban spaces have often been thought of in contrast to rural and more 'natural' spaces, where agriculture is normal. UC brings this 'naturalness' into the city (Hartig and Kahn, 2016) and is contributing to urban greenspace and infrastructure.

Cities as entities have often been considered not green, even anti-environmental, as places that consume much more than their share of resources (Vojnovic, 2013; Wigginton et al., 2016), but this is being reconsidered, as it is increasingly clear that cities may in fact be a more efficient way for humans to live, with a reduced individual footprint (Hartig and Kahn, 2016). Meyer describes this as cities' environmental opportunity (Meyer, 2013). UC is central to actualizing this environmental opportunity.

From a **social perspective**, it is clear that UC contributes social sustainability through engaging in social justice and bringing various people together who may otherwise not interact. As already pointed out in Section 20.2 above, UC is about much more than food, and it brings diverse people together, building trust and community by working together. A very specific example of social sustainability and community resilience through UC is highlighted in Chapter 19, which illustrated a specific case of UC's ability to help communities recover after Hurricane Sandy in New York City.

From an **economic perspective**, beyond the role crops play in greening the city and contributing to social justice, growing food in the city, even if it does not feed the city *in toto*, contributes to making a city more self-sufficient in terms of its food production. UC helps keep the food chain shorter, and forces a rethink of the idea of a 21st-century 'foodshed' and the relationship between urban and rural spaces of food production (Kloppenburg et al., 1996; Seto and Ramankutty, 2016). UC acts as a way of feeding the city without relying on long supply chains; it is part of an effort to shorten these chains, thereby creating more resilient regions. We also know from empirical work in Australia that proximity to community gardens can raise property values in surrounding areas (Chapter 10, this volume).

Although possibly seen as the antithesis of much of the UC analysed to date, there is a need to embrace the more commercially oriented forms of UC, and those systems that are technologically intensive (e.g. enclosed and climate-controlled, soil-less hydro- and aquaponics) in the future. These forms of UC are increasingly appearing in the urban landscape (e.g. Chapter 1 and 3, this volume; La Gorce, 2016; Martin, 2016), and although some of these cater to high-end restaurants, for many entrepreneurs the end goal is to make locally produced fresh food available to all, including people living in food 'deserts' around their urban farms. This style of urban farm is capable of producing year-round and possibly at volume, something the rationalist observers of UC are looking for. Although initially counterintuitive, the technology used can be efficient and help contribute to urban sustainability by limiting the distance for produce to be transported.

All in all, what makes UC such a valuable contribution to the sustainable city of the future is that it is a socio-ecological process. As such, it serves as a form of adaptive capacity because of its ability to enhance resiliency in an urban food system. A number of chapters in this volume (e.g. Chapters 2, 11 and 19) explicitly addressed the role of UC as adaptive capacity, and UC is a rich and varied field of interdisciplinary inquiry for deepening ideas and ideals for urban sustainability. This line of inquiry can benefit from much more attention, given the challenges of increased global urbanization and climate change, and where there is a need to increase societal resilience while decreasing its vulnerability.

20.5 Future Directions

This book has demonstrated that the practice of urban cultivation is much more than producing food and should be an accepted livelihood in cities with the appropriate land rights and regulatory environment guiding it. It helps make cities more pleasant, resilient and sustainable, and contributes in many ways to issues of social justice. As an accepted form of urban livelihoods and production capacity, UC is a component in a global agricultural system.

These aspects of UC should and will likely continue to attract research on its numerous expressions and practice. But there are further elements, less visible from the material in this book that also warrant attention by those interested in the topic. Here I echo a number of points made in Chapter 2 of this volume. Three strands of future research stand out as demanding more immediate attention: deeper work on theorizing urban cultivation; attention to the markets for UC; and engagement with the gender and power relations inherent in UC.

The first is **deeper work on the theorizing of UC**. Although I encouraged all authors in this volume to engage with theoretical frameworks (urban political ecology, critical urban studies/urban assemblage, sustainable livelihood framework) and many did, there is much more to be done. I feel that we stand at the precipice of a significant rethink of some fundamental theories in urban geography and urban sociology as topics such as foodsheds, urban–rural linkages, the city–hinterland dynamic, and others, all come into question with the rise of UC. Specifically, UC is forcing a reassessment of the spatial dynamics of the food system in and around a city, and this is an area rich in theorizing.

Although it is clear that UC is about much more than food, the cultivation of edibles is at the core of the practice and a better understanding of the actual and potential **markets for food and other products generated via UC** is needed, as this is poorly understood everywhere UC is practised. Given that much of the market for UC food is in the informal economy, this is understandable, but there is therefore an opportunity to analyse the pathway of UC-produced food from cultivation to consumption. This will permit a better understanding of the extent to which UC can contribute to a (sustainable) urban food system. As illustrated in Chapters 7 and 11, in the Global South, the marketing of urban produce is *ad hoc* and seemingly idiosyncratic. The destinations of food produced through UC in the GN is even less understood, assumed to be absorbed into a subsistence-oriented informal economy, but with little empirical evidence. As UC becomes more common and more regulated, the need to understand and plan for the marketing of its products becomes more pressing, as creative pathways will develop (e.g. Laterman, 2016). As more policy attention is given to UC, the fact that it is proximate to potential markets should be taken advantage of, as should its ability to form tight and personal connections between producers and consumers.

The issues of **gender and power dynamics** need much more attention. Gender permeates much of the literature on UC, including some of the work presented in this volume (e.g. Chapters 2 and 3) but is rarely its focus. However, it could be, especially when considered together with the power dynamics that surround UC. This is related to deepening theory on UC, especially the possibility of utilizing a feminist urban political ecology approach. Land and access rights are often differentiated by gender, and this needs to be looked at more systematically. Chapter 18, hints at the significance that power dynamics can wield regarding who feels comfortable participating in projects that focus on urban cultivation, due to their political affiliation. More investigations along those lines should be pursued; these would also help further the theorizing by using an urban political ecology framing. Understanding gender and other power dynamics can better provide directions for more targeted food-security planning and the building of resilient and sustainable cities.

In closing, UC as a form of multifunctional agriculture has significant potential to contribute to improving society on many levels. As I have tried to show with this volume, future work on UC will be richer if investigators work towards erasure of the GN/GS divide conceptually and empirically. Engaging with and reaching across the divide will enrich all future work. Finally, I encourage future research to focus more on the *processes* than the traits of UC, as these will demonstrate more similarity than difference between cities in the Global North and the Global South and truly embrace Roy's urban worlding theory (Roy, 2011).

Note

[1] Critique of the *Science* special issue noted that the terms 'urban' and 'city' are used interchangeably, yet in reality are quite different in their population density. This book has done as *Science* did, and I continue this semantic fluidity in this chapter, arguing that both 'urban' and 'city' refer to non-rural places.

References

Bassett, T.J. (1981) Reaping on the margins: a century of community gardening in America. *Landscape* 25(2), 1–8.
Covert, M. and Morales, A. (2014) Formalizing city farms: conflict and conciliation. In: Mukhija, V. and Loukaitou-Sideris, A. (eds) *The Informal American City: From Taco Trucks to Day Labor*. MIT Press, Cambridge, Massachusetts, USA, pp. 193–208.
Hartig, T. and Kahn, P.H. (2016) Living in cities, naturally. *Science* 352(6288), 938–940.
Hou, J. (2014) Making and supporting community gardens as informal urban landscapes. In: Mukhija, V. and Loukaitou-Sideris, A. (eds) *The Informal American City: From Taco Trucks to Day Labor*. MIT Press, Cambridge, Massachusetts, USA, pp. 79–96.
Kloppenburg, J., Hendrickson, J. and Stevenson, G. (1996) Coming in to the foodshed. *Agriculture and Human Values* 13, 33–42.
La Gorce, T. (2016) How does this garden grow? To the ceiling. *New York Times*, 22 July.
Laterman, K. (2016) New York buildings with communal gardens. *New York Times*, 5 August.
Martin, C. (2016) A ski town greenhouse takes local produce to another level. *New York Times*, 26 March.
McClintock, N. (2010) Why farm the city? Theorizing urban agriculture through a lens of metabolic rift. *Cambridge Journal of Regions, Economy and Society* 3, 191–207.
Meyer, W.B. (2013) *The Environmental Advantage of Cities: Countering Commonsense Antiurbanism*. MIT Press, Cambridge, Massachusetts, USA.
Mukhija, V. and Loukaitou-Sideris, A. (eds) (2014) *The Informal American City: From Taco Trucks to Day Labor*. MIT Press, Cambridge, Massachusetts, USA.
Opitz, I., Berges, R., Piorr, A. and Krisker, T. (2016) Contributing to food security in urban areas: differences between urban agriculture and peri-urban agriculture in the Global North. *Agriculture and Human Values* 33, 341–358.
Pearsall, H., Gachuz, S., Rodriguez Sosa, M., Schmook, B., van der Wal, H. and Gracia, M.A. (2017) Urban community garden agrodiversity and cultural identity in Philadelphia, Pennsylvania, US. *The Geographical Review*. DOI: 10.1111/j.1931-0846.2016.12202.x.
Redwood, M. (2009) *Agriculture in Urban Planning: Generating Livelihoods and Food Security*. Earthscan, London, UK.
Roy, A. (2011) Urbanisms, worlding practices, and the theory of planning. *Planning Theory* 10(6), 6–15.
Science (2016) Rise of the city. *Science* 352(6288), 906–907.
Seto, K.C. and Ramankutty, N. (2016) Hidden linkages between urbanization and food systems. *Science* 352(6288), 943–945.
Vojnovic, I. (2013) Advancing toward urban sustainability. In: Vojnovic, I. (ed.) *Urban Sustainability: A Global Perspective*. Michigan State University Press, East Lansing, Michigan, USA, pp. 1–34.
Wigginton, N.S., Fahrenkamp-Uppenbrink, J., Wible, B. and Malakoff, D. (2016) Cities are the future. *Science* 352(6288), 904–905.

Index

access 57–59, 61–62, 71, 136, 139–140, 220
adaptive capacity 135–137, 140–141, 143
    vulnerability 136, 141–143
alternative food networks (AFNs) 3, 29
Amaranthus (*Amaranthus* spp.) 84
apiculture 196–197, 203–204
    decline 196–197
    community 202–203
    government policy 201–202
    science 199–200
aquaponics 6
Austin, Texas 106–107, 114–115
    stakeholder conflicts 108–111
Australia
    community gardens 121–124
        characteristics 125
        environmental functions 127–129
        motivations 125–126
        stakeholder demographics 126–127
    political ecology 118–119, 129–131

bees 196–197, 203–204
    decline 196–197
    community 202–203
    government policy 201–202
    science 199–200
benefits 25, 28–29, 31, 33, 38, 40–42, 44–45, 173–174, 226, 229
    environmental 43, 99, 173–174
    health 40–41, 43–44, 125–126, 164, 236
    social 25, 28–29, 33, 44, 98–100, 119–121, 138–139, 151–153, 164, 172–173, 180, 229, 234–235, 238, 242–243, 246
Berlin 6

chickens 3, 109–111
China 207
    business enterprises 214–215
    E-commerce 215
    educators 215–216
    food security 209–210
    government policy 209–210, 214
        planning 214
    land 213–214
    motivations 211–212
    peri-urban agriculture 208–210, 213–214, 216–217
    production practices 210–211
    stakeholders 215
    urbanization 207
Colombia
    community gardens 160–161
    food security 159, 161–163
    health 164
    income 164
    land tenure 159, 160–161, 166–167
    social benefits 164
    social urbanism 164–166
    women 159–160, 167
common agricultural policy (CAP) 185–186
community development 53–55, 59–60, 61–62, 69, 87–88, 101, 103, 125–126, 231–232, 237–238, 243
community gardens 25, 39–41, 44–45, 130–131, 160–161, 224–225, 227, 229–230
    access 57–59, 61–62, 71, 112–114
    barriers 31, 226–227
    benefits 25, 28–29, 31, 33, 38, 40–42, 44–45, 173–174, 226, 229

community gardens (continued)
- environmental 43, 99, 173–174
- health 40–41, 43–44, 125–126, 164, 236
- social 25, 28–29, 33, 44, 98–100, 119–121, 138–139, 151–153, 164, 172–173, 180, 229, 234–235, 238, 242–243, 246
- characteristics 125
- community development 53–55, 59–60, 61–62, 69, 87–88, 101, 103, 125–126, 231–232, 237–238, 243
- environmental functions 127–129, 230–231, 233–234
- exclusionary practices 57–59, 61–62, 235
- food security 38, 50, 56, 76, 159, 161–163
- funding 74–75, 129, 164–165, 225
- government policy 53, 60–61
  - social urbanism 164–166
- harvest destination 74, 163
- labour 67, 75–76, 163
- marginalized groups 50–51, 52, 56–57, 61, 130, 225, 231–232, 235–236
- motivations 125–126
- political ecology 118–121, 129–131
- production practices 72–74, 225–226
- socio-economics 61, 94, 102–103
- stakeholder demographics 41–42, 55–57, 126–127, 225, 231–232
  - income 42–43
- study of 41, 121–124, 174–175, 232–233
- umbrella organizations 53–55

community supported agriculture (CSA) 187, 191, 194
cost, of food 88
Cuba 106–107, 111, 114–115
- food access 112–114
- state-led production 112
- sustainability 111–112

E-commerce 215
- REKO Circles 184, 189–191, 194
ecosystem services 5, 29–30, 197, 230–231, 233–234, 246
education 28, 69–71, 98, 125–126, 129, 153–156
environmental
- diversity 138–139, 230–231, 233–234
- functions 127–129, 230–231
- issues 87, 108–111, 175–178
- sustainability 106–108, 114–115, 120, 193–194, 245–247

Finland
- community supported agriculture (CSA) 187, 191, 194
- government policy 185–186
- urban agriculture 185, 191–192, 194
  - cultural values 192

- environmental benefits 193–194
- individual agency promotion 192–193
- nationalism 193–194
- sites 186–187
- study of 187
- urban food distribution networks 184–185
  - Herttoniemi Food Cooperative 184, 187
  - implications for farmers 191
  - REKO Circles 184, 189–191, 194

food
- access 57–59, 61–62, 71, 136, 139–140, 220
- cost 88
- environment 15
- justice 29, 93–95, 96–97, 243
- quality 87–88, 212
- security 12–14, 16–18, 27–28, 69, 96–97, 134–135, 136, 139–140, 159, 161–163, 209–210, 220, 222–223
  - urban planning 18–20
- systems 12–15, 89, 93–94
  - city food provisioning 17–18
  - food justice 29, 93–95, 96–97, 243
  - resiliency 20–21, 137

funding 74–75

Gardening the Community (GTG) 97, 100–102
Ghana 79, 89
- Amaranthus (*Amaranthus* spp.) 84
- food quality 87–88
- income generation 86
- jute mallow (*Corchorus olitorius*) 82–84
- livelihoods framework 80
- market characteristics 80–81
- production 81
- stakeholder relationships 83, 84–86
- study of 81
- urban infrastructure 88–89
- urban political ecology (UPE) 79–80
- vegetable marketing system 82–83, 84, 85–86, 89
- waste water irrigation 82–84, 86

Global North (GN) 1–5, 24
- alternative food networks (AFNs) 29
- Cuba 106–107, 111, 114–115
  - food access 112–114
  - state-led production 112
  - sustainability 111–112
- ecosystem services 5, 29–30, 197, 230–231, 233–234, 246
- Finland
  - community supported agriculture (CSA) 187, 191, 194
  - government policy 185–186
  - urban agriculture 185, 186–187, 191–194
  - urban food distribution networks 184–185, 187, 189–191, 194
- guerrilla gardening 27

home gardens 25–26
institutional gardens 27, 32–33, 125–126, 129
scale 25
Toronto, Canada 86–87
urban farms 27, 32
USA
    Austin, Texas 106–111, 114–115
    community gardens 31, 33, 38, 39–45, 50–62, 67, 69–76, 96, 98–100, 101–103
    Houston, Texas 66–77
    Lincoln, Nebraska 231–232
    Massachusetts 93–102
    New York, New York 232, 234–235, 237–238
    Portland, Oregon 87
    San Diego, California 50–62
    Santa Clara County, California 30–34
Global South (GS) 1–5
    Australia
        community gardens 121–129
        political ecology 118–119, 129–131
    China 207
        business enterprises 214–215
        E-commerce 215
        educators 215–216
        food security 209–210
        government policy 209–210, 214
        land 213–214
        motivations 211–212
        peri-urban agriculture 208–210, 213–214, 216–217
        production practices 210–211
        stakeholders 215
        urbanization 207
    Colombia
        community gardens 160–161
        food security 159, 161–163
        health 164
        income 164
        land tenure 159, 160–161, 166–167
        social benefits 164
        social urbanism 164–166
        women 159–160, 167
    food security *See* food security
    food systems 12–15
    Ghana 79, 89
        Amaranthus (*Amaranthus* spp.) 84
        food quality 87–88
        income generation 86
        jute mallow (*Corchorus olitorius*) 82–84
        livelihoods framework 80
        market characteristics 80–81
        production 81
        stakeholder relationships 83, 84–86
        study of 81
        urban infrastructure 88–89

        urban political ecology (UPE) 79–80
        vegetable marketing system 82–83, 84, 85–86, 89
        waste water irrigation 82–84, 86
    Kenya
        sack gardening 172, 178, 179–180
        slums 171–172, 174–181
    resiliency 20–21
    scale 13
    Senegal 134
        adaptive capacity 135–137, 140–141, 143
        food access 139–140
        food system characterization 137
        income 138–139
        nutrient recycling 140
        vulnerability 136, 141–143
    Tanzania
        pollution 148
        urban farming 146–147, 148, 149–156
    urban agriculture 12–13
    Zimbabwe
        community gardens 224–227
        food security 222–223
        government policy 223–224
        income 222
        planning 223–224
        poverty 222–223
        urban agriculture 220–221
government policy 33–34, 53, 112, 129, 142, 159, 209–210, 214, 223–224, 245
    common agricultural policy (CAP) 185–186
    political ecology 118–121, 129–131
    social urbanism 164–166
Green City Acres, British Columbia 89
guerrilla gardening 27

harvest destination 74, 139–140
health 99, 142
    benefits 40–41, 43–44, 125–126, 164, 236
    waste water irrigation 82–84, 86, 148
Herttoniemi Food Cooperative 184, 187
home gardens 25–26, 31–32
hydroponics 6

income 42–43, 138–139, 149–151, 164, 212, 222, 226
institutional gardens 27, 32–33, 125–126, 129
irrigation 73, 84, 225
    waste water irrigation 82–84, 86

justice
    food 29, 93–95, 96–97, 243
    social 107–108, 120, 173–174
jute mallow (*Corchorus olitorius*) 82–84

Kenya
    sack gardening 172, 178, 179
        benefits 179–180
    slums 171–172, 180–181
        environmental issues 175–178
        powerlessness 178–179
        study of 174–175

labour 67, 75–76
    women 16, 67, 84, 149–151, 247
        Colombia 159–160, 167
land tenure 71–72, 101–102, 119, 159, 160–161, 209, 244–245
Lincoln, Nebraska 231–232
livestock 3, 109–111

marginalized groups 50–51, 52, 56–57, 61, 130, 159, 161–164, 167, 225, 231–232, 235–236
    community gardens 160–161
    food security 159, 161–163
    health 164
    income 164
    land tenure 159, 160–161, 166–167
    social benefits 164
    social urbanism 164–166
    women 159–160, 167
markets 79, 139
    characteristics 80–81
Massachusetts 95–96
    food systems 93–94
    Gardening the Community (GTG) 97, 100–102
    Nuestras Raíces 97–100
motivations 125–126, 211–212

neoliberalism 51–53, 93, 120
new food equation (NFE) 19–20
Nuestras Raíces 97–100

planning 13, 18–20, 118–119, 160–161, 164–167, 214, 223–224, 244
political ecology 5–6, 79–80, 118–121, 129–131
Portland, Oregon 87
poverty 96, 101, 130, 156, 161–162, 220, 222–223
    slums 171–172, 180–181
        environmental issues 175–178
        powerlessness 178–179
        study of 174–175
prices 88
production practices 72–74, 140, 210–211, 225–226
    small plot intensive (SPIN) farming 89

REKO Circles 184, 189–191, 194
resiliency 20–21, 137, 230, 237–238, 242–243

San Diego, California
    community gardens 50–51
        community development 53–55, 59–60, 61–62
        exclusionary practices 57–59, 61–62
        food security 50, 56
        government policy 53, 60–61
        marginalized groups 50–51, 52, 56–57, 61
        neoliberalism 51–53
        San Diego community garden network (SDCGN) 53–55
        social benefits 51–53
        socio-economics 61,
        stakeholder demographics 55–57
Santa Clara County, California 30–31, 33–34
    community gardens 31, 33
    home gardens 31–32
    school gardens 32–33
    urban farms 32
school gardens 27, 32–33, 125–126, 129
Senegal 134
    adaptive capacity 135–137, 140–141, 143
    food access 139–140
    food system characterization 137
    income 138–139
    nutrient recycling 140
    vulnerability 136, 141–143
small plot intensive (SPIN) farming 89
social
    benefits 25, 28–29, 33, 44, 98–100, 119–121, 138–139, 151–153, 164, 172–173, 180, 229, 234–235, 238, 242–243, 246
    justice 107–108, 120, 173–174
    process 15–16
        city food provisioning 17–18
        variability 16–17
        women 16
    urbanism 164–166
socio-ecology 94–95, 97, 99, 135–137, 229–230, 238, 246–247
    adaptive capacity 135–137, 140–141, 143
    vulnerability 136, 141–143
socio-economics 24, 28, 61, 94, 102–103, 130, 220–221, 246–247
    income 42–43, 138–139, 149–151, 164, 212, 222, 226
    poverty 96, 101, 130, 156, 161–162, 220, 222–223
        slums 171–172, 174–181
socio-spatial 95, 97, 130, 146, 147–148, 156, 244
stakeholder 215
    backgrounds 149–151
    conflicts 108–111

demographics 41–42, 55–57, 68, 96–97,
    126–127, 148, 211–212, 225,
    231–232
  income 42–43, 138–139, 149–151, 164, 212,
    222, 226
  marginalized groups 50–51, 52, 56–57, 61, 96
subversive and interstitial food spaces (SIFS) 3
sustainability 106–108, 114–115, 120, 193–194,
    245–247
  politics of 111–112
  social justice 107–108
  stakeholder conflicts 108–111
sustainable livelihoods framework 7

Tanzania
  pollution 148
  urban farming 146–147, 148, 156
    education 153–156
    safety 152–153
    social benefits 151–153
    stakeholder backgrounds 149–151
    study of 149
theoretical frameworks 5
  sustainable livelihoods framework 7
  urban assemblage framework 6–7, 207–208,
    213–217
  urban political ecology (UPE) 5–6, 79–80, 94,
    118–121, 129–131, 197–199
Toronto, Canada 86–87

urban agriculture 1–2, 8–9, 66, 79, 93–95,
    134–135, 146–147, 159, 185,
    191–192, 194, 207, 220–221, 242–243,
    247–248
  alternative food networks (AFNs) 29
  apiculture 196–197
  benefits 25, 28–29, 31, 33, 38, 40–42, 44–45,
    173–174, 226, 229
    environmental 43, 99, 173–174
    health 40–41, 43–44, 125–126,
      164, 236
    social 25, 28–29, 33, 44, 98–100,
      119–121, 138–139, 151–153,
      164, 172–173, 180, 229,
      234–235, 238, 242–243, 246
  community gardens 25, 39–41, 44–45,
    130–131, 160–161, 224–225, 227,
    229–230
    access 57–59, 61–62, 71, 112–114
    barriers 31, 226–227
    benefits 25, 28–29, 31, 33, 38, 40–45,
      98–100, 119–121, 125–126,
      138–139, 151–153, 164,
      173–174, 226, 229, 234–236,
      238, 242–243, 246

characteristics 125
community development 53–55,
    59–60, 61–62, 69, 87–88,
    101, 103, 125–126, 231–232,
    237–238, 243
environmental functions 127–129,
    230–231, 233–234
exclusionary practices 57–59,
    61–62, 235
food security 38, 50, 56, 76, 159,
    161–163
funding 74–75, 129, 164–165, 225
government policy 53, 60–61, 164–166
harvest destination 74, 163
labour 67, 75–76, 163
marginalized groups 50–51, 52,
    56–57, 61, 130, 225, 231–232,
    235–236
motivations 125–126
political ecology 118–121, 129–131
production practices 72–74, 225–226
socio-economics 61, 94, 102–103
stakeholder demographics 41–43, 55–57,
    126–127, 225, 231–232
study of 41, 121–124, 174–175,
    232–233
umbrella organizations 53–55
definition of 2–5, 14–15
ecosystem services 5, 29–30, 197, 230–231,
    233–234, 246
food
  access 57–59, 61–62, 71, 136,
    139–140, 220
  cost 88
  environment 15
  justice 29, 93–95, 96–97, 243
  quality 87–88, 212
  security 12–14, 16–20, 27–28, 69,
    96–97, 134–135, 136, 139–140,
    159, 161–163, 209–210, 220,
    222–223
  systems 12–15, 17–18, 20–21, 29, 89,
    93–97, 137, 243
government policy 33–34, 53, 112,
    129, 142, 159, 209–210, 214,
    223–224, 245
  common agricultural policy
    (CAP) 185–186
  political ecology 118–121, 129–131
  social urbanism 164–166
guerrilla gardening 27
home gardens 25–26
institutional gardens 27, 32–33,
    125–126, 129
land tenure 71–72, 101–102, 119, 159,
    160–161, 209, 244–245
scale 13, 25

urban agriculture (*continued*)
    social 15–16
        benefits 25, 28–29, 33, 44, 98–100, 119–121, 138–139, 151–153, 164, 172–173, 180, 229, 234–235, 238, 242–243, 246
        justice 107–108, 120, 173–174
        process 15–18
        urbanism 164–166
    theoretical frameworks 5
        sustainable livelihoods framework 7
        urban assemblage framework 6–7, 207–208, 213–217
        urban political ecology (UPE) 5–6, 79–80, 94, 118–121, 129–131, 197–199
    urban farms 27, 32
        Senegal 134–143
        stakeholder conflicts 108–111
        Tanzania 146–156
urban assemblage framework 6–7, 207–208, 213–217
urban cultivation 3, 242
    definition of 3, 242
urban farms 27, 32
    Senegal 134
        adaptive capacity 135–137, 140–141, 143
        food access 139–140
        food system characterization 137
        income 138–139
        nutrient recycling 140
        vulnerability 136, 141–143
    stakeholder conflicts 108–111
    Tanzania 146–147, 148, 156
        education 153–156
        safety 152–153
        social benefits 151–153
        stakeholder backgrounds 149–151
        study of 149
urban food distribution networks 184–185
    Herttoneimi Food Cooperative 184, 187
    implications for farmers 191
    REKO Circles 184, 189–191, 194
Urban Harvest 68, 74
urbanization 172–173, 207, 238, 244
urban political ecology (UPE) 5–6, 79–80, 94, 118–121, 129–131, 197–199
USA
    Austin, Texas 106–107, 114–115
        stakeholder conflicts 108–111
    community gardens 39–41, 44–45
        access 57–59, 61–62, 71
        barriers 31
        benefits 31, 33, 38, 40–45, 69–71, 98–100
        community development 53–55, 59–60, 61–62, 69, 101, 103
        education 69–71, 98

    food security 38, 50, 56, 69, 76, 101
    funding 74–75
    government policy 53, 60–61
    harvest destination 74
    labour 67, 75–76
    land tenure 71–72, 101–102
    marginalized groups 50–51, 52, 56–57, 61, 96
    neoliberalism 51–53
    production practices 72–74
    social benefits 51–53
    socio-economics 61, 94, 102–103
    stakeholder demographics 41–43, 55–57, 68, 96
    study of 41
Houston, Texas 66, 67–68, 76–77
    food access 69, 76
    funding 74–75
    harvest destination 74
    labour 67, 75–76
    land tenure 71–72
    objectives 69–71
    production practices 72–74
    stakeholder demographics 68
    Urban Harvest 68, 74
Lincoln, Nebraska 231–232
Massachusetts 95–96
    food systems 93–94
    Gardening the Community (GTG) 97, 100–102
    Nuestras Raíces 97–100
New York, New York 232, 234–235, 237–238
Portland, Oregon 87
San Diego, California
    community development 53–55, 59–60, 61–62
    community gardens 50–51
    exclusionary practices 57–59, 61–62
    food security 50, 56
    government policy 53, 60–61
    marginalized groups 50–51, 52, 56–57, 61
    neoliberalism 51–53
    San Diego community garden network (SDCGN) 53–55
    social benefits 51–53
    socio-economics 61
    stakeholder demographics 55–57
Santa Clara County, California 30–31, 33–34
    community gardens 31, 33
    home gardens 31–32
    school gardens 32–33
    urban farms 32

Vienna, Austria 231
vulnerability 136, 141–143

women 16, 67, 84, 149–151, 247
    Colombia 159–160, 167

Zimbabwe
    community gardens 224–225, 227
        benefits 226
        challenges 226–227
        income 226
        management 225
        productivity 225–226
        stakeholders 225
    food security 222–223
    government policy 223–224
    income 222
    planning 223–224
    poverty 222–223
    urban agriculture 220–221